MACROCOGNITION METRICS AND SCENARIOS

Macrocognition Metrics and Scenarios

Design and Evaluation for Real-World Teams

Edited by

EMILY S. PATTERSON
Ohio State University, USA

&

JANET E. MILLER
Air Force Research Laboratory, USA

CRC Press
Taylor & Francis Group
Boca Raton London New York

CRC Press is an imprint of the
Taylor & Francis Group, an **informa** business

CRC Press
Taylor & Francis Group
6000 Broken Sound Parkway NW, Suite 300
Boca Raton, FL 33487-2742

First issued in paperback 2017

© 2010 by Emily S. Patterson and Janet E. Miller
CRC Press is an imprint of Taylor & Francis Group, an Informa business

No claim to original U.S. Government works

Version Date: 20160226

ISBN 13: 978-1-138-07208-4 (pbk)
ISBN 13: 978-0-7546-7578-5 (hbk)

**Visit the Taylor & Francis Web site at
http://www.taylorandfrancis.com**

**and the CRC Press Web site at
http://www.crcpress.com**

Contents

PART 1 THEORETICAL FOUNDATIONS

PART II MACROCOGNITION MEASURES FOR REAL-WORLD TEAMS

List of Figures

List of Tables

Acknowledgments

Support for book editing and a two-day conference on Macrocognitive Metrics in Columbus, Ohio, at The Ohio State University on June 14, 2007, was provided by the Air Force Research Laboratory (S110000012). The views expressed in this book are those of the authors and do not necessarily represent those of the Air Force. We thank Christopher P. Nemeth for sharing a book index from Healthcare Team Communication, Michael W. Smith for creating the book index, and Fernando Bernal for designing the book cover. We thank all of the chapter reviewers, and in particular Robert Hoffman, for his thoughtful and detailed reviews of many of the book chapters.

List of Contributors

Kyle Behymer is currently a Senior Researcher at 361 Interactive, LLC. Mr. Behymer received his M.S. in Human Factors Psychology in 2005 from Wright State University. From 2003–2006, he was a Research Psychologist at JST Applications in the defense and space industry. Mr. Behymer's research interests are human factors psychology and computer information systems.

C. Shawn Burke is a Research Scientist at the Institute for Simulation and Training at the University of Central Florida. Dr. Burke earned her doctorate in Industrial and Organizational Psychology from George Mason University. Her expertise includes teams and their leadership, team adaptability, team training, measurement, evaluation, and team effectiveness. Dr. Burke has published more than 60 journal articles and book chapters related to these topics and has had work accepted at more than 100 peer-reviewed conferences.

Nancy J. Cooke is a Professor of Applied Psychology at Arizona State University and is Science Director of the Cognitive Engineering Research Institute in Mesa, AZ. She is also Section Editor of *Human Factors*, a member-at-large of the Human Factors and Ergonomics Society's Executive Council, a member of the National Research Council's Committee on Human Systems Integration, and a member of the US Air Force Scientific Advisory Board. Dr. Cooke specializes in the development, application, and evaluation of methodologies to elicit and assess individual and team cognition.

April B. Courtice is a Research Psychologist at the Air Force Research Laboratory. Ms. Courtice received her M.S. in Psychology from Wright State University in 2009. She is a member of the Human Factors and Ergonomics Society and her main research interest is cognitive engineering and decision making.

Donald A. Cox is a Senior Scientist at Klein Associates Division of Applied Research Associates. Mr. Cox completed his M.Sc. in Computer Science at the University of Calgary. Mr. Cox has been working with computers for over 25 years in design, implementation, evaluation, and researching of systems and development methods with a special interest in qualitative methods. At Klein Associates, Mr. Cox is currently involved in projects evaluating advanced decision support for planning, making ethnographic insights more available to system developers, and characterizing the impact of new technology of submarine command team functioning.

Robert G. Eggleston is a Principal Research Scientist affiliated with the Air Force Research Laboratory with over 30 years of fundamental and applied research experience on human performance in complex work. He is currently leading the theoretical and applied development of work-centered support technology from a joint cognitive systems perspective, addressing the contextualized work needs over the range of orderly (simple, complicated), to unruly (complex and chaotic) military work problem situations.

John Elias is a Graduate Student in the Modeling and Simulation program at the University of Central Florida's Institute for Simulation and Training. His education includes studies in molecular biology, literature, and philosophy. His chief research interests lie in the interdisciplinary cognitive sciences, with focus specifically on intersubjectivity and its relation to interaction.

Stephen M. Fiore is an Assistant Professor of Cognitive Sciences at the University of Central Florida with the Department of Philosophy and Director of the Cognitive Sciences Laboratory at UCF's Institute for Simulation and Training. Dr. Fiore earned his doctorate in Cognitive Psychology from the University of Pittsburgh. His multidisciplinary research interests incorporate aspects of the cognitive, social, and computational sciences in the investigation of learning and performance in individuals and teams. Dr. Fiore has published more than 100 scholarly publications in the area of learning, memory, and problem solving at the individual and the group level.

John M. Flach is Professor and Chair of the Department of Psychology at Wright State University. He received his Ph.D. in Human Experimental Psychology from The Ohio State University in 1984. His current research interests are perceptual-motor skill, manual control, decision making, cognitive systems engineering, human-machine interface design (particularly in the domains of aviation, medicine, and virtual reality), and ecological psychology.

Brian Friel is an Associate Professor in the Psychology Department at Delaware State University. He holds a Ph.D. in Experimental Psychology from Kansas State University and a M.S. in Cognitive Psychology from the University of Oklahoma. His research interests include mental processes in decision making, measures of performance, reading and language processes, and bilingualism.

Jamie C. Gorman is a Postdoctoral Research Associate at Arizona State University and the Cognitive Engineering Research Institute. He completed his Ph.D. in Cognitive Psychology at New Mexico State University. His research interests are team cognition, communication, and the application of dynamical systems methods. Dr. Gorman is currently conducting research on team coordination and collaboration, real-time dynamics, and the deterministic structure of language.

Robert R. Hoffman is a Senior Research Scientist at the Institute for Human and Machine Cognition. Dr. Hoffman is recognized as one of the world leaders in the field of cognitive systems engineering and human-centered computing. He is a Fellow of the Association for Psychological Science and a Fulbright Scholar. His Ph.D. is in Experimental Psychology from the University of Cincinnati, where he received McMicken Scholar, Psi Chi, and Delta Tau Kappa Honors. Following a Postdoctoral Associateship at the Center for Research on Human Learning at the University of Minnesota, Dr. Hoffman joined the faculty of the Institute for Advanced Psychological Studies at Adelphi University. He began his career as a psycholinguist, and founded the journal *Metaphor and Symbol*. His subsequent research leveraged the psycholinguistics background in the study of methods for eliciting the knowledge of domain experts. Hoffman has been recognized internationally in disciplines including psychology, remote sensing, weather forecasting, and artificial intelligence, for his research on human factors in remote sensing, his work in the psychology of expertise and the methodology of cognitive task analysis, and for his work on HCC issues intelligent systems technology and the design of macrocognitive work systems. Dr. Hoffman is a Co-Editor for the Department on Human-Centered Computing in IEEE: Intelligent Systems. He is Editor for the book series, *Expertise: Research and Applications*. He is a co-founder and Track Editor for the *Journal of Cognitive Engineering and Decision Making*. His major current projects involve evaluating the effectiveness of knowledge management, and performance measurement for macrocognitive work systems.

Gary Klein is a Principal Scientist at Applied Research Associates. He was instrumental in founding the field of Naturalistic Decision Making. Dr. Klein received his Ph.D. in Experimental Psychology from the University of Pittsburgh in 1969. He was an Assistant Professor of Psychology at Oakland University (1970–1974) and worked as a Research Psychologist for the US Air Force (1974–1978). The R&D company he founded in 1978, Klein Associates, was acquired by Applied Research Associates in 2005. He has written: *Sources of Power: How People Make Decisions* (1998); *The Power of Intuition* (2004); *Working Minds: A Practitioner's Guide to Cognitive Task Analysis* (Crandall et al., 2006); and *Streetlights and Shadows: Searching for the Keys to Adaptive Decision Making* (2009). Dr. Klein developed a Recognition-Primed Decision (RPD) model to describe how people actually make decisions in natural settings. He also developed methods of Cognitive Task Analysis for uncovering the tacit knowledge that goes into decision making. He was selected as a Fellow of Division 19 of the American Psychological Association in 2006, and in 2008 he received the Jack A. Kraft Innovator Award from the Human Factors and Ergonomics Society.

Barbara Künzle is a Work and Organizational Psychologist at the Swiss Federal Institute of Technology, Zurich, Switzerland and holds a Ph.D. in Psychology. She did her Ph.D. research on anesthesia teams with the Department of Management,

Technology. She published several articles on leadership and coordination modes. Her current research is focused on leadership, coordination, and teamwork in healthcare teams.

John D. Lee is a Professor in the Department of Industrial and Systems Engineering at the University of Wisconsin, Madison. Previously he was with the University of Iowa, and was the Director of Human Factors Research at the National Advanced Driving Simulator. Before moving to the University of Iowa, he was a Research Scientist at the Battelle Human Factors Transportation Center for six years. He received his Ph.D. from the Department of Mechanical and Industrial Engineering at the University of Illinois. Professor Lee's research focuses on the safety and acceptance of complex human-machine systems by considering how technology mediates attention. Specific research interests include trust in technology, advanced driver assistance systems, and driver distraction.

Michael P. Letsky is currently Program Officer for the Collaboration and Knowledge Interoperability Program at the Office of Naval Research (ONR), Arlington, VA. He manages a research program of academic grants and innovative small business projects seeking to understand team cognition and team performance. He previously worked for the Army Research Institute of the Behavioral and Social Sciences where he developed their long-range strategic research plan and also served on the Army Science Board on Highly Maneuverable Forces. Dr. Letsky's education includes a B.S. in Electrical Engineering (Northeastern University), an MBA and DBA in Operations Research (George Washington University).

Colin Mackenzie is a Physician, graduating from Aberdeen University, Scotland in 1968. He is currently Professor of Anesthesiology, Associate Professor of Physiology, University of Maryland School of Medicine. His research interests include human factors in emergencies and trauma resuscitation. He has been continuously funded by federal grants for the past 18 years, and has over 120 peer-reviewed publications.

Anne M. Miller is an Assistant Professor in the School of Nursing at Vanderbilt University. She received her R.N. (Royal Melbourne Hospital) and her Ph.D. in Psychology at the University of Queensland. She has numerous scientific publications and has served as Principal Investigator on a number of funded grants. Her current research interests are resilience engineering for healthcare environments, clinical decision making and continuity of clinical care, and clinical information design.

Janet E. Miller is a Senior Electronics Engineer at the Air Force Research Laboratory, Sensors Directorate at Wright-Patterson Air Force Base, Dayton, Ohio. Dr. Miller received a Ph.D. in Industrial and Systems Engineering from The Ohio State University in 2002. Currently, Dr. Miller is advancing the state-of-the-art in macrocognition research and application in military situations as well as in trust in complex, analytic technologies.

Emily S. Patterson is an Assistant Professor in the Health Information Management and Systems Division at The Ohio State University Medical Center, School of Allied Medical Professions. Dr. Patterson received a Ph.D. in Industrial and Systems Engineering from The Ohio State University in 1999. Her professional focus is in applying human factors knowledge methods to improve the design of complex, socio-technical settings, and particularly in human-computer interaction and computer-supported cooperative work. She has served as Principal Investigator on a number of federal grants and contracts. In 2004, Dr. Patterson received the Alexander C. Williams, Jr., Design Award from the Human Factors and Ergonomics Society, and in 2008 she received the Lumley Research Award from The Ohio State University. Dr. Patterson currently serves as Associate Editor for *IEEE Systems, Man, and Cybernetics Part A*, on the advisory board for the *Joint Commission Journal on Quality and Patient Safety*, and formerly on the editorial board for the *Human Factors* journal. She currently serves as an advisory board member to the Center for Innovation of the National Board of Medical Examiners and formerly to the Joint Commission International Center for Patient Safety.

Scott S. Potter is a Principal Scientist with Charles River Analytics. He completed his Ph.D. in Cognitive Systems Engineering at The Ohio State University. He has 15 years experience in applying cognitive work analysis as the underlying basis for the design of decision-support tools for complex, dynamic, high-risk environments. Before joining Charles River Analytics in April of 2007, he was Principal Engineer at Aegis Research Corp. for seven years; for the six years prior, he was a Principal Engineer with Carnegie Group.

John Raacke is the Associate Dean in the College of Arts & Sciences and an Assistant Professor of Psychology at the University of North Carolina at Pembroke. He completed his Ph.D. in Experimental/Cognitive Psychology at Kansas State University with emphasis in Judgment and Decision-Making. Dr. Raacke's research has examined decision making in a number of applied areas of psychology. Most recently, Dr. Raacke's studies have dealt with investigating expertise, team development in dynamic environments, juror decision making, the use of conditional probabilities, and social networking.

Michael A. Rosen is a Doctoral Candidate in the Applied Experimental and Human Factors Psychology program at the University of Central Florida and since fall 2004 has been a Graduate Research Assistant at the Institute for Simulation and Training, where he won the Student Researcher of the Year in 2006. His research interests include individual and team decision making and problem solving, human-computer interaction, team performance, and training in high-stakes domains such as healthcare and the military.

Emilie M. Roth is the Owner and Principal Scientist of Roth Cognitive Engineering. She has been involved in cognitive systems analysis and design in a variety of domains including nuclear power plant operations, railroad operations, military command and control, medical operating rooms, and intelligence analysis. She received her Ph.D. in Cognitive Psychology from the University of Illinois at Champaign-Urbana in 1980. She currently serves on the editorial board of the journals *Human Factors* and *Le Travail Humain*, and is Editor of the Design of Complex and Joint Cognitive Systems track of the *Journal of Cognitive Engineering and Decision Making*.

Robert Rousseau is an Emeritus Professor at Université Laval and currently Director of C3, a firm specializing in Research and Development (R&D) in cognitive systems engineering. As a Senior Defence Scientist at DRDC Valcartier between 2003 and 2006, he led projects on cognitive issues in evaluation of Decision Support Systems (DSS), decision making in command and control (C2), and situation awareness. He has extensive expertise on cognitive aspects of work automation, information systems implementation, change management and team decision making. He was for many years Chair of the School of Psychology at Université Laval.

Eduardo Salas is Trustee Chair and Professor of Psychology at the University of Central Florida where he also holds an appointment as Program Director for the Human Systems Integration Research Department at the Institute for Simulation and Training (IST). Before joining IST, he was a Senior Research Psychologist and Head of the Training Technology Development Branch of Naval Air Warfare Center Training Systems Division for 15 years. Dr. Salas has co-authored over 300 journal articles and book chapters and has co-edited 19 books. His expertise includes teamwork, team training strategies, training effectiveness, decision making under stress, and performance measurement tools. Dr. Salas is a Fellow of the American Psychological Association, the Human Factors and Ergonomics Society, President-Elect of the Society for Industrial and Organizational Psychology, and a recipient of the Meritorious Civil Service Award from the Department of the Navy.

Daniel Schwartz is a Consumer Psychologist at Procter & Gamble. He received his Ph.D. in Experimental Psychology from Wright State University. From 2005 until 2008, he was a Research Fellow at the Air Force Research Laboratory. Dr. Schwartz's current interests are consumer understanding, product research and development, and social media.

James Shanteau is a University Distinguished Professor of Psychology and Commerce Bank Distinguished Graduate Faculty. Professor Shanteau received his Ph.D. in Experimental Psychology from University of California, San Diego. Since joining the faculty at Kansas State University, he has held visiting appointments at Universities of Michigan, Oregon, Colorado, Cornell, Toulouse, and National

Science Foundation. His research interests include studies of expertise (especially medical decision making) and studies of consumer healthcare choices (especially organ donation and transplantation). Professor Shanteau has received over $5.9 million from agencies such as National Institute of Mental Health, Division of Organ Transplantation, National Institute of Child Health and Human Development, and National Science Foundation. His publications include over 75 articles in referred journals, 10 books, 65 book chapters, 6 encyclopedia entries, 7 monographs, 21 proceedings papers, 14 technical reports, and 3 computer programs. He served on grant-review panels for the National Institute of Health, Environmental Protection Agency, and National Science Foundation. He served on a National Research Council Committee to redefine mental retardation standards, and on a DHHS Advisory Committee on Organ Transplantation (ACOT). He is a Fellow of the American Psychological Association and a Charter Fellow of the American Psychological Society.

Lawrence G. Shattuck is a Senior Lecturer and Director of the Human Systems Integration Program at the Naval Postgraduate School in Monterey, CA. Dr. Shattuck received his Ph.D. in Cognitive Systems Engineering from the Ohio State University in 1995. He is a retired Army Colonel with 30 years of service. During his last assignment on active duty (1995–2005) he was a Professor and Director of the Engineering Psychology Program at the United States Military Academy, West Point, NY. His research interests include communication of intent, military command and control, human-computer interaction, and human error.

Wayne Shebilske is a Professor in Psychology at Wright State University. Dr. Shebilske received his Ph.D. in Psychology at the University of Wisconsin, Madison, in 1974. After nine years on the faculty of the University of Virginia Psychology Department, Dr. Shebilske served as Study Director for the Committee on Vision at the National Academy of Sciences. Dr. Shebilske was a Professor at Texas A&M University where he was the Graduate Program Advisor. In 1992, he was awarded a National Research Council Senior Research Associateship at the Armstrong Laboratory Intelligent Training Branch. Dr. Shebilske has worked closely with government, military, and private agencies identifying critical issues for study in many areas, including design and development of aerospace systems, visual display equipment, medical devices, standards for pilots and drivers, automated instruction for complex skills, and virtual reality systems.

Kimberly Smith-Jentsch is an Assistant Professor in Industrial and Organizational Psychology at the University of Central Florida. She received her Ph.D. in Industrial and Organizational Psychology from University of South Florida in 1994. She has earned a number of awards, including the M. Scott Myers Award for Applied Research in the Workplace (2001), the Dr. Arthur E. Bisson Award for Naval Technology Achievement (2000), and the NAVAIR Senior Scientist

Award (2000). Her research interests include mentoring, teamwork, leadership, assertiveness, training and development, and performance measurement.

Rick P. Thomas is an Assistant Professor in Psychology at the University of Oklahoma. His primary research focus is in the emerging field of cognitive decision theory. This field integrates theory and methods from cognitive and ecological psychology with behavioral decision theory. Most of Professor Thomas's research in cognitive decision theory relies on computational models of memory and knowledge acquisition to account for judgment and decision-making phenomena. For example, he is currently investigating how a modified version of MINERVA-DM, HyGene, accounts for the effects of hypothesis generation on probability judgments, confidence judgments, and hypothesis testing (Thomas, Dougherty, Springler & Harbison, 2008; http://www.thehygeneproject.org/). Professor Thomas also does work concerning the study of expertise (Thomas et al., 2001; http://www. k-state.edu/psych/cws/team.htm), primarily in the areas of performance evaluation and the development of decision-support tools. For example, he has conducted both basic and applied research to develop and evaluate decision-support tools to aid pilot navigation through convective weather (Burgess & Thomas, 2004; Elgin & Thomas, 2004).

Norman W. Warner is a Senior Scientist at the Naval Air Systems Command, Patuxent River, MD. His Ph.D. is from the University of South Dakota in Human Factors Engineering. His expertise is in automated decision support, human decision making, and team collaboration. Over the past 28 years, he has conducted research with a variety of organizations (for example, Naval Air Systems Command, Office of Naval Research, Air Force and Army). Dr. Warner was awarded two patents in decision-support technology and published over 25 peer-reviewed articles. Currently, his research focus is on understanding team decision making and the cognitive mechanisms used during asynchronous, distributed team collaboration.

Robert L. Wears is an Emergency Physician, Professor in the Department of Emergency Medicine at the University of Florida, and Visiting Professor in the Clinical Safety Research Unit at Imperial College London. He serves on the board of directors of the Emergency Medicine Patient Safety Foundation, and on the editorial board for *Annals of Emergency Medicine*. He is also on the editorial board of *Human Factors and Ergonomics*, the *Journal of Patient Safety*, and the *International Journal of Risk and Safety in Medicine*. Dr. Wears has been an active writer and researcher with interests in technical work studies, joint cognitive systems, and particularly the impact of information technology on safety and resilient performance. His work has been funded by the Agency for Healthcare Research and Quality, the National Patient Safety Foundation, the Emergency Medicine Foundation, the Society for Academic Emergency Medicine, the Army Research Laboratory, and the Florida Agency for Health Care Administration.

David J. Weiss, the co-developer of the CWS index, received his bachelor's degree from the University of Pennsylvania and his doctorate from University of California, San Diego. He has spent his entire teaching career in the Psychology Department at California State University, Los Angeles. He has published extensively in the area of judgment and decision making, and also has developed several new statistical procedures. Much of his empirical work has employed functional measurement methodology, and his textbook *Analysis of Variance and Functional Measurement* (Oxford University Press, 2006) includes the only elementary exposition of that approach. The text also includes the CALSTAT suite of user-friendly computer programs he developed. Together with spouse Jie W. Weiss, he edited *A Science of Decision Making: The Legacy of Ward Edwards* (Oxford University Press, 2009).

Sterling L. Wiggins is a Principal Scientist at the Klein Associates Division of Applied Research Associates. He completed his M.A. in Education with a focus on Learning, Design, and Technology at Stanford University. Mr. Wiggins has 15 years experience in research and development at the convergence of humans, technology, and work in high-consequence environments. Mr. Wiggins is currently active in projects advancing human sensemaking through system design, and developing augmented reality applications for small unit leaders in the military.

David D. Woods is a Professor at Ohio State University in the Department of Integrated Systems Engineering. He received his Ph.D. in Psychology from Purdue University in 1979. From his initial work following the Three Mile Island accident in nuclear power, to studies of coordination breakdowns between people and automation in aviation accidents, to his role in today's national debates about patient safety, he has studied how human and team cognition contributes to success and failure in complex, high-risk systems. He is author of *Behind Human Error*, received the Jack A. Kraft Innovator Award from Human Factors and Ergonomic Society for advancing cognitive engineering and its application to safer systems, and received a Laurels Award from *Aviation Week and Space Technology* (1995) on the human factors of highly-automated cockpits. He has served on a National Academy of Engineering/Institute of Medicine Study Panel applying engineering to improve healthcare systems and on a National Research Council panel on research to define the future of the national air transportation system.

Yan Xiao is directing patient safety research at Baylor Health Care System, Dallas, TX. After finishing doctoral training in Human Factors from University of Toronto in 1994, Dr. Xiao joined University of Maryland School of Medicine to conduct research in communication, coordination, information technology, team performance, and clinical alarms. Most recently he was a Tenured Professor of Anesthesiology at University of Maryland School of Medicine in Baltimore, MD. Dr. Xiao is the author of more than 130 journal and proceeding articles and book chapters. His current research has been supported by National Science

Foundation, National Patient Safety Foundation, and Agency for Healthcare Research and Quality. His current research interests include team communication and technology-supported coordination.

Daniel J. Zelik is a Doctoral Candidate in the Cognitive Systems Engineering program at The Ohio State University. He has experience analyzing and modeling human performance as a co-op student at NASA Johnson Space Center. He currently investigates how professional intelligence analysts cope with data overload and infuse rigor into their analytical processes, as well as strategies for creating insight into computer-supported information analysis activity.

Preface
Macrocognition: Where Do We Stand?

Emily S. Patterson, Janet E. Miller, Emilie M. Roth, and David D. Woods

Real-world settings require effective teamwork to accomplish the objectives of an organization. Drastic improvements in performance are anticipated by many over the next decade via the implementation of technological innovations that facilitate teamwork. With many of the anticipated innovations, the concept of a team will be substantially broadened to include physically and temporally distributed members, ad hoc groups that form and then disband once an objective has been met, and anonymous interactions with others who have common or interacting goals without any sense of shared identity.

A daunting challenge is accurately predicting which innovations will truly improve team performance in a real-world setting, particularly for primarily cognitive (as opposed to physical, which are usually directly observable) activities. Most of the traditional human performance and cognitive process measures were developed for individuals working on relatively simple tasks without sophisticated computerized decision aids or communication networks. Without suitable measures and evaluation methodologies, how can we predict in advance whether the benefits of a proposed innovation justify the costs of implementation and maintenance? This book advances the state-of-the-art in team process and performance measurement, specifically for real-world teams in complex settings, to evaluate the usefulness of a proposed innovation and compiles recent exciting advances by prominent macrocognition researchers.

What is Macrocognition?

The short definition of macrocognition is 'the way we think in complex situations' (Klein et al., 2003; Chapter 4). The 'macro' in macrocognition emphasizes that cognition in real-world settings occurs across multiple individuals using tools and shared resources, as contrasted with 'micro'cognition done 'between the ears' of an isolated individual at a discrete point in time (Woods, 1987; Woods & Roth, 1986; Roth et al., 1986).

Complexity in modern work settings is likely to increase in the future due to a number of interrelated drivers. Technological and communication innovations are the primary motivation for explicitly considering the 'macro' level of cognition for study in its own right. For instance, recent technological innovations have provided new opportunities to extend cognitive capacity, including advanced sensors, automation, alerting algorithms, digital display formats ('visual analytics'), and massive digital data storage capacity. Advances in communications technologies are already transforming how cognitive work is conducted, including new communication formats such as Instant Messaging, Chat, and Twitter; dedicated portable handheld devices like Kindles and BlackBerries; additional features on existing devices such as cellular phones with embedded digital cameras, Global Positioning Systems (GPS), and email and Internet connectivity; shared information repositories that can be accessed and updated real-time simultaneously by multiple users like Wikipedia and blogs; and access to specialized expertise that is not within organizational or personal networks, such as Ask-An-Expert services. Due to the technological and communication innovations as well as other factors, the complexity of real-world environments is anticipated to increase, with more interconnections ('coupling'), a faster pace of change, more integration of diverse software packages and hardware, and higher expectations for productivity ('faster better cheaper').

Macrocognition is a relatively new term—in fact, new enough that some still question whether the term is needed—and the associated theoretical frameworks, concepts, terms, and definitions are a rapidly moving target. This chapter takes stock of the macrocognition research in this book and answers the question: *Where do we stand?* An important contribution of the creation of this book is clarifying the scope of macrocognition and identifying areas of consensus and disagreement among the community.

Historical Influences on Macrocognition Research

The term macrocognition can be viewed as a natural evolution and aggregation of several lines of research that have mutually influenced each other for the last few decades:

- Cognitive Systems Engineering (CSE) was introduced as a field in the 1980s with a focus on 'cognitive adaptations to complexity' (Rasmussen & Lind, 1981; Schraagen et al., in press) and 'coping with complexity' (Woods, 1988).
- Ethnographics observations of 'cognition in the wild' (Hutchins, 1995) revealed a reliance on specialized tools to 'externalize cognition' and make aspects of cognition public in order to aid coordination across individuals in well-defined roles.
- 'Situated action' research (Suchman, 1987) emphasized how events in the world ('event-driven' as opposed to 'self-paced' worlds) challenge the implementation of plans, thereby suggesting that plans are better viewed as 'resources for action' rather than hard constraints on activity.

- A substantial body of research on 'naturalistic decision making' and decision making under uncertainty has revealed a number of phenomena, including that decision making in complex settings differs substantially from decision making in traditional psychology laboratory-based studies with simpler tasks (Klein et al., 1993; Orasanu & Connelly, 1993), that expert practitioners take 'shortcuts' from recognizing a situation to implementing default procedural responses without evaluating alternative plans (Rasmussen et al., 1991), and that commitments to irrevocable courses of action are often strategically delayed to increase flexibility, particularly in overconstrained situations (Pearl, 1982).
- Research on anomaly response in high-consequence worlds such as NASA Johnson Space Center revealed that 'safing' activities were often performed in parallel with activities aimed at diagnosing problems and that low-probability, high-consequence explanations for data were often treated as the actual explanation in parallel with the highest probability explanation (Chapter 8 in Woods & Hollnagel, 2006).
- Research on human expertise has emphasized the importance of differences between novice and expert strategies in problem solving (Feltovich et al., 1997) and the importance of knowledge building to foster shared understandings ('shared mental models', 'team situation awareness') across team members (Letsky et al., 2008; Salas & Fiore, 2004).
- One way to frame shared resources in a distributed setting is as potential bottlenecks in cognitive processing, such as attention, memory, workload, goals, and coordination bottlenecks (Roesler & Woods, 2007).
- An important distinction between macrocognition and more classic cognitive models (for example, the observe-orient-decide-act (OODA) loop frequently used by the US military, which was developed by John Boyd) is that there is nearly always historical cognitive activity that shapes how cognitive work is framed. For example, problems have already been framed, hypotheses already exist as likely explanations for situations, plans are already in progress, and actions are already being executed. Therefore, the emphasis of the macrocognitive functions is on 'RE' doing the cognitive work as a result of triggers to reconsider, revise, reframe, or otherwise change the default activity space. For example, although there are occasions where new plans need to be generated that are different than plans in progress and do not match the typical procedures employed, this is not the dominating case for (re-)planning as a macrocognitive function (Klein et al., 2003).

In macrocognition, everything relates to everything, and simplified theoretical frameworks cannot reduce the complexity of the phenomena under study. Therefore, while focusing on a particular activity like executing a procedure can be important for gaining traction in conducting research, there will still be a need to consider interactions with other activities, such as monitoring the environment for changes, adapting a procedure to fit the dynamic situation, interpreting the guidance of distant supervisors as to the intent of an original procedure that cannot be executed in order to adapt it, assessing whether a plan in progress can be done within the anticipated window of opportunity, coordinating the implementation of multiple procedures, and timing activities so that they do not negatively interact with other activities (Woods & Hollnagel, 2006; Hollnagel & Woods, 2005).

In specifying the scope of macrocognition, there is a clear consensus that the core value of the community of researchers is that cognitive work is complex (note that this definition is inclusive of research conducted both in an actual, naturalistic work setting as well as a humans-in-the-loop laboratory setting where complex problems are simulated). Common elements for the research compiled in this book are:

- complex environment;
- embedded in a social context;
- artifacts require technical expertise;
- event-driven;
- dynamic demands;
- conflicting goals;
- organizational constraints;
- high consequences for failure.

There is similarly consensus among the researchers that macrocognition can be simultaneously considered at three interacting (macro) levels of analysis, even when there is usually an emphasis on the middle layer:

1. growth and decay of effective *expertise* and associated knowledge in individuals;
2. joint *activity* conducted in teams; and
3. factors that facilitate or inhibit organizational adaptability and *resilience*.

Finally, one of the main contributions of this book has been to advance the debate on what are the core functions in (macro)cognitive work. Although there is unlikely to be a final, comprehensive list mutually agreed upon by the community anytime soon, we believe that these functions and definitions represent an advance in clarity and consensus from pioneering work published in Klein et al. (2003); (see Chapter 4, Figure 4.1 for an updated figure summarizing macrocognitive functions and supporting processes):

- *Detecting:* Noticing that events may be taking an unexpected (positive or negative) direction that require explanation and may signal a need or opportunity to reframe how a situation is conceptualized (sensemaking) and/or revise ongoing plans (planning) in progress (executing). (Related terms: detecting problems, monitoring, observe, anomaly recognition, situation awareness, problem detection, reframing).

- *Sensemaking:* Collecting, corroborating, and assembling information and assessing how the information maps onto potential explanations; includes generating new potential hypotheses to consider and revisiting previously discarded hypotheses in the face of new evidence. (Related terms: orient, analysis, assessment, situation assessment, situation awareness, explanation assessment, hypothesis exploration, synthesis, conceptualization, reframing.)
- *Planning:* Adaptively responding to changes in objectives from supervisors and peers, obstacles, opportunities, events, or changes in predicted future trajectories; when ready-to-hand default plans are not applicable to the situation, this can include creating a new strategy for achieving one or more goals or desired end states. (Related terms: replanning, flexecution, action formulation, means-ends analysis, problem solving.)
- *Executing:* Converting a prespecified plan into actions within a window of opportunity (Related terms: adapting, implementation, action, act); this includes adapting procedures based on incomplete guidance to an evolving situation where multiple procedures need to be coordinated, procedures which have been started may not always be completed, or when steps in a procedure may occur out of sequence or interact with other actions.
- *Deciding:* A level of commitment to one or more options that may constrain the ability to reverse courses of action. Decision making is inherently a continuous process conducted under time pressure that involves re-examining embedded default decisions in ongoing plan trajectories for the predicted impact on meeting objectives, including whether to sacrifice decisions to which agents were previously committed based on considering trade-offs. This function may involve a single 'decision maker' or require consensus across distributed actors with different stances toward decisions. (Related terms: decision making, decide, choice, critical thinking, committing to a decision.)
- *Coordinating:* Managing interdependencies across multiple individuals acting in roles that have common, overlapping, or interacting goals. (Related terms: collaboration, leadership, resource allocation, tracking interdependencies, communication, negotiation, teamwork.)

In summary, our view of the scope of macrocognition is:

- joint activity distributed over time and space;
- coordinated to meet complex, dynamic demands;
- in an uncertain, event-driven environment with conflicting goals and high consequences for failure;
- made possible by effective expertise in roles;
- shaped by organizational (blunt end) constraints;
- that produces emergent phenomena.

How This Book is Organized

The book is divided into three sections.

Part 1, Theoretical Foundations lays the groundwork for how measurement theory in general is relevant to developing macrocognitive measures. In Chapter 1, Robert R. Hoffman provides an overview of terminology from measurement theory literatures, including providing an example of moving from a theoretical concept to an operational definition of measurement. He also argues that there is a fundamental difference between the terms 'measures' and 'metrics' in that 'metrics express policy that is shaped by goals and value judgments—*things that are external to the measurement process itself*' (Chapter 1, italics in original). In Chapter 2, Robert R. Hoffman describes challenges for macrocognitive measurement. He provides a review of lessons learned from prior attempts at measurement, including Hits, Errors, Accuracy, Time (HEAT) measures, Human Reliability Analysis (HRA), Cognitive Reliability and Error Analysis Method (CREAM), and usability and learnability questionnaires. He provides a list of desirable characteristics for macrocognitive measures, and encourages us to do a broad exploration of possible measurement approaches: 'Measurement for complex cognitive systems, and for macrocognition, will certainly involve totally new approaches, new kinds of scales, and new analytical techniques.' (Chapter 2). In Chapter 3, Eduardo Salas and colleagues provide a number of insights based on their wealth of experiences measuring teamwork processes. They start by providing guidelines for developing any measure, including: '(1) what to measure, (2) what tools to use (that is, how to measure it), (3) when and from whom to collect the information, (4) how to get useful responses (for example, maximize validity, while limiting obtrusiveness), and (5) how to index the responses' (Chapter 3). In developing macrocognition measures in particular, their insights include clear distinctions between macrocognition levels, unit of analysis, processes, emergent states, dimensions, moderators, and outcomes.

Part 2, Macrocognition Measures for Real-World Teams details recent advances in macrocognition measures (see Table P.1 for an overview of the nine chapters that provide macrocognition measures). We believe that all of the measures detailed in these chapters could be immediately applied in evaluations of proposed technological innovations, or at least provide a significant jumpstart in developing measures appropriate for a particular evaluation.

Part 3, Scenario-Based Evaluation Approaches describes advances in methodologies well suited for enhancing team performance at a macro-level of analysis in complex, dynamic settings. In Chapter 13, Emilie M. Roth and Robert G. Eggleston propose a work-centered evaluation to address mismatches between traditional evaluation paradigms, including usability testing and controlled laboratory experimentation, and evaluation goals and pragmatic constraints for studies done with a macrocognitive focus. In Chapter 14, Emily S. Patterson, Emilie M. Roth, and David D. Woods provide guidance on how to design evaluation scenarios that are sufficiently complex to represent the problem

Table P.1   Overview of macrocognition functions and measures

Ch.	Title	Macrocognitive Function(s)	Measure(s)
4	Macrocognitive Measures for Evaluating Cognitive Work	All	Detecting (3), Sensemaking (4), Planning (3), Executing (4), Deciding (3), Coordinating (3)
5	Measuring Attributes of Rigor in Information Analysis	Sensemaking	Rigor
6	Assessing Expertise When Performance Exceeds Perfection	Detecting	CWS ratio
7	Demand Calibration in Multitask Environments: Interactions of Micro and Macrocognition	Detecting	Distraction
8	Assessment of Intent in Macrocognitive Systems	Executing	Intent
9	Survey of Healthcare Teamwork Rating Tools: Reliability, Validity, Ease of Use, and Diagnostic Efficacy	Coordinating	Teamwork: 10 instruments
10	Measurement Approaches for Transfers of Responsibility During Handoffs	Coordinating	Handoff quality: Outcomes, Content of interactions, Interaction processes, Learning
11	The Pragmatics of Communication-based Methods for Measuring Macrocognition	Coordinating	Communication: Flow/Timing (4), Content (4)
12	From Data, to Information, to Knowledge: Measuring Knowledge Building in the Context of Collaborative Cognition	Sensemaking	Knowledge Building: Content, Process

space likely to be encountered in real-world operations. A substantial set of 'complicating factors' organized by macrocognitive function are provided along with examples from multiple domains that can be used to efficiently 'scale up' the complexity of existing scenarios. In Chapter 15, Scott Potter and Robert Rousseau describe and provide an example of applying a Decision-Centered Testing (DCT) methodology that assesses the resilience of a proposed innovation. A key element of their methodology is explicitly designing the study to 'stress the edges' based on a deep understanding of the cognitive demands of the work environment. In Chapter 16, John M. Flach and colleagues describe how using synthetic task environments from a macrocognitive perspective alter conceptualizations of the system, situation, constraints, awareness, performance, work activities, movement through a state space, and fidelity as compared to microcognitive research. In Chapter 17, Sterling Wiggins and Donald Cox present detailed, practical guidance on how to perform a heuristic evaluation of a proposed innovation using Cognitive Performance Indicators (CPI). This chapter is particularly interesting in that the authors describe how an evaluation can be used to steer design directions early in the design process to more promising system conceptualizations, as opposed to performing a backend evaluation of a completed system prior to implementation.

Conclusion

> There are powerful regularities to be described at a level of analysis that transcends the details of the specific domain, but the regularities are not about the domain specific details, they are about the nature of human cognition in human activity (Hutchins, 1992).

Although the research described in these chapters was conducted in disparate domains, ranging from healthcare to driving to military settings, the emerging consensus on core macrocognitive functions and levels enables a common language and level of abstraction that facilitates sharing and lays the foundation for future work. This book captures and disseminates exciting new developments in macrocognition research that significantly advance our ability to measure 'messy complexity', and thus improve our ability to predict which innovations will truly improve the performance of real-world teams.

References

Feltovich, P. J., Ford, K. M., & Hoffman, R. R. (1997). *Expertise in context: Human and machine.* Cambridge, MA: MIT Press.

Hollnagel, E. & Woods, D. D. (2005). *Joint cognitive systems: Foundations of cognitive systems engineering.* Boca Raton. FL: Taylor & Francis.

Hutchins, E. (1992). Personal Communication with David Woods.

Hutchins, E. (1995). *Cognition in the wild.* Cambridge, MA: MIT Press.

Klein G., Ross K. G., Moon B. M., Klein D. E., Hoffman, R. R., & Hollnagel E. (2003). Macrocognition, *IEEE Intelligent Systems, 18*(3) 81-85.

Klein, G. A., Orasanu, J., Calderwood, R., & Zsambok, C. E. (Eds.) (1993). *Decision making in action: Models and methods.* New York, NY: Academic Press.

Letsky M. P., Warner N. W., Fiore S. M., Smith C. A. P. (Eds.) (2008). *Macrocognition in teams: Theories and methodologies*. Aldershot: Ashgate Publishing.

Orasanu, J., and Connolly, T. (1993). The reinvention of decision making. In G. Klein, J. Orasanu, R. Calderwood, & C. Zsambok (Eds.), *Decision making in action: Models and methods* (pp. 3-20). Norwood, NJ: Ablex.

Rasmussen, J., Brehmer, B., & Leplat, J. (Eds.) (1991). *Distributed decision making: Cognitive models for cooperative work.* New York, NY: Wiley.

Rasmussen, J., & Lind M. (1981). Coping with complexity. In H. G. Stassen (Ed.), *First European annual conference on human decision making and manual control*. New York, NY: Plenum.

Roesler, A. & Woods, D.D. (2007). Designing for Expertise. In: R. Schifferstein, R. & P. Hekkert, *Product experience – A multidisciplinary approach*; Oxford, UK: Elsevier.

Roth, E. M., Woods, D. D., & Gallagher, J. M. Jr. (1986). Analysis of expertise in a dynamic control task. In *Proceedings of the Human Factors Society. 30th Annual Meeting*, (pp. 179-181).

Salas E., & Fiore S. M. (2004). Team cognition: Understanding the factors that drive process and performance. Washington, DC: American Psychological Association.

Schraagen, J. M., Klein, G., & Hoffman, R. (in press). The macrocognitive framework of naturalistic decision making. In J. M. Schraagen, L. Militello, T. Ormerod & R. Lipshitz (Eds.), *Naturalistic decision making and macrocognition*. Aldershot, UK: Ashgate Publishing.

Pearl, J. (1982) GODDESS: A Goal Directed Decision Supporting Structuring System, *IEEE Trans. Pattern Analysis and Machine Intelligence, 4*(3); 250.

Suchman, L. (1987). *Plans and situated actions: The problem of human-machine communication*. Cambridge, UK: Cambridge University Press.

Woods, D. D. (1987). Commentary: cognitive engineering in complex and dynamics worlds. *International Journal of Man-Machine Studies, 27*(5-6), 571-585.

Woods, D. D. (1988). *Coping with complexity: the psychology of human behaviour in complex systems. Tasks, errors, and mental models: a festschrift to celebrate the 60th birthday of Professor Jens Rasmussen.* Edited by L.P. Goodstein, S.E. Olsen, & H.B. Andersen. London; New York: Taylor & Francis.

Woods, D. D. & Hollnagel, E. (2006). *Joint cognitive systems: Patterns in cognitive systems engineering.* New York. NY: Taylor & Francis.

Woods, D. D. & Roth, E. M. (1986). *The role of cognitive modeling in nuclear power plant personnel activities.* Washington DC: U. S. Nuclear Regulatory Commission, (NUREG-CR-4532).

PART 1
Theoretical Foundations

Chapter 1

Theory $\nrightarrow$ Concepts $\nrightarrow$ Measures but Policies $\nrightarrow$ Metrics

Robert R. Hoffman

What may be considered to be a dream, fantasy, or ideal... is that of achieving complete automaticity of inference (Bakan, 1966, p. 430).

Introduction

The word 'measure' comes into English via Old French from words in Sanskrit, ancient Greek, and Latin meaning to apportion, as in 'to dole out portions of some commodity.' Derivatives include such words as 'moon' and 'month.' In fact, there is a plethora of derivative words denoting or relating to the moon, attributed to the fact that the phases of the moon were the first means of measuring time periods longer than a day. The word 'measure' also traces back to an Indo-European word meaning to acquire 'exact knowledge,' and from that derived words including 'cunning' (Partridge, 1958).

There is of course no such thing as *the* theory of measurement (see Mitchell, 1986, 1999). There are numerous treatments of measurement and measurement issues in all the sciences, and in the philosophy of science. These range from the basic to the esoteric, from simple to mathematically brain-numbing (cf. Krantz et al., 1971). Much of the literature in philosophy of science reflects that discipline's historical obsession with the physical sciences, as in Karl Popper's discourse (1959) on the question of how precise a set of measurements has to be in order for measurements alone to refute a theory. Topics of interest include the problems related to uncertainty, the conundrums of measurement in quantum and relativistic physics, and the meaning of psychophysical measurements as attributes of discriminatory responses (for example, Bergmann & Spence, 1944).

Measurement procedures can become substitutes for thought instead of aids to thought (Bakan, 1966, p. 436). In recent years, the procurement of advanced information processing systems has been accompanied by calls for 'metrics.' But this call is shouted in a partial vacuum in which metrics seem to be understood apart from any foundations of measurement theory. Only when such notions are clear can one begin to consider ways in which measurement postulates might be formed regarding macrocognitive phenomena and the evaluation of cognitive work.

Some Terminology

Outside the ticket booth to a rollercoaster is a painted sign showing the figure of a small child in outline. Above the figure it reads:

'You must be this high (4 feet 8 inches) to ride the *Rollercoaster of Doom!*'

- Height is a *conceptual measurable*, a frame of reference for a certain kind of length. (Its conceptual nature is revealed by pointing out that we would not refer to the width of a person who is lying down when we mean their height.)
- While we could measure the concept of height using category assignments (for example, tall, short, and so on), we seek a *numerical scale.*
- Numbers representing feet and inches are adopted as the numerical scale, rather than, say meters or furlongs.
- The numerical scale, itself a framework for positioning a set of numbers of a particular kind, is to be interpreted as *measurement scale,* specifically a scale of height.
- 4'8" is a particular *measurement* on that measurement scale, which might be used, as in the present example, to divide individuals into classes: those who are shorter in height than the given height, those who are taller in height, and those who are exactly 4'8" (within a certain tolerance).
- Height is measured by having the customer stand up against the outline. If the top of the customer's head is higher than the topmost part of the outline, their height is greater than 4'8" and the customer gets to ride the ride. This is the *operational definition*, a specification of a replicable, dependable procedure for making individual measurements.
- 4'8" is a measurement that is then used as a *metric*, that is, a specific value that serves as a decision gate or threshold.

'Why 4 feet 8 inches, Mommy?' moans the disappointed 4'6" child to his mother.

One must have a theory of the (complex) relation of height to such things as age and safety. This is then combined with some policy that drives the threshold to 4'8", as opposed to any other value.

This example shows that metrics do not arise either immediately or automatically from measures. Metrics express policy that is shaped by goals and value judgments—*things that are external to the measurement process itself*—in this case, such things as not wanting to kill the paying customer, not wanting to get sued, and so forth. Metrics are thresholds or decision criteria that are used in an evaluation. This requires, or presupposes, value judgments based on a policy. 'Systolic blood pressures over 130 mm Hg are not normal for the average person weighing . . .' It can take a good while to establish such metrics in addition to the time and effort needed to determine and routinize the operations used to make measurements, often because the systems that are being measured are complex.

Theory → Concepts → Measures

Theories consist of distinguishable sets of assertions or postulates. Some of the assertions in a theory are laws. The Laws refer to the subject matter—the phenomena the theory is about and is supposed to explain; the 'things' that are identified, classified, and counted. Definitions of concepts in the subject matter form what is called the Ontology. Yet another set specifies the Methodology. A scientific or empiricist methodology would assert, for instance, that knowledge that is pertinent to the theory must come from experience and observation. The postulates of a methodology specify what counts as acceptable method, including means for making measurements (see Quine, 1951).

The parts of a theory (ontology, metatheory, laws, methodology) are interconnected by entailment relations. For instance, an entity in the ontology must appear in more than one law in order for the theory to 'hang together.' But, at least to some extent, specifications of measures and methods of measurement can be pulled out from a theory and recycled, as it were. You can alter or even kill a methodology without killing the theory of which it is a part. Conversely, you can kill a theory and yet retain its methodology. For instance, in a study of a complex sociotechnical system, one might use the method of Hierarchical Task Analysis, which originated in the 1960s (Annett & Duncan, 1967), or the method of Goal-Directed Task Analysis, which originated in the 1980s (for example, Endsley, 1988), The two methods are quite similar, but each would come wrapped in a different theory.

Yet, methodologies carry theory along with them at least some bits of a theory. Note, for instance, that to use hierarchical task analysis in a modern study of sociotechnical systems, one is presupposing some notion of 'task,' which is a concept from an ontology. *Measures and measurements are always theory-laden.*

The definition of measurement offered by the operationalist philosophy (see Bergmann & Spence, 1944) is a strictly 'behavioral' one that attempted to reduce all measurement to metrology: the measure defines the concept. This would make physics the science of meter reading. In psychology, the naïve operationalist view resulted in the old saw that intelligence is what an intelligence test measures. (There is always some danger of confusing the measure with the thing being measured.) Fortunately, psychology proper followed the lead of physics and quickly left this operationalist notion behind (Krantz, 1972, p. 1435). Science is, after all, about understanding the invisible. There is more to the meaning of theoretical concepts than what is given by their measures. There *has* to be. Take temperature, for example. A common thermometer can only measure temperatures up to the boiling point of mercury. How do we know that the temperature of a star, measured using a spectrometer, is the same thing as the temperature measured using a mercury thermometer? In order to bridge the gap, one must rely on theory (in this case, thermodynamics) and one must make certain theoretical assumptions (in this case, assumptions involving Planck's constant).

This dependence of measurement on theory carries over to the study of cognitive work. Both errorful and correct performance (whether of an individual of or a work system) are defined relative to the goals of the work activity, so even for a 'behavioral' measure, the activities that constitute the work will be carved up according to theoretical categories that are believed to define the work and its goals. This inevitable and tight linkage of measures to theory is one of the themes of this chapter: *You cannot measure something without having some theory of it. Measurement is not the same thing as simply performing operations that generate numbers, or marking boxes in a checklist.*

Policies → Metrics

> The goal of measurement is to find the numbers that fit the world. The purpose of metrics is to shape the world to match desired numbers (Anonymous, 2008).

Though the immediate goal of measurement is to classify measurables and predicate names and numbers to the classes, measurement always has broader goals and purposes. While measurement is tightly linked to theory, metrics are tightly linked to policy and valuation. This shows in discussions of human factors standards and guidelines (for example, Karowski, 2006). Actually, what shows is the gap between guidelines and measures, or even to conceptual measurables. For human-computer interaction one finds usability guidelines such as 'use plain language to express errors' and 'use consistent sequences of actions and terminology whenever appropriate' (Stewart, 2006). There are concepts here, but where are the conceptual measurables? Of all the dozens and dozens of listed international and governmental standards, one has to work to find discussions of actual measures. One might argue that the listing of measures is not the goal for listings of standards, but this falls flat given the prominent discussions of performance measurement that accompany discussions of standards and guidelines. What starts out promising to be a meaningful analysis gets reduced to subjective evaluations using checklists (for example, 'Does the dialog sequence match the logic of the task? Yes or no?' and 'Can the user choose how to restart an interrupted dialog? Yes or no?'). Clearly, standards and guidelines represent value judgments that stem from policy, driven by and justifying some sort of investment.

Assuming that one has successfully gone from a conceptual measurable to one or more reasonable operational definitions, and has successfully gone from operational definitions to specific measurement procedures, and has then linked the measurements to one or more meaningful measurement scales, one cannot then assume some magic leap to a 'metric' without having some sort of policy or goal. Without some policy to specify what is desired (or good), how is one to determine what a decision threshold should be? In one context, '85 percent correct' might represent a useful metric. In another context, it might be totally misleading and genuinely dangerous. In one context, '35 percent better than before' might appear to be a significant gain whereas in some other context it might not. One might hope to compare the performance of a work system to some historical baseline, on the view that any improvement will serve as a metric for making value judgments. But even that is often difficult for practical reasons, such as the difficulty of finding or getting access to meaningful historical data.

So if someone says '*I need a metric for x,*' the answer must be a question: *'OK, what's the policy?'* If the answer to that rejoinder does not specify the policy, and has no reference to conceptual measurables, but instead refers nebulously to things such as 'performance improvement' or 'efficiency' or 'human-system integration,' then the researcher is completely free to come up with conceptual measurables. A consequence of the abrogation of responsibility—leaving the policy tacit—is the sponsor might end up hearing things they do not want to hear when measures and metrics are delivered.

Conclusion

Key points of this chapter are presented in the Concept Map in Figure 1.1.

Challenges to macrocognitive measurement include 'moving target' complications. Cognitive work in sociotechnical systems is a moving target (Ballas, 2007; Dekker, et al., 2003), because of changes in the world and changes in the technology. This requires resilience and adaptability in the work (Hollnagel & Woods, 1983). This entails a need for adaptatiliby in measurement. One complication is that the time frame for technology insertions sometimes does not allow researchers to apprehend much change in process (Scholtz, 2005). Another complication is that of the 'Fundamental Disconnect,' which is the mismatch of the rate of change in sociotechnical systems versus the time frame for effective controlled experimentation (Ballas, 2007; Dekker, et al., 2003). These challenges call into question something that seems to be implied by how people sometimes use the word 'metric.'

I can explain by using the analogy to the game of darts. The game has rules (rules for what is allowed in the play of the game, and rules for keeping score) and a measurement scale (the rings and sectors on the dart board). These specify the measurements to be made (for example, the 'bull's eye' constitutes a high-scoring throw). But the game of darts is fixed; cognitive work is not. Metrics for cognitive work systems must be open to refinement and adaptation, even if the driving policies are stable or invariant. Yet, as we look at the research and

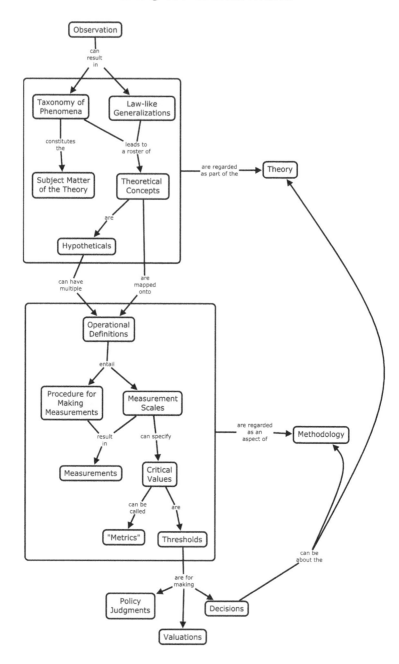

Figure 1.1 This concept map explains the derivation going from observations to classifications, to conceptual measurables, to measures, to measurements, and finally to metrics

procurement communities, we apprehend a tacit justificationist view that sought-for metrics are bull's eyes.

Acknowledgment

The author's contribution was through participation in the Advanced Decision Architectures Collaborative Technology Alliance, sponsored by the US Army Research Laboratory under Cooperative Agreement DAAD19-01-2-0009.

References

Annett, J., & Duncan, K. D. (1967). Task analysis and training design. *Occupational Psychology, 41,* 211-222.

Bakan, D. (1966). The test of significance in psychological research. *Psychological Bulletin, 66,* 423-437.

Ballas, J. A. (2007). Human centered computing for tactical weather forecasting: An example of the 'Moving Target Rule.' In R. R. Hoffman (Ed.), *Expertise out of context: Proceedings of the Sixth International Conference on Naturalistic Decision Making* (pp. 317-326). Mahwah, NJ: Erlbaum.

Bergmann, G., & Spence, K. W. (1944). The logic of psychophysical measurement. *Psychological Review, 51,* 1-24.

Dekker, S. W. A., Nyce, J. M., & Hoffman, R. R. (2003). From contextual inquiry to designable futures: What do we need to get there? *IEEE Intelligent Systems,* March-April, 74-77.

Endsley, M. R. (1988). Design and evaluation for situation awareness enhancement. In *Proceedings of the 32nd Annual Meeting of the Human factors and Ergonomics Society* (pp. 97-101). Santa Monica, CA: Human Factors and Ergonomics Society.

Hollnagel, E. & Woods, D. D. (1983). Cognitive systems engineering: New wine in new bottles. *International Journal of Man-Machine Studies, 18*(6), 583-600.

Karwowski, V. (Ed.) (2006). *Handbook of standards and guidelines in ergonomics and human factors.* Mahwah, NJ: Erlbaum.

Krantz, D. H. (1972). Measurement structures and psychological laws. *Science, 175,* 1427-1435.

Krantz, D. H., Luce, R. D., Suppes, P., & Tversky, A. (1971). *Foundations of measurement, Volume 1: Additive and polynomial representations.* New York, NY: Dover.

Mitchell, J. (1986). Measurement scales and statistics: A clash of paradigms. *Psychological Bulletin, 100,* 398-407.

Mitchell, J. (199). *Measurement in psychology: A critical history of a methodologcical concept.* Cambridge, UK: Cambridge University Press.

Partridge, E. (1958). *Origins: A short etymological dictionary of modern English.* New York, NY: Macmillan.

Popper, K. R. (1968/orig. 1959). Degrees of testability. Chapter 7 in *The logic of scientific discovery* (pp. 112-135). New York, NY: Harper Torchbooks.

Quine, W. V. O. (1951). Two dogmas of empiricism. *The Philosophical Review 60*, 20-43.

Scholtz, J. (2005). Metrics for the evaluation of technology to support intelligence. In *Proceedings of the Human Factors and Ergonomics Society 49th Annual Meeting* (pp. 918-921). Santa Monica, CA: Human Factors and Ergonomics Society.

Stewart-Brown S. (2006). *What is the evidence on school health promotion in improving health or preventing disease and, specifically, what is the effectiveness of the health promoting schools approach?* Copenhagen: WHO.

Chapter 2

Some Challenges for Macrocognitive Measurement

Robert R. Hoffman

> We tend to think that certain procedures are required by the subject matter when in fact the situation is only that those procedures are the ones most readily available to us. It is for this reason that the development of new scales and techniques of measurement can be so valuable... the range of possibilities is basically limited only by our imagination and ingenuity (Kaplan, 1964, pp. 190, 197).

Introduction

Measures enable all of the activities of cognitive systems engineering. They support the process of proficiency scaling and the study of apprentice-journeyman-expert development. Measures make possible the study of the information requirements of decision makers. Measures make reliability evaluation possible when multiple judges analyze protocols. Measures set criteria for the creation of scenarios (for example, representativeness, degree of difficulty, and so on). Measurements of one sort or another culminate nearly all procedures of Cognitive Task Analysis (Crandall et al., 2006, Chapter 14). Measures will enable the study of macrocognition.

In the context of complex cognitive systems, significant issues and challenges arise with regard to measurement. Many of these were apprehended decades ago by scientists including Alphonse Chapanis (see Chapanis, 1999), Erik Hollnagel (for example, Hollnagel & Woods, 1983) Jens Rasmussen (for example, 1986) Thomas Sheridan (1974), Harold Van Cott (for example, Van Cott & Chapanis, 1972; Van Cott & Warrick, 1972), David D. Woods (1988), and others (see for example Sinaiko, 1951). This includes discussion of issues of fidelity in simulations, field studies, or 'real-world' observations, and so forth. As was becoming clear over 50 years ago, some of the challenges that emerge at the system level call into question the extensibility of comfortable notions of measurement when what one really needs to study are macrocognitive phenomena and what one really needs to evaluate is the goodness of cognitive work in complex sociotechnical systems.

Hits, Errors, Accuracy, and Time (HEAT) Measurement

Discussions of measurement in the literature of human factors present approaches to evaluating validity and reliability, and depend uncritically on distinctions such as that between subjective and objective measurement. Standard treatments (O'Neill, 2007; Stanton et al., 2005; Wilson & Corlett, 2005) focus on measurement of the performance of individuals (for example, response correctness, errors, reaction time, frequencies of behaviors or behavior types, and so on). Gawron (2000) is an excellent compendium, with examples of many human factors measures and measurement scales of performance. The introduction of measures of mental workload (Hart & Staveland, 1988; Moray, 1979) and measures of situation awareness (Endsley & Garland, 2000) represent advances in human factors measurement.

Most measures of human performance are measures of HEAT: Hits, Errors, Accuracy, and Time. These enter into calculations of efficiency, effort, or other concatenations (for example, relations of speed and accuracy).

Take for example the NAVSEA 2003 Human Performance Technical Metrics Guide. As we summarize it in Table 2.1, it presents HEAT measures. Measures such as heart rate and ratings of perceived exertion can be extremely important when one considers the full range of human factors (for example, response time while wearing protective clothing).

The NAVSEA Guide, like other presentations, points to some measures that seem related to cognitive work. In many rosters of measures, the appearance of system-level considerations is just a cameo. To illustrate, we can synopsize Table 7.1 in Meister (2004). It gives the impression that human factors measurement *is* human performance measurement. The categories include: Reaction time measures (for example, time to complete an activity); Accuracy and correctness measures (of detection, identification, and so on); Error measures (frequency, type, and so on); Response characteristics (for example, magnitude, reliability); and Effort measures (for example, resources consumed).

Characteristic of HEAT measurement, Meister relegates anything that is 'subjective' to a category called 'Behavior Categorization by Self and Others.' This includes measures such as 'estimation of the degree of behavior displayed' and 'analysis of operator/crew behavior characteristics' (p. 172). He also includes interviews and self reports. These measures are listed last, and barely discussed. In Meister's categories one finds only a hint that the humans who are performing are thinking and feeling, and are part of a larger cognitive work system: Measurements such as frequency of communications, diagnostic checks, and requests for information. Alas, such things are described antiseptically (for example, the 'frequency of observing responses,' p. 172).

Macrocognition needs a measurement approach that goes beyond the 'John Henry versus the steam hammer' mentality. Measures of the individual worker's raw performance (speed, efficiency, accuracy, output) have important uses, but are

Table 2.1   Measures listed in the NAVSEA 2003 Human Performance Technical Metrics Guide

Work Accomplished Per Unit Time (productivity, efficiency)
Probability of success Percent of goals achieved by the user Control actions per unit of time Control actions per operator task Time to complete while wearing protective equipment Error recovery time Percent of time spent correcting errors Time to respond Time to detect Time to recognize Time to diagnose Time to respond while wearing protective equipment
Task Accuracy
Sustained performance accuracy over time Error rates Probability of success Percent of goals achieved by the user Percent of errors corrected successfully by the user Task accuracy while wearing protective equipment
Decision Accuracy
Detection accuracy Recognition accuracy Diagnostic accuracy
Cognitive Workload
NASA Task Load Index (TLX) Subjective Workload Assessment Technique (SWAT) Modified Cooper-Harper Scale Performance on secondary tasks
Situational Awareness
Objective Measures of Situational Awareness Situational Awareness Global Assessment Technique (SAGAT) Subjective Measures of Situational Awareness
Physical Workload
Heart rate Rating of Perceived Exertion (RPE) Revised NIOSH lifting equation

not adequate to the understanding of complex cognitive systems, which is where the payoff for investment really comes.

The HEAT Runs Cold

'... it is a pointless, though widespread practice to use a physical measure or a count as a 'definition' of a psychological variable; this practice obscures the fact that all one has done is measured a physical variable, or counted' (Krantz, 1972, p. 1435).

One of the most common HEAT measures is 'number of sub-goals (tasks or activities of a certain kind) accomplished per unit time.' Suppose that using a new software system and its associated work method, a team of two people (navigator and payload manager) is able to simultaneously execute the missions of two Unmanned Aerial Vehicles (UAVs), whereas using the current work method (and its technology), each three-person UAV team can execute only one mission at a time. Chalk one up to efficiency. A procurement contract is just a signature away.

Suppose further that at the end of the day the workers confess that they feel particularly drained. The new work method leaves them feeling helpless when the mysterious automation limits their opportunity to exercise their expertise. In their legacy work, they had felt they were largely in control of the missions. One can see here why we refer to John Henry versus the steam hammer.

It is often the things that are easy to measure that are the things that get measured. HEAT measures pass the scientific purity test, often without much comment. Sometimes all it takes is for the researcher to assert that the measures used were 'objective' or 'behavioral.' Often it can be relatively easy to measure performance in terms of the primary task goals, for example by counting the number of documents analyzed, the number of targets identified, and so forth. It can be easy to evaluate software usability by measuring response time on well-defined and regularly-occurring action sequences, or by collecting practitioner self-ratings of mental workload. All these are relatively easy to measure or count, and quench the human thirst for holy numerosity.

As Scholtz (2005) argued, what is critical and more difficult is to develop methods for measuring things at a meaningful, system level. We need to measure such things as the goodness of the technology, the learnability of work methods, and the resilience of the work system. Argument on this point should not be necessary. But merely achieving this understanding does not make measurement easy.

In his presentation on human factors measurement, Meister (2004) relied heavily on the words 'behavior,' 'objectivity' (and predictably, studies involving 'subjects'), echoing the language of human performance measurement and its behaviorist legacy. Meister then, also predictably, has to stretch the language to system-level analysis. This is impossible, the puddinged proof being the Common

Waffling Error, which is to use the word 'system' sometimes to mean the work system (humans-machines) and sometimes to refer to just the technology (the automation). Then the logic falls completely apart:

> . . . one way of tracing the effects of a behavioral action like a command decision on the system is to measure its effects in relation to the accomplishment of the system goal, which is the only fixed element in the process.. It is not easy to decompose the system-level measure . . . into its constituent elements (p. 194).

By definition, the goals in complex cognitive work are never fixed. Work cannot be adequately understood in terms of simplistic causal chain decompositions. Only if one decides, *as a matter of policy*, to hold the human hostage to the goals of the machine can one even attempt a reduction of system-level measurement to human performance measurement. And even then, it is a pretense that the analysis will be sufficient.

A clear case of an effort to move beyond HEAT measurement is the Cognitive Reliability and Error Analysis method (CREAM). In his critique of Human Reliability Analysis (HRA) methods, Hollnagel (1998) argued that HRA relies on simple models of human performance, relies on sequential/hierarchical decomposition that is blind to system-level features and effects, relies on simple distinctions (that is, success versus failure) that also mask system-level features, and is, in general, insensitive to context (Serwy & Rantanan, 2007). Using multiple tables, including one representing the context ('common performance conditions'), CREAM allows one to chart multiple paths describing errors and how and why they occur, both retrospectively and predictively.

It comes as little surprise that methods such as CREAM seem tedious when applied, since the goal is to model complexity. However, advances are being made in creating software support for the conduct of such methods (Serwy & Rantanan, 2007). Such CREAM tabular representations show why it is that system-level analysis is necessary. Hollnagel (1998) used the example of a collision in a New York subway. The braking system failed to stop the train, and this could be traced to an earlier failure to properly inspect the brakes. In addition, the operator missed seeing a red light. However, in this portion of track the light signals were spaced too closely; the devices were placed back in 1918. On top of all this, the light was itself not working reliably and was located in a region of track that was brightly illuminated, making it difficult to discern even when working reliably. A traditional 'blame game' would attribute this accident to human error. But clearly, the error is systemic.

We've Known it All Along

The importance of systems-level analysis has been known for many years, for instance by engineers and others who have attempted to model complex and dynamic systems (Bar-Yam, 2003; Jagacinski & Flach, 2003; Senge, 1990), and industrial and human factors psychologists whom we mentioned earlier, who witnessed the emergence of complex cognitive systems. Here is a passage from the 1951 paper that introduced the notion of the 'sociotechnical system,' a report on a study of changes wrought in the coal mining industry by the introduction of powered machines:

> The mining engineer might write a simple equation: 200 tons of coal equals 40 men over 200 yards over 24 hours. But there are no solutions of equivalent simplicity for the social and psychological difficulties raised. For psychological and social difficulties of a new order appear when the scale of a task transcends the limits of simple spatio-temporal structure (Trist & Bamforth, 1951, p. 14).

Sociotechnical systems considerations have recently emerged as important in technology procurement and acquisition (Goguen, 1994; Neville et al., 2007):

Reference has been made to the disappointing levels of performance of many new systems, in particular to the common failures of investments in new technology and new working practices. Organizations so rarely undertake systematic evaluations of their investments against their original goals. The original estimates of performance may have been over-optimistic. The estimates may have been political statements to persuade senior managers to release capital for investment. There may have been little expectation that the goals would actually be met. Project champions and managers may not want to have subsequent evaluation since they do not wish to have any failures exposed. Key people may also have moved on. There are many good human, political, and organizational reasons why evaluation appears not worth the effort. A sociotechnical perspective explicitly assumes a commitment to evaluating the performance of new systems against the goals of the organization and the people in it, and includes the explicit inclusion of social, technical, operational, and financial criteria (Clegg, 2000, p. 473).

This broad perspective emerged at different times in a number of communities of practice (see Hoffman & Militello, 2008), but the communities are now converging, as exemplified by discussions at meetings of the Human Factors and Ergonomics Society on such topics as how to integrate human-systems analysis into systems engineering (Deal, 2007) and determining the costs of integrating (or costs of not integrating) cognitive systems engineering in to systems engineering (Neville et al., 2007; Zachary et al., 2007).

HEAT is inadequate for evaluating for the goodness of cognitive work. Therefore it is inadequate for the study of macrocognition.

The Shape of Measures to Come

> Doctrine has barred certain scientists from joining the mainstream by artificially erected walls, by conceptual injunctions against admixtures from sources

suspected as contaminated because they failed to pass the orthodox test of purity, namely, that one plus one must be made to equal two (Weiss, 1967, p. 801).

Measures always have to be interpreted. Measure X gets interpreted as a measurement scale of Y. In some cases (for example, using a ruler to measure length), this process is so familiar it is tacit. Kaplan (1964) makes this explicit in the example of measuring group morale by measuring group performance effectiveness, on the supposition that groups with higher morale will perform more effectively. Measurement scales are common in psychology, and in cognitive systems engineering. An example is the Winner et al. (2007) measure of the extent to which individuals understand the commander's intent: the actual measure was rankings of alternative courses of action.

The value of a measurable on some dimension is almost always understood as the result of other factors or variables, which can themselves sometimes be measured. In these ways, measurement scales are derived from measures. For example, the rate of increase in performance (on primary task goals) might be concatenated with a measure of the understandability of the interface and the result might be interpreted as a single measurement scale that reflects the goodness of an interface. The study of macrocognitive phenomena will hinge on multiple measures and compound measures that are interpreted so as to support convergence on statements about the cognitive work (for example, that an interface promotes high levels of situation awareness).

The process of using multiple measures to form measurement scales is referred to as conjoint or derived measurement. There are many kinds of conjoint measurement structures depending on the independence or non-independence of the measures (Krantz, 1972; Krantz, et al. 1971). This hinges on suppositions about the relations of the individual measures, which can also expressed as a measurement structure or set of axioms (see Krantz, 1972). For instance, an evaluation of a new interface might involve measuring:

- Usability based on a rating administered after an initial practice period; and
- Learnability measured by the number of practice trials it takes for participants to reach a level of 85 percent correct across the trials in the practice period.

Clearly, these two are related in meaningful ways, and one might try out different combinations, such as trying to transform them to a common numerical scale and then simply actually adding together the usability rating (higher number means greater usability) and the inverse of the trials-to-criterion result (higher number means greater learnability), to form a scale of goodness. With regard to cognitive work systems, one might want to evaluate them for resilience and look at performance for tough cases, or performance when a mission (or activity) gets derailed. One might want to evaluate a cognitive work system with regard to teamwork functions, and look at whether team members can describe other team members' goals, anticipate other team members' needs, or cope with goal conflicts. But in all such cases, precisely what would one measure? How would one forge a meaningful set of measures and then create a measurement structure that allows one to conjoin the measurements into a meaningful scale that maps onto the policies that might be used to set metrics? Let's look at some specific challenges.

The 'Pleasure Principle' of Human-Centered Computing (Hoffman & Hayes, 2005; Hoffman & Woods, 2005) asserts that: *The good cognitive work system instills in the human a sense of joyful engagement.* The worker feels that they are living in the problem rather than fighting with the technology. How can one evaluate the extent to which new technology supports enhanced immersion or enhanced direct perception? How can a cognitive work system be designed such that effects on intrinsic motivation can be measured (Hoffman et al., 2008)? How can we evaluate the extent to which new technology accelerates achievement of proficiency, or the ability of workers to cope with rare or tough cases (Hoffman & Fiore, 2009; Hoffman et al., 2004)?

How might it be possible to track changes in work such as 'the discovery of toolness.' What is it about some new software that workers find valuable? Does software move workers toward new ways of working, even ways not anticipated by the designer? We know that new technologies typically force people to create work-arounds and kluges, and that there are different kinds of kluges and work-arounds (Koopman & Hoffman, 2003). That taxonomy points to things that might be counted and measured. But the step from conceptual measurable to an operational definition of a measure is neither direct nor easily come-by in the case of complex cognitive systems. For example, what would a 'team kluge' be?—it cannot merely be a kluge that was created by an individual. How do researchers notice team kluges now? Can we formalize some methods for finding instances, that is, can we specify one or more operational definitions? What then? Do we just count them? Or do we measure their attributes, and if so, which attributes?

Sometimes one can readily apprehend a possible measure, but then a conceptual gap comes in working back from that to the theory. For example, we could count Post-It Notes® on the assumption that more notes mean a less usable system, but how can you be certain of what such counts, by themselves, might tell you?

To ask such questions is to take a first step in refining a measure. This is what it is all about. Measures and measurement methods are not frozen things. They *always* get refined.

Measurement and Macrocognition

A critical outstanding need for the study of cognitive work is to extend conceptual definitions of macrocognitive concepts (functions and supporting functions) to operational definitions. To date, sets of known or familiar measures are used to 'get at' macrocognitive phenomena (see Chapter 4). Thus, for instance, we can use measures of situational awareness and interpret them as measures of sensemaking. But how do we operationally define sensemaking such that anyone could use the procedures to identify instances of it? Such questions must be asked of all the macrocognitive functions and processes.

Each macrocognitive function and process will be the basis for a set of measures. In developing the measures, one searches for domain-specific or appropriate aspects of macrocognitive functions that can be used to evaluate hypotheses (for example, about how an intervention in the work might enhance system performance, remaining open to the possibility that it might end up degrading system performance in some respects). For example, let's take 'common ground' (Klein et al., 2004). Members of a team must share some knowledge in order to coordinate, replan, and so forth. Achieving common ground requires establishing a set of shared goals, and assigning roles and responsibilities. Maintaining common ground is a continual process of communicating and coordinating, of updating knowledge and beliefs, of anticipating needs and activities. This requires shared beliefs about what each team member believes and knows, about each team member's intent, capabilities, and so forth.

Such a roster of features can be taken as the *conceptual definition* of sensemaking. The challenge is then to link the conceptual definition to an operational definition.

There are some things we can say about macrocognitive measurement with confidence, precisely because of the premise that measurement must be meaningful at a system level. For instance, measurement for macrocognition must be sensitive to trade-offs, that is, comparisons of increases or decreases of one sort (for example, expanding the expert's range of 'the familiar') against increases or decreases of another sort (increased likelihood of surprise). This is not just a matter of using measures that can be placed in ratios, but using measures in the context of experimental designs that allow for the discovery that there are trade-offs, and what their magnitude is. Here are examples.

Wulf et al. (2002) found that training programs that provided rapid and accurate feedback significantly improved the learning curve of the trainees. Metrical guidance would lead to a decision to provide rapid and accurate feedback. However, it turned out that the feedback reduced performance when trainees moved into the actual work context. The reason is that the group getting rapid and accurate feedback never developed skills in generating their own feedback, and so on the job they were handicapped.

A study of training Morse coders found that trainees who were good right from the beginning actually flunked out of the program at higher rates than their classmates. The reason was that the slower learners had to build up memory strategies. The fast learners never needed those strategies, but as the rate of Morse code transmission increased, metacognitive memory strategies became essential, and the fast learners couldn't keep up anymore. A metrical decision based on initial performance would not have reflected a good of policy.

Another thing that is certain is that macrocognitive measurement will involve all manner of comparisons: increases and decreases, changes in range or variance, change in rate or sign, and so forth. An interface that is easy to learn might better support the achievement and maintenance of high levels of situational awareness. Alternatively, an interface that is harder to learn might result in lower levels of performance initially, but in the longer term be better at supporting the achievement and maintenance of high levels of situational awareness. The alternative hypotheses suggest differing relations between the measures and the conceptual measurables that the measures are believed to reflect.

Table 2.2 illustrates how the measurement of multiple kinds of things and predictions of decreases can be involved in a single study. This study was of a redesign for the workstation of the Weapons Control Officer on board AEGIS cruisers (Kaempf et al., 1992). In all cases, the measures involve comparison, that is, the measures were used as indicators of beneficial change, but the measurements were of decreases. In this table we refer to *system* performance, and not human performance. One will see that the measures are not entirely different from the general kinds of measures used in HEAT measurement, however the focus is on the achievement of the primary system goals.

Table 2.2   Measures and predictions from a study of performance using a new work method

Measure	Anticipated/Desired Indication
Improved system performance at preventing hostile air strikes	Decrease in relative frequency (number of hostile air strikes that were successful divided by the total number of hostile air strikes)
Improved system performance at avoiding the shoot-down of friendlies	Decrease in relative frequency (number of friendly assets shot down divided by the total number of friendlies that were identified)
Improved system performance at targeting	Decrease in relative frequency (number of missiles fired that missed the target divided by the total number of missiles fired)
Improved utilization of aircraft resources	Decrease in relative frequency (number of times that deployed aircraft had to return to refuel divided by the total number of times aircraft were deployed)

Table 2.3 presents some examples of measures that were generated for the domain of intelligence analysis. In a study of analytical work (Patterson et al., 2001), it became apparent that expert analysts know how and when to use a variety of databases. They know how to access and manipulate the data within them,

Table 2.3   Multiple macrocognitive measures for intelligence analysis performance

Work Aspect	Measure	Expectation
Quality of models/ arguments/theories	Understanding of changes in over time	Increase
	Accuracy in prediction	Increase
	Comprehensiveness of knowledge capture	Increase
	Sensitivity to context	Increase
Level of validation of models/ arguments/theories	Percentage of properly vetted sources and data	Increase
	Appropriateness of sources	Increase
	Skepticism about data and collection procedures	Decrease
Analytic dexterity	Ability to manipulate disparate data	Increase
	Number of databases used	Increase
	Use of knowledge shields	Decrease
Amount and nature of collaboration	Time spent synchronizing knowledge bases	Decrease
	Effectiveness of pass-ons	Increase
	Availability of and visibility to back channel information exchange (i.e., knowledge sharing)	Increase
	Costs of coordination (shared mental models, situation understanding)	Decrease

important characteristics of the data (for example, credibility, age), and when it is important to compare them. It was also found that skilled analysts actively search for explanations of data rather than seeking to explain away data that does not fit their frame, a process that Feltovich et al. (2001) describe as using 'knowledge shields.' Thus, these two findings from a cognitive task analysis suggested a compound measure of 'analytic dexterity,' having these three measures:

- ability to manipulate disparate data;
- number of databases used; and
- use of knowledge shields.

One would expect that interventions to support intelligence analysts would enable increases on the two former measurements, and decreases on the latter measurement. This shows how investigation of the effects of an intervention can rely on multiple measures, some of which involve comparisons with an expectation of finding an increase, and some with an expectation of seeing a decrease.

Measurement and Complexity

Measurement for the design and evaluation of sociotechnical systems is an exercise in complexity:

1. The set of potential measurables is not bounded and includes many different *kinds* of things and events: attributes, distinctions, dimensions, causal factors, categorizations, complex and often subtle phenomena, and so on.
2. The scope over which the measurables can range is not bounded. Distinctions or dimensions might pertain to cognitive phenomena, that is, the work activities of individuals, or the work of larger collectives. They might also apply to the ways in which the work involves changing and adapting the methods of work.
3. Appraisals of the sociotechnical system, before and after an intervention, can vary greatly among different stakeholders. For example, the usabilty of a system can vary across different perspectives, both with regard to the operation of the workplace itself (for example, its profitability, degree of stress, efficiency, quality) and from the point of view of the values of different stakeholders (for example, management, the workers, the unions, the shareholders, the Internal Revenue Service, and so on).
4. The collectives of humans (cognitive systems engineers) and machines who attempt to create or improve sociotechnical work systems are *themselves* a complex sociotechnical system engaged in macrocognitive activities. Thus, the cognitive systems engineers are subject to the *reductive tendency*. Research conducted under the rubric of Cognitive Flexibility Theory (for example, Spiro et al., 1988) identified characteristics of learning material and performance situations that cause cognitive difficulty. A key finding is that people often deal with complexity through oversimplification, leading to misconception and faulty knowledge application. For instance, people often think of things that are dynamic using a language of structures and statics. Additional dimensions of difficulty, and their implications for the measurement of macrocognition and the design of sociotechnical systems, are presented in Table 2.4.

Challenges to system-level measurement also include demand characteristics effects. Measurement has consequences for the work. For formative design—that is, the creation of new cognitive work systems—the system performance must be observable, revealing how the humans adapt as the technology matures. This is 'design for observability' (Gualteri et al., 2005). An Army program was developed to measure processes of command and control (see Chapter 4). The researchers used a measure of time as a metric for judging success in planning — specifically, how long a plan remained in place before it had to be modified was to be suggestive of the 'goodness' of the plan. However, in observing military planners in exercises, it was found that using this measure created an unintended

Table 2.4   Some of the dimensions of difficulty and their implications for macrocognitive measurement

Mechanism versus Organicism
Is it assumed that important and accurate understandings can be gained by attempting to isolate and study parts of the system, when in fact the entire system must be understood if any hypothetical parts are to be understood? Measurement must be sensitive to complex, nonlinear, interactive, self-organizing characteristics that can emerge because the sociotechnical system has qualities and processes that are more than the sum of its component parts (for example, Waldrop, 1992).
Discrete versus Continuous
Are processes and events measured as if they consist of discernable steps when they may be continua? Are attributes described by a small number of categories (for example, dichotomous classifications like large/small), when it may be necessary to utilize continuous dimensions or large numbers of categorical distinctions?
Separable versus Interactive
Are processes described as occurring independently or with only weak interaction, when there might be strong interaction and interdependence? The measurements might be insensitive to interdependency of effects across components of the sociotechnical system.
Sequential versus Simultaneous
Are processes thought of as occurring one at a time, when in fact multiple processes occur at the same time? Operation of the reductive tendency on this dimension emphasizes measuring the work as a linear set of steps, as in an assembly line, rather than a matter of interactiveness and simultaniety.
Homogeneous versus Heterogeneous
Are components or explanatory schemes thought of as being uniform (or similar) across a system—or are they really diverse? The reductive assumption is that the processes, values, ways of doing things, cultural norms, abilities, loyalties, and so forth, are pretty much uniform across the many diverse subunits of the sociotechnical system.
Single versus Multiple Representations
Do descriptions afford single (or just a few) interpretations, functional uses, categorizations, and so on, when they should afford many? Are multiple representations (for example, multiple perspectives, schemas, analogies, models, case precedents, and so on) and therefore multiple measures required to capture and convey the meaning of a process or situation? Can a single line of explanation convey a concept or account for a phenomenon, or are multiple overlapping lines of explanation required for adequate coverage?
Linear versus Nonlinear
Are functional relationships understood as being linear when in fact they are nonlinear? The reductive assumption is that changes, effects of interventions, and perturbations of various kinds to the sociotechnical system will have effects that can be measured as increments. This would not account for the possibility of cascades and other nonlinear responses (for example, 'butterfly effects' in which very small input changes can propagate large output changes; Resnick, 1996). Possible implications in measurement would include insensitivity to automation surprises.

Table 2.4   *Concluded*

Regular versus Irregular
In calling things by the same name, does the measurement characterize the domain by a high degree of routinizabilty or prototypicality across cases, when in fact cases differ considerably from each other even when they are called by the same name? Is it believed that there are strong elements of symmetry and repeatable patterns in concepts and phenomena, when in fact there is a prevalence of asymmetry and absence of consistent pattern?
Universal versus Conditional
Is it believed that measures are applicable in much the same way (without the need for substantial modification) across different situations, or is there considerable context-sensitivity in their application? The reductive assumption is that one or another principle for design has the same applicability and measurable effects throughout the many different and changing contexts of work and practice.

consequence. Commanders tried to hold on to mediocre plans, showing how good they were at planning when they should have been replanning as quickly as appropriate.

Codicil

What might be the fate of cognitive systems engineering if, like psychology, it is resigned to a never ending plea bargain, hoping to be blessed as having scientific status? Cognitive systems engineering confronts the challenge of understanding very high orders of cognitive work, matched by very high levels of importance to society. Is the science of complex cognitive systems up to the task of studying 'moving target' work systems? Is the science capable of dealing with highly contextualized events, that is, events that are essentially unique? Measurement will illuminate the path. As the saying goes, there are the hard sciences, and then there are the difficult sciences. To paraphrase Kaplan (1964, p. 181) whether one or another particular measure applies is not for us to decide, but only to discover, for it depends on the facts of the case, not on our whim. Measurement for complex cognitive systems, and for macrocognition, will certainly involve totally new approaches, new kinds of scales, and new analytical techniques. My goal for this chapter has been to empower people to think outside of the box when it comes to measurement and metrics. Measures, and metrics, can be created to do whatever you need them to do. Whether they do it or not should be a matter of empirical fact and logic. On the main, we expect that advances in measurement for complex sociotechnical systems, and macrocognition, must generate novel, and even uncomfortable, ideas and methods.

Acknowledgment

The author's contribution was through participation in the Advanced Decision Architectures Collaborative Technology Alliance, sponsored by the US Army Research Laboratory under Cooperative Agreement DAAD19-01-2-0009.

References

Bar-Yam, Y. (2003). *The dynamics of complex systems*. Boulder, CO: Westview Press.

Bergmann, G. (1943). Outline of an empiricist philosophy of physics. *American Journal of Physics, 11*, 248-258.

Chapanis, A. (1999). *The Chapanis chronicles*. Santa Barbara, CA: Aegean Publishing Company.

Clegg, C. W. (2000). Sociotechnical principles for system design. *Applied Ergonomics, 31*(5) 463-477.

Crandall B., Klein G., Hoffman R. R. (2006). *Working minds: A practitioner's guide to cognitive task analysis*. Boston: The MIT Press.

Deal. S. (Organizer) (2007). Human Systems Integration (HSI) Interest Meeting. *Held at the Human Factors and Ergonomics Society 51st Annual Meeting*, Baltimore, MD.

Endsley, M. R., & Garland, D. L. (2000). *Situation awareness analysis and measurement*. Hillsdale, NJ: Erlbaum.

Feltovich, P., Coulson, R., & Spiro, R. (2001). 'Learners' (mis)understanding of important and difficult concepts' in K. D. Forbus & P. J. Feltovich (Eds.), *Smart machines in education: The coming revolution in educational technology*. Menlo Park, CA.

Gawron, V. J. (2000). *Human performance measures handbook*. Mahwah, NJ: Erlbaum.

Goguen, J. (1994). Requirements engineering as the reconciliation of social and technical issues. In M. Jirotka & J. Goguen (Eds.), *Requirements Engineering: Social and Technical Issues* (pp. 165-200). Burlington, MA: Elsevier.

Gualtieri, J. W., Szymczak, S., & Elm, W.C. (2005). Cognitive System Engineering-Based Design: Alchemy or Engineering. In *Proceedings of the Human Factors and Ergonomic Society 48th Annual Meeting* (pp. 254-258). Santa Monica, CA: Human Factors and Ergonomics Society.

Hart, S. G., & Staveland, L. E. (1988). Development of NASA-TLX (Task Load Index): Results of empirical and theoretical research. In P. A. Hancock and N. Meshkati (Eds.), *Human mental workload* (pp. 239-250). Amsterdam: North Holland.

Hoffman, R. R., Fiore, S. M., Klein, G., Feltovich, P. & Ziebell, D. (2009). Accelerated learning (?). *IEEE: Intelligent Systems*, 18-22.

Hoffman, R. R., & Hayes, P. J. (2005). The Pleasure Principle. *IEEE Intelligent Systems, 1*(January/February), 1) 86–89.

Hoffman, R. R., Lintern, G., & Eitelman, S. (2004). The Janus Principle. *IEEE Intelligent Systems*, *2*(March/April), 78-80.

Hoffman, R. R., Marx, M., Amin, R., McDerrmott, P., & Hancock, P. A. (2008). The metrics problem in the study of cognitive work: A proposed class of solutions. *Technical Report*, Institute for Human and Machine Cognition, Pensacola, FL.

Hoffman, R. R. & Militello, L. G. (2008). *Perspectives on cognitive task analysis: Historical origins and modern communities of practice*. Boca Raton, FL: CRC Press/Taylor and Francis.

Hoffman, R. R., & Woods, D. D. (2005). Steps toward a theory of complex and cognitive systems. *IEEE: Intelligent Systems, 1*(January/February), 76-79.

Hollnagel, E. (1998). *Cognitive reliability and error analysis method*. New York, NY: Elsevier.

Hollnagel, E. & Woods, D. D. (1983). Cognitive systems engineering: New wine in new bottles. *International Journal of Man-Machine Studies*, *18*(6), 583-600.

Jagacinski, R. J., & Flach, J. M. (2003). *Control theory for humans: Quantitative approaches to modeling performance*. Mahwah, NJ: Erlbaum.

Kaempf, G. L., Wolf, S., Thordsen, M. L., & Klein, G. (1992). Decision making in the AEGIS combat information Center. *Report under contract N66001-90-C-6023, Control and Ocean Surveillance Center, Naval Command*, San Diego, CA.

Kaplan, A. (1964). Power in perspective. In R. L. Kahn & E. Boulding (Eds.), *Power and conflict in organizations*. London: Tavistock .

Klein, G., Woods, D. D., Bradshaw, J. D., Hoffman, R. R., & Feltovich, P. J. (2004). Ten challenges for making automation a "team player" in joint human-agent activity. *IEEE Intelligent Systems*, *6*(November/December), 91-95.

Koopman, P., & Hoffman, R. R., (2003). Work-arounds, make-work, and kludges. *IEEE Intelligent Systems*, *6*(November/December), 70-75.

Krantz, D. H. (1972). Measurement structures and psychological laws. *Science, 175*, 1427-1435.

Krantz, D. H., Luce, R. D., Suppes, P., & Tversky, A. (1971). *Foundations of measurement, Volume 1: Additive and polynomial representations.* New York, NY: Dover.

Meister, D. (2004). *Conceptual foundations of human performance measurement.* Mahwah, NJ: Erlbaum.

Moray, N. (Ed.) (1979). *Mental workload: Its theory and measurement*. New York, NY: Plenum.

NAVSEA. (2003). NAVSEA 03 Human Performance Technical Metrics Guide. Dahlgren, VA: Naval Sea Systems Command, Human Systems Integration Directorate.

Neville, K., Hoffman, R. R., Linde, C., Elm, W. C., & Fowlkes, J. (2008). The procurement woes revisited. *IEEE Intelligent Systems, 1*(January/February), 72-75.

O'Neill, M. J. (2007). *Measuring workplace performance*. Boca Raton, FL: Taylor and Francis.

Patterson, E. S., Roth, E. M., & Woods, D. D. (2001). Predicting vulnerabilities in computer-supported inferential analysis under data overload. *Cognition, Technology, and Work, 3*(4), 224-237.

Rasmussen, J. (1986). *Information processing and human–machine interaction: An approach to cognitive engineering*. New York, NY: North-Holland.

Resnick, M. (1996). Beyond the centralized mindset. *Journal of the Learning Sciences, 5*,(1), 1-22.

Scholtz, J. (2005). Metrics for the evaluation of technology to support intelligence. In *Proceedings of the Human Factors and Ergonomics Society 49th Annual Meeting* (pp. 918-921). Santa Monica, CA: Human Factors and Ergonomics Society.

Senge, P. M. (1990). *The fifth discipline: The art and practice of the learning organization*. New York, NY: Doubleday.

Serwy, R. D., & Rantanan, E. M. (2007). Evaluation of a software implementation of the Cognitive Reliability and Error Analysis Method (CREAM). In *Proceedings of the Human Factors and Ergonomics Society 51ˢᵗ Annual Meeting* (pp. 1249-1253). Santa Monica, CA: Human Factors and Ergonomics Society.

Sheridan, T. B., & Ferrell, W. R. (19764). *Man-machine systems: Information, control, and decision models of human performance*. Cambridge, MA: MIT Press.

Sinaiko, H. W. (Ed.) (1951). *Selected papers on human factors in the design and use of control systems*. New York, NY: Dover.

Spiro, R. J., Coulson, R. J., Feltovich, P. J., & Anderson, D. K. (1988). Cognitive flexibility theory: Advanced knowledge acquisition in ill-structured domains. In *Proceedings of the 10th Annual Conference of the Cognitive Science Society* (pp. 375-383). Hillsdale, NJ: Lawrence Erlbaum. Also appears in R. B. Ruddell, M. R. Ruddell and H. Singer (Eds.) (1994). *Theoretical models and processes of reading* (pp. 602-615). Newark, DE: International Reading Association.

Stanton, N. A., Salmon, P. M., Walker, G. H., Baber, C., & Jenkins, D. P. (2005). *Human factors methods: A practical guide for engineering and design*. Burlington, VT: Ashgate Publishing.

Van Cott, H. P. & Chapanis, A. (1972). Human Engineering tests and evaluation. In H. P. Van Cott & R. G. Kinkade (Eds.), *Human engineering guide to equipment design* (pp. 701-728). Washington, DC: American Institutes for Ressearch.

Van Cott, H. P., & Warrick, M. J. (1972). Man as a system component. In H. P. Van Cott & R. G. Kinkade (Eds.), *Human engineering guide to equipment design* (pp. 17-40). Washington, DC: American Institutes for Research.

Waldrop, M. M. (1992). *Complexity: the emerging science at the edge of order and chaos*. New York, NY: Simon & Schuster.

Weiss, P. (1967). One plus one does not equal two. In G. Quarton, T. Melnechuck & F. Schmidt (Eds.), *The neurosciences* (pp. 801-821). New York, NY: Rockefeller University Press.

Wilson, J. R., & Corlett, N. (Eds.) (2005). *Evaluation of human work* (3rd edn). Boca Raton, FL: Taylor and Francis.

Winner, J. L., Freeman, J. T., Cooke, N. J., & Goodwin, G. F. (2007). A metric for the shared interpretation of commander's intent. In *Proceedings of the Human Factors and Ergonomics Society 51st Annual Meeting* (pp. 122-126). Santa Monica, CA: Human Factors and Ergonomics Society.

Woods, D. D. (1988). Coping with complexity: the psychology of human behavior in complex systems. In L. P. Goodstein, H. B. Andersen, & S. E. Olsen (Eds.), *Tasks, errors, and mental models: A Festschrift to celebrate the 60th birthday of Professor Jens Rasmussen* (pp. 128-148). London: Taylor and Francis.

Wulf G., McConnel N., Gärtner M., & Schwarz A. (2002) Enhancing the learning of sport skills through external-focus feedback. *Journal of Motor Behavior. 34*(2):171-82.

Zachary, W., Neville, K, Fowlkes, J., & Hoffman, R. R. (2007). Human total cost of ownership: The Penny Foolish Principle at Work. *IEEE Intelligent Systems, 2*(March/April), 22-26.

Chapter 3

Measuring Macrocognition in Teams: Some Insights for Navigating the Complexities

C. Shawn Burke, Eduardo Salas, Kimberly Smith-Jentsch, and
Michael A. Rosen

Introduction

Twenty-first century organizations are constantly adapting and changing in order
to remain competitive. The root driver of this change in many organizations is the
complexity and dynamism of the operating environment. This trend of increasing
complexity is not expected to slow down in the coming years or decades. In fact,
increased technological sophistication and the need to be responsive to global issues
are likely to increase the need for organizations to be adaptive. A primary strategy
used by organizations to meet these demands involves shifting from individual to
team-based work arrangements. Therefore, it has become essential for researchers
and practitioners alike to better understand how to facilitate team performance
within such environments. In response to this need, there has been much research
over the past two decades that has begun to examine the cognitive processes and
resulting emergent states that contribute to successful team performance within
such environments (for example, Cooke et al., 2004; Lewis, 2003; Mathieu et al.,
2000).

This increasing focus on understanding cognition within teams has resulted
in great interest in the higher-order construct of macrocognition. Macrocognition
refers to the cognitive processes and emergent states that occur when entities are
operating within complex, dynamic environments where there is often no one
clear solution path to goal attainment (see Letsky et al., 2008; Schraagen et al.,
2008). While macrocognitive processes have been examined in the context of
teams, much more common has been the examination of macrocognitive emergent
states. Where processes have been defined as 'members' interdependent acts that
convert inputs to outcomes through cognitive, verbal, and behavioral activities
directed toward organizing taskwork to achieve collective goals' (Marks et al.,
2001, p. 357), emergent states have been defined as 'constructs that characterize
properties of the team that are typically dynamic in nature and vary as a function
of team context, inputs, processes, and outcomes' (Marks et al., 2001, p. 357).
Emergent states tend to be cognitive, motivational, or affective in nature rather than
describing the nature of team interaction as processes do (Marks et al., 2001).

The importance of macrocognition lies in its ability to be used as a higher-order organizing construct for much of the various cognitive components and provide an explanatory mechanism for how teams operating in dynamic, complex environments are able to collectively coordinate their actions in an interdependent and adaptive manner, thereby facilitating team effectiveness. Additionally, the bulk of team cognition research to date has dealt with behavior coordination. Macrocognition in teams extends the types of team performance considered by addressing collaborative knowledge work (for example, problem solving, planning, sensemaking; Rosen et al., 2008). The recognition of the importance of many of these macrocognitive processes and corresponding states has spurred the development of measurement methodologies and techniques. Measurement tools are needed for both research and practical purposes. Theories and frameworks of macrocognition in teams are relatively new and in need of empirical evaluation and refinement. Additionally, being able to measure macrocognition in teams is essential for developing training systems to improve these process and designing technology to support them.

Despite the growth of techniques and specific metrics within the literature there has been a lack of guidance for practitioners. Due to the lack of guidance, efforts to effectively create such measures or implement existing measures often create confusion. Therefore, the purpose of this chapter will be threefold. First, to inform practitioners as to the importance and complexity of macrocognition measurement. Macrocognition will be defined and a few of the complexities of measuring this cognitive process will be delineated. Second, to provide practitioners with a brief set of initial guidelines to assist in navigating the complexity of performance diagnosis, development, and training of teams operating in complex problem solving environments. Finally, by illustrating the importance, complexity, and challenges inherent in measuring macrocognition we hope to highlight the need for additional research in this area.

The Importance of Measurement

The design and implementation of systematic, reliable, and valid measurement tools is often ignored or given only superficial attention in team training systems despite its importance in efficiently and effectively developing high-performance teams. It is only through systematic measurement, often at multiple times across a team's developmental lifespan, that a true picture can be gained concerning the types of interventions needed in order for the team to reach its maximum potential. Without such information the practitioner is often left to assess future needs based on 'gut reactions.' However, for the practitioner, the creation and implementation of macrocognitive measures are often time consuming and confusing. While much is known about the qualities of good measurement, the 'research to practice' gap remains a large challenge to the field. That is, researchers often do not translate or communicate their findings in a way that is of use to practitioners. For example, research has shown that effective team performance measurement systems must: (a) possess good psychometric qualities, (b) capture relevant competencies and performance across time, and (c) capture processes and emergent states (Cannon-Bowers & Salas, 1997). To do so, the measurement system, as a whole, must employ multiple methods and approaches to assess the performance of teams, subteams, and individuals through the use of methods which facilitate near real-time diagnosis (Salas et al., 2007). This latter point is even more important when measurement tools are being used to assess teams operating in real contexts or field settings (that is, outside a laboratory or classroom environment) as often instructors or leaders may have limited time in which to conduct and deliver feedback from the assessment.

While much is known about the qualities of good performance measurement systems, ensuring that measurement tools meet the above specified qualities is not always easy. Moreover, in a field environment where teams are operating in context it may not be possible to achieve each of the above characteristics. However, practitioners should strive to negotiate the many practical constraints within an organization to the fullest degree possible. Due to the complexity of measuring macrocognition and its various components it will not be possible to address the entire spectrum of components that comprise the higher-order construct of macrocognition within the scope of the current chapter. Instead this chapter will focus on a limited, but representative, subset of key issues for practitioners. The measurement challenges discussed below pose the most difficulty for practitioners. First, however, the focus of measurement will be more specifically delineated.

Conceptualizing the Measurement Target: Macrocognition

One of the initial steps in developing a measurement system is always clearly defining what to measure. This issue will be dealt with in more specific terms in the next section, but first we will define macrocognition in a general sense. While there have been several definitions of macrocognition (for example, Klein et al., 2003; Schraagen et al., 2008; Letsky et al., 2008) for the purposes of the current chapter, macrocognition will be defined as 'the process of transforming internalized team knowledge into externalized team knowledge through individual and team knowledge building processes' (Smith-Jentsch et al., 2008, p. 9). Framing macrocognition in this manner implies several bounding criteria. First, macrocognition is a multi-level phenomenon. Because the knowledge-building process involves individual as well as team knowledge building, the unit of analysis must include both the individual team members as well as the team as a whole. Second, this conceptualization of macrocognition is not defined by where it can be studied (that is, laboratory versus field; cf Klein et al., 2003), but instead by the processes involved, which may appear within either laboratory or field settings provided that the right level of complexity is present in the task environment and knowledge requirements of team members (that is, a rich, collaborative problem-solving scenario). Macrocognition in this perspective '... is collaboratively mediated occurring within and across individuals during team interaction; is influenced by artifacts in the environment and/or created by the team; is an emergent cognitive property; develops and changes over time; is domain dependent and collaboration environment dependent' (Smith-Jentsch et al., 2008, p. 7).

The conceptualization put forth by Smith-Jentsch et al. (2008) proposes that macrocognition is comprised of five dimensions (that is, internalized knowledge, externalized team knowledge, individual knowledge-building processes, team knowledge-building processes, and team problem-solving outcomes). Internalized team knowledge refers to 'the collective knowledge held in the individual minds of team members.' (Smith-Jentsch et al., 2008, p. 14). Internalized team knowledge reflects an emergent state and includes two subdimensions, specifically, team knowledge similarity (that is, task mental model similarity, team interaction knowledge similarity, teammate knowledge similarity, shared situation awareness), and team knowledge resources (that is, task knowledge stock, interpositional knowledge, recognition of teammate expertise, and individual situation awareness).

Externalized knowledge, also reflecting an emergent state, is defined as, 'facts, relationships, and concepts that have been explicitly agreed upon, or not openly challenged or disagreed upon, by factions of the team' (Smith-Jentsch et al., 2008, p. 16). Correspondingly it is composed of the following subdimensions: externalized cue strategy associations, pattern recognition and trend analysis, and uncertainty resolution (see Table 3.1).

Reflecting the process side of macrocognition are the latter two dimensions: individual and team knowledge-building processes. Individual knowledge building, as the name implies, occurs within individual team members and refers to the 'process which includes action taken by individuals in order to build their own knowledge' (Smith-Jentsch et al., 2008, p. 12). Individual knowledge building comprises three subdimensions: individual information gathering, individual information synthesis, and knowledge object development (see Table 3.1). Finally, team knowledge building is a process that occurs at the team level and '... includes actions taken by teammates to disseminate information and to transform that information into actionable knowledge for team members' (Smith-Jentsch et al., 2008, p. 13). It comprises five subdimensions: team information exchange, team knowledge sharing, team solution option generation, team evaluation and negotiation of alternatives, and team process and plan regulation (see Table 3.1).

Integrating the four high-level macrocognitive dimensions predicts team problem-solving outcomes (for example, plan quality, efficiency of planning process, and plan execution). Internalized team knowledge, individual and team knowledge-building processes, and externalized team knowledge all impact team problem-solving outcomes, albeit through different mechanisms (Smith-Jentsch et al., 2008). For example, internalized team knowledge serves to directly impact the team knowledge-building process which, in turn, directly impacts the individual

Table 3.1   Specification of macrocognitive dimensions and subdimensions

Construct		Definitions
Dimension #1: Internalized Team Knowledge		
Subdimension #1a	Team Knowledge Similarity	'…the degree to which differing roles understand one another…, or how well the team members' understand the critical goals and locations of important resources' (Smith-Jentsch et al., 2008, p. 14).
Subdimension #1b	Team Knowledge Resources	'"Team members" collective understanding of resources/ responsibilities associated with the task.' (Smith-Jentsch et al., 2008, p. 15). Specific forms include: task knowledge stock (Austin, 2003), interpositional knowledge (Volpe et al., 1996), recognition of teammate expertise, individual situation awareness (Endsley, 1995).
Dimension #2: Externalized Team Knowledge		
Subdimension #2a	Externalized Cue-Strategy Associations	'The team's collective agreement as to their task strategies and the situational cues that modify those strategies (and how)' (Smith-Jentsch et al., 2008, p. 16).
Subdimension #2b	Pattern Recognition and Trend Analysis	'The accuracy of the patterns or trends explicitly noted by members of a team that is either agreed upon or unchallenged by other team members' (Smith-Jentsch et al., 2008, p. 16).
Subdimension #2c	Uncertainty Resolution	'The degree to which a team has collectively agreed upon the status of the problem variables' (Smith-Jentsch et al., 2008, p. 17).
Dimension #3: Individual Knowledge Building Process		
Subdimension #3a	Individual Information Gathering	'Involves actions individuals engage in to add to their existing knowledge such as reading, asking questions, accessing displays, etc.' (Smith-Jentsch et al., 2008, p. 12).
Subdimension #3b	Individual Information Synthesis	'Involves comparing relationships among information, context, and artifacts to develop actionable knowledge' (Smith-Jentsch et al., 2008, p. 12).

Table 3.1   *Concluded*

Construct		Definitions
Subdimension #3c	Knowledge Object Development	'Involves creation of cognitive artifacts that represent actionable knowledge for the task' (Smith-Jentsch et al., 2008, p. 12).
Dimension #4: Team Knowledge Building Processes		
Subdimension #4a	Team Information Exchange	'Involves passing relevant information to the appropriate teammates at the appropriate times' (Smith-Jentsch et al., 2008, p. 13).
Subdimension #4b	Team Knowledge Sharing	'Involves explanations and interpretations shared between team members or with the team as a whole. These explanations may be augmented by graphic visualizations or shared workspaces' (Smith-Jentsch et al., 2008, p. 13).
Subdimension #4c	Team Solution Option Generation	'Offering potential solutions to a problem' (Smith-Jentsch et al., 2008, p. 13).
Subdimension #4d	Team Evaluation & Negotiation of Alternatives	'Clarifying and discussing the pros and cons of potential solution options' (Smith-Jentsch et al., 2008, pp. 13–14).
Subdimension #4e	Team Process & Plan Regulation	'Discussing or critiquing the team's knowledge building process or plan following feedback on its effectiveness' (Smith-Jentsch et al., 2008, p. 14).

development of externalized knowledge and team problem-solving outcomes. Individual and team knowledge-building processes interact in a reciprocal manner to impact one another and serve as feedback loops back into the team's internalized knowledge stocks. Finally, internalized and externalized knowledge serve as moderators of key relationships.

While it is only recently that the higher-order construct of macrocognition has begun to receive attention within the team literature, many of the processes and corresponding emergent states in Table 3.1 have received considerable attention in their own right. The majority of measurement work that has been conducted within the team domain involves team knowledge-building processes and internalized team knowledge. Therefore we focus on these two categories in the following sections.

Measuring Macrocognition in Context: Challenges and Guidelines

In developing or implementing any measure designed to assess a component of macrocognition in context, five fundamental challenges are often faced by practitioners: (1) what to measure, (2) when and from whom to collect the information, (3) what tools to use (that is, how to measure it), (4) how to get useful responses (for example, maximize validity, while limiting obtrusiveness), and (5) how to index the responses (that is, decision whether to aggregate responses, and if so, how). This latter challenge will only be briefly mentioned in the context of the other challenges as there is currently not much guidance to offer on the best manner for aggregation as dictated by context and theory.

Each of these challenges is fundamental to developing good measurement systems. They are particularly difficult in the context of macrocognition in teams because of the complex environments in which macrocognition emerges. Next, each of the challenges identified above will be briefly described followed by a set of related insights regarding potential mitigation strategies.

Challenge #1: Deciding what to measure—the content of macrocognition metrics

Often the decision as to what to measure is a challenging one for practitioners. For teams that are operating in context, often the recognition of the need to assess the team does not arise until there is a performance problem that is realized by the team members themselves or their supervisors. This is unfortunate as it often gives assessment a negative connotation, but often many organizations do not take the time to conduct systematic repeated assessment because it is not a 'sexy' use of their limited resources.

Once there is a recognition of the need for assessment, practitioners are often still shooting in the dark trying to determine what to assess. For example, is it a problem with a lack of knowledge, or lack of skill, or is it attitudinal? Even once this initial level is decided through the application of a needs analysis (see Burke, 2005; Goldstein, 1993), there is often a secondary question of what component of the knowledge, skill, or attitudes needs to be assessed. Within the current chapter our focus is on macrocognition so we begin there, yet the question remains, what is the specific content that should be included in measures of macrocognition? Within the current framework (that is, Smith-Jentsch et al., 2008) the answer is—it depends on whether one is interested in internalized team knowledge, externalized team knowledge, individual knowledge building, or team knowledge building. Below are a set of guidelines which may assist the practitioner with the subcomponents of this challenge.

Insight #1a: The purpose of the measurement system should drive decisions about content Within the realm of macrocognition, perhaps one of the earliest decisions regarding measurement content at a high level is whether to assess emergent states, processes, or a combination. There are many purposes for which measurement systems can be utilized (for example, assessment, development, selection; formative versus summative evaluation). However, a formal or informal assessment of needs and gaps is often used to assist in determining the purpose of measurement (that is, what questions is the measurement system designed to answer). For example, is the measurement system supposed to drive formative or summative evaluations (for example, are data going to be compared to a criterion or used to generate feedback)? If the question focuses on summative evaluation, emergent state components of macrocognition versus the actual processes involved will likely be the focus of evaluation. This might be done in the case where the practitioner is interested in determining the level of macrocognition for the purposes of something other than a developmental nature, such as evaluating the effects of different collaborative technologies.

Conversely, if the assessment is being conducted for developmental purposes or so that diagnostic feedback can be delivered it is often best to capture both process as well as emergent states. Emergent states provide information pertaining to where the team or team member is currently at with regards to the specific state, but will not provide information on the dynamic processes that lead to the current level of the state. Using the example of team member knowledge, if knowledge of team member expertise (emergent state) was assessed, one would understand the degree to which team members were knowledgeable concerning where specific expertise lay within the team. However, there would be no information to provide the practitioner with a picture of how they achieved the current state as that is reflected by the process measures—in the case of macrocognition, individual, and team knowledge building. Additionally, the processes of performance are the focus of training systems, and therefore, information about processes will be needed to generate feedback. Process is sometimes more challenging to measure due to its dynamic nature, but measurement is most powerful when both are measured as it takes both pieces together to provide the information that will most efficiently assist in targeting an intervention and delivering diagnostic feedback.

Insight #1b: Use needs analysis to develop the content of metrics   Specific measurement content will be determined by the macrocognitive components identified as important in the gap analysis or needs assessment (see Burke, 2005; Goldstein, 1993). Once the needs assessment reveals the specific components of macrocognition that the practitioner is interested in, the operational definitions of those constructs (see Table 3.1) will assist in determining the content of the specified measure. For example, if one is interested in the individual knowledge-building process and more specifically in individual information gathering, one would want to ensure that the content included would capture the various methods by which information could be gathered in the specific context in which the team was operating. For example, for a team operating in a command and control type of environment, one might envision content including markers assessing the degree to which information was being sought from outside sources, team members, system data, collaborative workspaces, and displays.

Challenge #2: Deciding where the data comes from

Another decision that must be made is from whom to collect the information on cognitive processes and the corresponding states. While some components of macrocognition solely reside within the minds of the individual team members, other components are manifested through communication, system interaction, and other behavioral indicators. These data must be collected at a level of analysis consistent with its conceptual definition. This leads to the following guidelines.

Insight #2a: Assess internalized team knowledge primarily by using the individual team member as the source of information   As internalized knowledge represents the 'collective knowledge held in the individual minds of team members' (Smith-Jentsch et al., 2008, p. 14) it is best and most commonly assessed via direct querying of individuals about their own knowledge.

Insight #2b: Assess externalized team knowledge or team knowledge building by collecting information from the team as a whole as well as outside observers   When the goal of measurement is to assess externalized team knowledge, there are more options in terms of deciding from whom to collect the data. More specifically, as externalized knowledge refers to 'facts, relationships, and concepts that have been explicitly agreed upon, or not openly challenged and disagreed upon, by factions of the team' (Smith-Jentsch et al., 2008, p. 16) it can be assessed not only from the team members themselves, but also from those who have had a chance to observe the team in action. Its externalized nature also makes it possible that it can be collected from system information such as time stamped audio, digital communication logs, and external artifacts that may be generated within shared workspaces during the team's interaction (Smith-Jentsch et al., 2008).

Similarly, as team knowledge building represents 'a process which includes actions taken by teammates to disseminate information and transform that information into actionable knowledge for team members' (Smith-Jentsch et al., 2008, p. 13), data concerning it can be gathered from a myriad of sources. Given that it is a process consisting of components that are primarily externalized as knowledge is being built, it can be assessed by examining the actions of the team either through direct queries or the observations of others. Components such as team information exchange, team knowledge sharing, team solution option generation, team evaluation and negotiation of alternatives, and plan regulation are all dynamic components whose presence is primarily observed through externalization of behavior.

Insight #2c: Assessment of individual and team knowledge building can be augmented through the collection of objective system data   In describing team knowledge building in the context of macrocognition, the prior guideline alludes to how data could be collected from the system with which the team is interacting. Specifically, system data such as digital communications or graphic visualizations that are captured by the system may assist in collecting this type of information. With recent advances in technology, individual knowledge building, a primarily internalized process, may also be able to be assessed through an examination of system-collected data. For example, for the portions of the individual knowledge-building process that are internalized, methods such as eye tracking and keystroke analysis may be leveraged. While it is noted that eye tracking may be currently prohibitive for teams operating in real contexts; depending on the nature of the tasks, keystroke analysis can augment data collected from other sources. Furthermore, for those aspects of individual knowledge building that are externalized—they involve overt actions—data may be collected by observers or by the system.

Challenge #3: Deciding on the best measurement format

Another challenge for practitioners that is expected to increase as a greater breadth of measures are created for the various components of macrocognition, is deciding on the best measurement format for a particular purpose. While the choice of content will assist in narrowing down the selection of available formats, the context of the measurement environment and population must also be taken into account. When possible it is best to assess the same content through multiple measurement methods/formats so that obtained results are not due to the specific format but reflect the actual process or state of interest. In some cases the choice of format is not difficult because there is little variation in the methods available to assess a specific cognitive process or cognitive state. In addition, for practitioners charged with assessing teams in context the choices are often further narrowed down as many of the methods by which cognitive process and states are measured are cumbersome outside of a laboratory environment (see Challenge #4).

Insight #3a: Use multiple methods   Do not rely solely on any one format or method. There is often no one best measurement format, but multiple formats that work. The key to conquering the format challenge is to be aware of the different formats available for the specific component of macrocognition that is of interest and use them in combination when possible. Below is a brief description of some of the methods available for the various components of macrocognition.

The measurement of internalized team knowledge is an area that has received increased attention for the past several years. This type of knowledge has most often been assessed through the use of card sorts, concept mapping, paired comparison ratings, scenario probes, and questionnaires (see Table 3.2). These are methods that individual team members complete and then responses are often aggregated to examine knowledge of team member responsibilities and corresponding resources and the degree to which the knowledge held amongst team members is similar (see Challenge 2 for more information). When assessing externalized team knowledge and individual or team knowledge building, slightly different methods are used because these components reflect externalized processes or states. With respect to

Table 3.2   Potential methods used to assess macrocognition (adapted from Smith-Jentsch et al., 2008)

Construct	Methods
Dimension #1: Internalized Team Knowledge	Card sorts, concept mapping, paired comparison ratings, scenario probes, questionnaires
Dimension #2: Externalized Team Knowledge	Communication and analysis of artifacts on shared workspaces
Dimension #3: Individual Knowledge Building Process	Eye tracking, communication, system queries, concept maps, verbal or written probes during task, debrief questionnaires, notes, diagrams, sketches, artifacts
Dimension #4: Team Knowledge Building Processes	Communication analysis and analysis of artifacts on shared workspaces

externalized team knowledge, cue-strategy associations, pattern recognition, and uncertainty resolution are often measured through an examination of the team's communication or artifacts that are generated in any shared workspace the team is using (Smith-Jentsch et al., 2008, see Table 3.2). Team knowledge building is often assessed using the same methods. Conversely, because individual knowledge building reflects a process that has internalized and externalized components, externalized components may be captured by looking at team communication transcripts where the focus is on the individual's communication as compared to the team as a unit. For internalized components, methods such as keystroke data, eye tracking, and verbal protocols may be used (see Table 3.2), although the latter is fairly intrusive.

Challenge #4: Getting quality responses

While getting quality information from the individuals completing the measurement tools is always important, it is often more challenging when information is being collected in context versus within a laboratory environment. Specifically, when collecting information from teams operating in context there are many constraints that are more easily manipulated within laboratory environments as compared to field environments. Perhaps two of the largest roadblocks in getting quality responses in a field environment pertain to the motivation of those completing the measure and the time constraints that are often present given that there is a real job to do with goals and deadlines outside of the measurement task.

Insight #4a: Ensure the measure has face validity   When conducting measurement in context, ensuring that the measure has face validity is an important consideration.

While face validity is not directly related to the psychometric properties of the measure (for example, reliability or formal validity of measurement), it is important from a motivational standpoint. Face validity refers to the degree to which the measure appears to assess the content in question. When individuals do not perceive the measure to be related to the construct of interest they are less willing to spend quality time on completing the measure as they do not see its relevance. For example, pair-wise comparison ratings require team members to make many ratings and because the reason for making these ratings is not apparent to most team members in operational settings, there can be 'push back' when collecting these data in the field.

Insight #4b: Limit the obtrusiveness of the measure   Measure obtrusiveness has commonly been discussed in two different veins. One manner in which researchers talk about obtrusiveness pertains to the degree to which completing the measure causes members to perform differently on the task. This type of obtrusiveness can be seen in measures of internalized knowledge such as situation awareness, where the method used involves stopping the team in the middle of the task and inserting queries that tap members' current levels of situation awareness. A second way that researchers talk about obtrusiveness is the degree to which the measure is cumbersome to deliver in a field environment or the degree to which it disrupts the task, regardless of whether it causes those filling it out to do the task differently. When talking about obtrusiveness in this way many of the measures of macrocognition are obtrusive, with those typically used to measure internalized team knowledge perhaps being the largest offenders. However, current work is being conducted to develop mechanisms which allow the embedding of some of the more obtrusive measures into the context of the task environment. Most recently this has been conducted with measures of shared situation awareness in which the queries are embedded in the tasks themselves so they appear as a natural part of the ongoing flow of events.

Insight #4c: Ensure the measure has predictive validity   While the last guideline does not necessarily impact the motivation levels or time constraints of the individual completing the measure, it is important from the practitioner point of view. Given the value of time for teams operating in context, practitioners want to ensure that the measures that are chosen are actually able to predict performance. Practitioners should attempt to ensure that not only are the measures used predictive, but that, when using a combination of measures, each one offers a unique contribution in order to make the most efficient use of the time that is allotted for measurement purposes. Using measures that are not predictive or that offer no diagnostic capability will distally impact the motivation of those who are completing them as they realize that nothing useful has come of the time that they have taken to complete the measures. Unfortunately while this is an extremely important aspect of measurement and many of the measurement tools used to assess the various components of macrocognition have been shown to be able to predict performance, research is just emerging which specifically compares the predictive value of different measurement techniques (for example, aggregation methods, content, and methods combined).

Concluding Comments

The interest in cognition within teams has witnessed a dramatic increase in the past 20 years. Most recently macrocognition has emerged as a higher-order construct which is comprised of many cognitive processes and emergent states. As with cognition in general, measuring macrocognition in teams is a challenging endeavor for many reasons, not the least of which is a lack of identification of the challenges and mitigation strategies for practitioners prior to stepping into the quagmire known as macrocognition. In this chapter we have attempted to highlight a few of the more prominent challenges that practitioners must overcome in attempting to successfully devise measurement systems which capture the complexity of this cognitive construct. In addition, a few corresponding mitigation strategies have been identified in the form of insights that may assist in navigating the complexity.

While we have highlighted some of the most prominent challenges, most of which revolve around a recognition of the complexity of developing a systematic measurement system, there is much work that needs to be done. At this point in time, work in the area of macrocognition is just in its infancy. Thus, many of the insights put forth in the current chapter are leveraged from what has been learned regarding the measurement of a variety of the cognitive constructs that have been argued to be subsumed under the macrocognition (see Letsky et al., 2008). Yet the explanatory power of macrocognition as a higher-order construct over those cognitive components which have been examined in the team's literature in the past remains to be empirically verified. If the construct is deemed to offer incremental predictive power then empirical efforts, both in the laboratory and the field, need to be conducted to verify the insights put forth in the context of macrocognition. Additionally, there are potentially additional insights that can be gathered as the construct is increasingly evoked as an explanatory mechanism by those within the laboratory as well as the field.

Acknowledgements

The views, opinions, and findings contained in this article are the authors and should not be construed as official or as reflecting the views of the University of Central Florida or the Department of Defense. Writing this chapter was supported by a Multidisciplinary University Research Initiative Grant from the Office of Naval Research (Contract #: N000140610446).

References

Austin, J. R. (2003). Transactive memory in organizational groups: The effects of content, consensus, specializations, and accuracy on group performance. *Journal of Applied Psychology, 88,*(5), 866-878.

Burke, C. S. (2005). Team task analysis. In N. Stanton, A. Hedge, K. Brookhuis, E. Salas, E. & H. Hendrick (Eds.), *Handbook of human factors and ergonomics methods* (pp. 56.1-56.8). Boca Raton, FL: CRC Press.

Cannon-Bowers, J. A. & Salas, E. (1997). A framework for developing team performance measures in training. In M. T. Brannick, E. Salas, & C. Prince (Eds.), *Team performance assessment and measurement: Theory, methods, and applications* (pp.45-62). Mahwah, NJ: Lawrence Erlbaum Associates.

Cooke, N. J., Salas, E., Kiekel, P. A., & Bell, B. (2004) Advances in measuring team cognition. In E. Salas, & S. M. Fiore (Eds.), *Team Cognition*, (pp. 83-106). Washington DC: APA.

Endsley, M. R. (1995). Toward a theory of situation awareness in dynamic systems. *Human Factors, 37*(1), 32-64.

Goldstein, I. L. (1993). *Training in organizations: Needs assessment, development, and evaluation* (3rd edn). Pacific Grove, CA: Brooks/Cole Publishing.

Klein, G., Ross, K. G., Moon, B. M., Klein, D. E., Hoffman, R. R., & Hollnagel, E. (2003). Macrocognition. *IEEE Intelligent Systems, 3*(May/June), 81-85

Letsky, M. Warner, N., Fiore, S. M., & Smith, C. (Eds.) (2008). *Macrocognition in Teams: Theories and Methodologies.* London, UK: Ashgate Publishers.

Lewis, K. (2003). Measuring transactive memory systems in the field: Scale development and validation. *Journal of Applied Psychology, 88*(4), 587-604.

Mathieu, J. E., Heffner, T. S., Goodwin, G. F., Salas, E., & Cannon-Bowers, J.A. (2000). The influence of shared mental models on team process and performance. *Journal of Applied Psychology, 85,* 273-283.

Marks, M. A., Mathieu, J. E., & Zaccaro, S. J. (2001). A temporally based framework and taxonomy of team processes. *Academy of Management Review, 26*(3), 356-376.

Rosen, M. A., Salas, E., Fiore, S. M., Letsky, M., & Warner, N. (2008). Tightly coupling cognition: Understanding how communication and awareness drive coordination in teams. *International Journal of Command and Control* (online), *2*(1).

Salas, E., Rosen, M. A., Burke, C. S., Nicholson, D., & Howse, W. R. (2007). Markers for enhancing team cognition in complex environments: The power of team performance diagnosis. *Aviation, Space, and Environmental Medicine Special Supplement on Operational Applications of Cognitive Performance Enhancement Technologies., 78*(5),B77-B85.

Schraagen, J. M., Militello, L. G., Ormerod, T., & Lipshitz, R. (2008). *Naturalistic decision making and macrocognition.* Aldershot, UK: Ashgate Publishing.

Smith-Jentsch, K., Fiore, S. M., Salas, E., Warner, N., & Letsky, M. (2008). Theory and measurement development   for research in macrocognition. *Technical report submitted as part of the ONR CKI MURI SUMMIT Project,* University of Central Florida.

PART II
Macrocognition Measures for Real-world Teams

Chapter 4

Macrocognitive Measures for Evaluating Cognitive Work

Gary Klein

Introduction

Schraagen et al. (2008) defined macrocognition as the study of cognitive adaptation to complexity. Figure 4.1 shows the primary macrocognitive functions and processes. Although these six functions can be distinguished, they are also related to each other. No activity in cognitive work would draw on just a single function or process.

The primary macrocognitive functions in Figure 4.1 are decision making, sensemaking, detecting problems, (re)planning, adapting, and coordinating. These are activities, not states of knowledge. They are interrelated but each involves a different set of strategies. They are the key functions that need to be performed in natural settings, whether at the individual, the team, or the organizational level. The bottom half of Figure 4.1 shows the macrocognitive processes that support the functions on the top.

The purpose of this chapter is to describe a range of measures that can be used to assess the six macrocognitive functions shown in Figure 4.1. To date, most macrocognitive studies have been descriptive, using cognitive field research methods. As we learn more about the functions and processes, we may want to collect more quantitative data. For example, the investigation that resulted in the Recognition-Primed Decision (RPD) model (Klein et al., 1986) was fueled by incident accounts, but one of the reasons that people found the RPD model credible was that the types of decision strategies were coded and showed the predominance of the RPD strategy over option-comparison strategies. One of the reasons that several researchers replicated the RPD finding (see Klein, 1998 for a summary) was that they too could collect and analyze quantitative data. Scientific inquiry depends on measurement. Therefore, to advance the scientific inquiry into macrocognition we will need measures for the functions and processes.

The primary interest in suggesting a set of measures is to encourage researchers to more systematically collect and report data as they perform naturalistic studies of macrocognition. Discipline in data collection and analysis will allow us to inject more rigor in our research, and to increase the pace of scientific progress in our field. Measurement procedures will also help us evaluate how

Figure 4.1   Macrocognition model

effective training programs and information technologies are in supporting the functions; and they will enable us to contrast different conditions and different populations.

This chapter only addresses the macrocognitive *functions*. Measures can be developed for the macrocognitive *processes* by researchers who study them in the future.

Macrocognitive Paradigms

Before considering any measures, we first need to examine the types of paradigms available to study the different macrocognitive functions. The systematic collection of quantitative data generally requires the use of a controlled scenario that can be presented to participants, as opposed to observation or to interviews about specific incidents (but see Klein et al., 1986 for an example of a project that relied on coding of qualitative interview data). This in turn depends on paradigms for observing the macrocognitive phenomena. This section identifies a set of these paradigms. Most studies of macrocognitive functions have relied on just a small number of paradigms.

Decision Making

Decision-making exercise (DMX)   The concept of a Decision-Making Exercise (DMX) (also known as a Tactical Decision Game) is a scripted scenario, usually with an unexpected twist at the end, that enables the researchers to see how participants will make some difficult choices (Klein, 2004).

Scripted scenario   This is a DMX played out over time, with new developments emerging. The participant is faced with tough decisions (these may be tightly scripted in the scenario), along with information requests along the way. Baxter & Phillips (2004) used such a scripted scenario to collect more quantitative data from participants.

Sensemaking

Garden path   The basic Garden Path paradigm attempts to lead participants down the wrong road in interpreting a situation. The initial cues strongly suggest one explanation, which is incorrect. Subsequent cues will enable the participant to realize what is actually happening. The measure is how quickly people recover and arrive at a correct interpretation (Feltovich et al., 1984; Rudolph, 2003).

Focus   Data is dribbled in, while the participant tries to see connections, patterns, and the story. This can be done with think-aloud protocols. The analog is the Bruner & Potter study (1964) where a photograph is gradually brought into focus while the subjects try to guess what the photograph represents. By the end of the session, the photograph is completely specified. The measure is how much in advance of this end point will participants make the correct identification. For macrocognitive studies, the Focus paradigm can be used with stories. Klein et al. (2002) used this paradigm with information operations specialists. The study did not attempt to mislead the participants (as would have happened with a garden path paradigm) but, instead, provided seemingly unrelated cues that actually were connected to each other, to see how long it took the participants to discover this connection.

Starter kit   The participants are asked how they understand some complex phenomenon and, for purposes of illustration, they are given several different possible mechanisms of how the phenomenon may be working. The mechanisms might be diagrams, text descriptions, and so on. Their task is to select the mechanism that is closest to their beliefs about what is going on, and then explain where it still misses the point. Crandall et al. (2006) give the example of cartoon representations of the way laundry pretreatments work. The participants, who had been unable to articulate the mechanism of action, were able to identify the illustration that came closest to what they believed was happening.

Sherlock Holmes   A situation is described. Participants have to imagine what led up to it—they have to form explanations. This paradigm differs from the focus paradigm in that the cues are not introduced over time but rather are presented all at once and the participants are asked to speculate about where it started from, as opposed to where it is leading to.

Prediction paradigm   In this turn-taking exercise, participants generate predictions of the next state. This works very well with adversarial exercises (for example, Klein & Peio, 1989, used it with chess). It can also work with single-person exercises, to see how accurate the predictions are and their basis. This paradigm is consistent with Vicente & Wang's (1998) account of experts as being able to take better advantage of redundancies in the situation.

  The paradigm can also be used to study anticipatory thinking. Instead of asking for predictions, participants are studied to see how they anticipate the likely developments in a situation based on whether they direct their attention toward critical cues. What is measured here is attentional tracking, given the trajectory of the situation, rather than predictions about future states. For example, Pradhan et al. (2005) tracked the eye movements of participants in a driving simulator. As the scene developed, predefined risky situations played out. The measure was whether the drivers looked in the right places, which would indicate that they appreciated what the risks were.

Lights out   A situation is presented to participants and then, after a suitable period of time to study the elements, they are removed without warning. The task is to recreate the situation. This paradigm was originated by de Groot (1946) to show that skilled chess players could readily recreate coherent positions but not positions made up of randomly assigned pieces. Endsley (1995) has used this paradigm in the Situation Awareness Global Assessment Technique (SAGAT).

(Re)Planning

PlanEx   This is a simple planning exercise, basically a DMX that requires the participant to prepare a plan. The measures are the quality of the plan, the ability of the participant to see the 'sweet spot' for leveraging resources, and so on.

PlanEx scenario   In this version, the participant prepares a plan (or is given a plan) and then is engaged in implementing the plan. The scenario includes perturbations that require goal shifts, prioritization, replanning, and so forth.

Challenge round   The participant develops a plan, based on the guidance and intent from higher command levels, and is then faced with probe questions: 'If X happened, what would you do?' 'If you had a choice of accomplishing Y, would you make that choice?' These probe questions are designed to get at the goal priorities and tradeoffs (for example, Shattuck & Woods, 2000). In some ways they resemble the DMX 'Taking a Stand' presented in Klein (2004).

Detecting Problems

It is very difficult to study problem detection in a controlled environment. Most studies present the problem to the participants, rather than waiting for the participants to discover what is going wrong. Therefore, problem detection may best be studied through Cognitive Task Analysis interviews; particularly Critical Decision Method (CDM) interviews (Klein et al., 1989; Crandall et al., 2006) in which the problem detection is part of the story being told.

Garden path   The one controlled paradigm that does allow for problem detection is the garden path scenario. Researchers can study the point at which the participants become suspicious that the initial hypothesis may be wrong.

Adapting

Curve ball   Most studies of how people, teams, and organizations adapt employ some form of curve ball scenario in which the prepared plan or mission is rendered semi-obsolete by unexpected circumstances, and the participants have to determine that the original plan will no longer work. The participants then reprogram their resources, and possibly even revise their goals. Sometimes the participants have to make the discovery for themselves (which requires problem detection); other times they are informed of the new direction by their superiors.

Coordinating

Prediction paradigm   The Prediction paradigm was discussed earlier (Klein & Peio, 1989), as a means of studying sensemaking. It is also a way to study teams. The hypothesis is that teamwork depends in part on making oneself predictable, and being able to predict the others. Therefore, members of more effective teams should show a higher ability to predict each other's actions. Blickensderfer (1998) demonstrated this for tennis doubles, giving them DMXs that consisted of diagrams showing the location of all four players and the position and direction of the tennis ball, and asking how their partner would react. The teams that had done better in a tournament also showed higher prediction accuracy.

Macrocognitive Measures

To describe measures for each macrocognitive function, each subsection is organized into typical aspects of the function, and some potential variables that can be measured. Sensemaking is described first rather than Decision Making because a Sensemaking aspect will be included under Decision Making and it is easier to cover the topics in this order.

Sensemaking

It is difficult to study sensemaking directly. Therefore, researchers may prefer to examine different aspects of sensemaking. Four potential aspects of sensemaking are: seeking information, forming and revising mental models, noticing anomalies/ detecting problems, and applying frames/mental models. These four are intended as illustrations, and are consistent with the Data/Frame model of sensemaking (Klein et al., 2007). Other aspects could be added as appropriate.

Seeking information   Seeking information can be measured by the time spent on a document, rating the quality or the perceived information value of the document being studied, or the perceived quality of documents selected for review (perceived by the person reading the article). Such measures were used in the research of Patterson et al. (2001) on experienced intelligence analysts. To present intelligence analysts with an unclassified but naturalistic task, Patterson et al. (2001) used the Ariane 5 incident, in which a European Space Agency rocket had to be blown up shortly after launch. The explanations for the failure changed over time; the analysts had to use open source data to figure out what had happened. The analysts had more data available to them (approximately 2000 sources) than they could review in a three to four hour session, so they had to prioritize, hoping that they wouldn't miss a critical article. Some of the data analysis was qualitative, such as differentiating the strategies used to search through the documents, and the strategies used to resolve conflicts. Patterson et al. (2001) also used some quantitative measures, such as the number of high-profit documents (predefined) retrieved and the degree of convergence between high-profit documents and the documents that the study participants identified as key for their understanding.

Seeking information also requires one to differentiate relevant from irrelevant data. The worse this discrimination, the more time wasted. The same measures apply here as above: time spent on a document, rating of the value of a document, and quality of the document selected for further scrutiny. Objective performance standards can be set by the document ratings generated by experts (Patterson et al., 2001).

Forming and revising mental models   One variable is the ability to notice connections among events, people, and places. Measurements can come from 'think-aloud' verbal protocols, looking for linkages between messages in a stream of messages. Q-sorts, concept maps, and other methods can reflect these connections. Klein et al. (2002; 2007) studied Army information operations specialists, relying on a think-aloud task to collect data (see Crandall et al., 2006) to capture their interpretations as they read through a stream of messages.

Another variable that can be measured is the quality of the frames/mental models selected. Verbal protocols can be coded to reflect requests for information. Critical decision method probes can capture the frames that are used. (If a simulation exercise is conducted, the probes can be inserted at the appropriate points in the exercise.) Smith et al. (1986) captured this type of data in a study of aviation troubleshooting. The pilots, working in a simulator, had to diagnose malfunctions. Their verbalizations and actions during simulated trials enabled the researchers to categorize which mental models they were using, and to measure whether their mental models accurately matched the malfunction they were facing.

Another quality of forming and revising mental models that can be measured involves recovery from fixation, and escaping from a garden path. Measures of this include the time spent on the garden path, the amount of discrepant information that has to be received before the participant leaves the garden path, the number of participants in a condition who escape the garden path, and even the corrective actions taken. These kinds of measures were used by Feltovich et al. (1984) in their study of pediatric cardiologists who needed to escape from a garden path scenario, by Rudolph (2003) who used a garden path paradigm to study anesthesiologists, and by Baxter & Phillips (2004) who studied the difficulty Marine lieutenants had in escaping from fixation.

Noticing anomalies and detecting fixation   Measures include the way weak signals are handled, success in detecting the problem, and success in noticing negative cues (events that should have happened but didn't). Snowden et al. (2007) used these measures in studying teams of military officers and intelligence analysts who were confronted with garden path scenarios. This study tabulated the frequency with which the weak signal was mentioned in personal notes generated by each team member, and by the number of times a weak signal was mentioned in the group discussion. A related measure is the ability to discern the 'problem of the day.' Pliske et al. (1997), using the critical decision method, studied the way weather forecasters identified what they needed to watch carefully during their shift. How quickly was the problem of the day identified (versus, in hindsight, the earliest it might have been detected)? It is easier to measure latency when a simulation exercise is used because the initial cues are injected under the researcher's control.

Applying frames/mental models   The use of frames to form explanations can be assessed on the basis of the quality of the explanations offered. In a simulation scenario, the explanations can be elicited through requests from the researcher. Or participants might be asked to produce verbal protocols as they perform the task, and these can be rated by experts.

The use of frames such as mental models to anticipate events can be studied in terms of actions taken or the patterns of information gathered (in comparison to a predefined standard in a simulated exercise, or as rated by subject-matter experts in a critical incident study), or even eye movements. Pradhan et al. (2005) tracked eye movements to show how drivers with more experience were twice as effective at anticipating threats as new drivers. The experienced drivers understood what might turn into a problem and directed their attention to these potential threats, as shown by their eye movements.

Team Sensemaking   Klein et al. (in press), building on the data/frame model of sensemaking, have distinguished several sensemaking strategies that emerge at the team level. These include strategies for identifying a frame (for example, using simple rules, autocratic assertions, consensus), questioning a frame (using a form of ritualized dissent, such as a devil's advocate), and comparing competing frames (voting, consensus, and autocratic assertions by the leader). Measures of team sensemaking could distinguish which strategy a team was using, as well as the team's effectiveness and efficiency in applying that strategy.

Decision Making

Studies of decision making in natural settings have shown that the difficult part is usually figuring out what is going on; people don't usually need to wrestle with choices once they are clear about the nature of the situation. Therefore, sensemaking is a key aspect of naturalistic decision making. The other aspects we will cover are option generation and option evaluation. As with all of these subsections, the aspects covered are not meant to be exhaustive or comprehensive. Rather, they are intended to illustrate ways to measure the macrocognitive functions.

Sensemaking   One variable that can be measured is how quickly and effectively people can identify a prototype, pattern, or frame. Critical decision making data (for example, Klein et al., 1989), a form of cognitive task analysis (CTA) (Crandall et al., 2006), has been used to obtain data from retrospective accounts, as in the Klein et al. (1986) study of firefighters. Other examples that used the critical decision method of cognitive task analysis are Klein et al. (2004a), who studied Unmanned Aerial Vehicle operators, and Kaempf et al. (1996), who studied Navy commanders of Aegis cruisers. Each of these efforts used cognitive task analysis to determine how decision makers identified the nature of situations, how they used their interpretation to make decisions, and how they modified their assessments.

   Another variable is the way decision makers identify relevant goals and how their goals shift during a dynamic event. One way to capture these shifts is to use a Situation Awareness Record, described by Klein et al. (1989) as a way of representing shifts in interpretations of the situation during an incident that is studied using a critical decision method interview. The situation awareness record can be used to study both sensemaking and decision making. Each shift in the way the situation is understood is coded as an elaboration (building on the previous interpretation and adding more detail) or a shift (changing the previous interpretation in some important ways), and reflects shifts in the goals pursued, the expectancies, and the relevant cues. Decisions typically follow the way the situation is interpreted. The situation awareness record can guide the interview process and structure the kinds of data being collected.

Option generation and evaluation   Under time pressure and uncertainty, decision makers may only generate and evaluate a single option (Klein, 1998). The decision-making strategy used by participants can be suggested by qualitative interview data but then coded into categories with relatively clear criteria. The coding can be assessed using inter-rater agreement measures.

In some situations, such as chess, people do need to generate new options or generate several options. One measure is the quality of the first option considered. Klein et al. (1995) used think-aloud protocols for chess players examining a complex position to identify the first option they mentioned. The positions had been pre-rated by a panel of chess grandmasters, and their ratings were used to assess the quality of the moves generated by the study participants.

Team decision making   At the team level, we find emergent decision strategies. These include voting, recruiting supports, forming a consensus, collaborating to synthesize an option from the ideas presented by different team members, autocratic decision by the leader, and so forth. Measures of team decision making would distinguish which type of strategy was being used, the benefits/costs of applying that strategy along with the efficiency and effectiveness the team showed in applying it.

Detecting Problems

This macrocognitive function has primarily been discussed by Klein et al. (2005). Klein (2006) examined problem detection at a team level. We can distinguish at least two aspects of problem detection—noticing that a situation is different than expected, and noticing that the margin of error for a plan or activity has become uncomfortably small.

Noticing Anomalies

The first aspect, noticing that the situation may be different, has subsequently been incorporated into the data/frame model (Klein et al., 2007) in the form of questioning a frame and noticing anomalies. It has already been discussed earlier in the section on sensemaking. The garden path paradigm is useful for studying problem detection, measuring the point at which the participants realize that the situation is different than they thought, or, in a dynamic scenario, that the fault has propagated and is detected. One elaboration of the garden path scenario would be to dribble in cues that are themselves innocuous, but in concert reveal a threatening pattern. As Klein et al. (2005) discuss, the noisiness of the background and other factors can be manipulated to make it harder to detect the problem. The measure is still the point at which the participant notices that the situation has changed or requires a different strategy.

Noticing anomalies depends on having clear expectancies. Therefore, an additional measure can be the sharpness and accuracy of expectancies. Endsley's (1995) description of situation awareness methods is relevant here, particularly for generating predictions. A related approach is the prediction paradigm (Klein et al., 1989), which measures the accuracy of predictions in an ongoing exercise.

Reduced margin of error   The second aspect of detecting problems is noticing that the margin of error has gotten smaller. This aspect is more difficult to study and measure. Eye movements (for example, Pradhan et al., 2005) and other measures of information search may be sensitive to the actions of participants to track factors that may become important. The measures might address whether people begin to identify plausible risks and sequences of events that could result in their failure, or began to lose confidence that the sequence of actions was going to proceed as planned.

Team problem detection   Klein (2006) has examined barriers to problem detection in teams, and these offer some suggestions for measures in studies of team performance. For example, the latency of the first team member to spot a weak signal or new fault, compared to when it was first introduced into the exercise or when it was first available, can be measured, but a full measure would include the time course of the spread of the alarm through the team and the time at which the team reacted. Complications can be introduced such as having the fault or problem be a function of different cues that form a pattern but are presented to different members of the overall team.

Planning

We can distinguish at least three main aspects of planning: Generating a Course of Action (COA), allocating resources, and replanning.

Generating a course of action   The most impressive feat of planning occurs when a leader looks over the situation and quickly sees what to do. In military contexts, this is referred to as 'coup d'oiel.' Napoleon is often identified as the exemplar of an intuitive commander who relied on this experience to see when, where, and how to strike, depending on the circumstances. The essence of coup d'oiel is to see the sweet spot, the small lever that can result in a large impact. In one informal study we measured this in a military context—infantry exercise—by presenting a map-based situation. The three experienced regimental commanders saw what they had to do in ten minutes or less. The three less-experienced S-3 officers took 45 minutes to an hour to calculate the layout of forces; at best, their solutions were as good as those of their commanding officer.

Therefore, the paradigm would rely on a prepared exercise with a predetermined sweet spot. The measures would be the time needed to locate the sweet spot in the situation, or to find another sweet spot that is judged to be of equivalent importance. Expert ratings would be used to construct the scenario and materials and to pre-rate different responses.

Allocating resources   Performance here depends on the ability to identify the primary resources and to see new and more effective ways of using these resources. Measures here would consist of ratings of the way resources were allocated. In studying actual performance, using retrospective accounts, the evaluation would be done by Subject-Matter Experts (SMEs) using quality rating scales. In a simulated exercise, a measure could be the number of participants who select different resource configurations, which would then be rated by SMEs.

Replanning

We can view replanning as an aspect of planning, although some plans are executed as drawn up and require little or no replanning, and sometimes replanning takes on a different dynamic than planning itself (see Klein et al. 2003 for a description of replanning). Replanning usually begins when people detect problems with a plan and diagnose these as inherent—they are unlikely to go away. Thus, replanning is often initiated by problem detection. In a CDM study, the data can be reviewed and rated as to whether the problems could have been noticed earlier. In a simulation, with controlled inputs, researchers can measure the time needed to appreciate the significance of early warning signs. These measures can also be used in situations where the replanning is triggered by opportunities rather than by problems. Replanning also depends on success in discovering important goal properties (measures can include the speed of discovering new properties and the accuracy of noticing relevant properties and the proportion of participants who made the necessary discovery) and in performing goal tradeoffs. To evaluate the way people perform goal tradeoffs, a scenario could be presented, unexpected events could be introduced, and performance in prioritizing competing goals would be observed and subsequently rated.

Replanning also presents some critical decisions that allow measurement. How long will it take to make the changes—and will there be enough time to make the changes and disseminate the revised plan? In a simulated task, a confederate, acting as a supervisor, would request this information which would then be rated. Is it better to stick with the original, flawed plan or to shift to a new one? In an exercise, the measures would be the speed of shifting to a new direction, and the proportion of respondents who stick versus those who shift.

There is no subsection on team planning or team replanning because planning and replanning are typically accomplished by teams, rather than by individuals.

Adapting

The macrocognitive function of adapting covers a wide range of activities. We can distinguish three of these—adapting a plan, adapting goals, and adapting mental models (Klein & Pierce, 2001).

Adapting a plan   This aspect of adapting has already been discussed in the previous section, as replanning. Measures would include the time needed to detect the problem with the initial plan, the time needed to formulate a revision, and the quality of the revision, as rated by SMEs.

Adapting goals   The military has a saying that no plan survives the first encounter with the adversary. This saying emphasizes the need for adapting the plan in order to reach the goals of the mission. However, another form of adapting is to change the goals themselves. Klein (2007) has described why it is so critical to adapt the goals when faced with complex conditions and wicked problems. Klein (2009) has also discussed the concept of goal fixation, whereby people and teams resist changing their goals even when these goals have become obsolete or counterproductive. Therefore, in a simulated exercise, one measure is the proportion of individuals/teams who actually do abandon counterproductive goals. In a study using retrospective accounts of plans exposed to turbulent and complex situations, goal revisions can be coded and tabulated.

Adapting mental models   Here we come to the central aspect of macrocognition. If we define macrocognition as the cognitive adaptation to complexity, the core of adaptation is to reformulate one's understanding of events and their connections. Reformulating the way one understands a situation is often necessary for problem detection because it is only when people begin to see a situation differently that they can spot the problem. Klein & Hoffman (2008) have described a range of strategies for investigating different kinds of mental models. The measures would cover differences in the nature of beliefs as a function of new messages or types of feedback. The measures need to reflect whether people repaired flaws in their beliefs, such as beliefs about causal relations.

Team adaptation   Klein & Pierce (2001) defined adaptive teams as ones that are able to make the necessary modifications to meet new challenges. Based on their description, measures of team adaptation might focus on external adaptations (for example, changes in the plans and goals) and internal adaptation (changes in the organizational structure of the team). Measures of external adaptation would assess the team's ability to find solutions for a wide variety of conditions, relying on improvisation rather than careful and detailed planning. Measures of internal adaptation would reflect changes in the way the team parsed difficult tasks, and kept coordination costs down. Klein & Pierce suggest several additional measures of a team's adaptive capacity: level of awareness of mistakes and shortcomings, preparation for workarounds, ability to calibrate common ground, mindset for adaptation, and appropriate balance of centralized control and decentralized control.

Coordinating

The last macrocognitive function covered in this chapter is specifically focused on teams and organizations—coordinating the activities of different participants (including intelligent agents as team members). We'll consider three aspects of supporting coordination: adhering to the basic compact of responsibilities that come with being part of a team or organization, increasing predictability, and making it easier to direct other team members.

Adhering to the basic compact   Klein et al. (2004b), building on the work of Clark (1991), postulated that effective teams depend on a basic compact to work together, and to maintain and repair common ground, particularly if there seems to be some risk of confusion. To measure the strength of adherence to a basic compact, team members in a simulation can be exposed to competing demands to determine the degree of counter-pressure needed to abrogate the basic compact. Similarly, teams need to coordinate competing goals, and the quality of their coordination would be reflected by the amount of time needed to successfully negotiate these differences. The basic compact, an obligation on all team members, will require coordination costs for detecting and repairing common ground breakdowns. Measures here include efficiency and effectiveness in performing central and secondary tasks despite the coordination costs, as a result of reducing those costs. Measures should also reflect the time/effort spent on coordination. Researchers might obtain a process trace of team activity and use that to derive a ratio: the time on primary/secondary tasks versus time on coordination. Following Hoffman (this volume), macrocognitive approaches are likely to include more of these kinds of compound measures.

Predictability   Team members need to be predictable in order to coordinate smoothly. In order to be predictable, team members need to accurately model their colleagues' intentions and actions. In studying actual team interactions, researchers can request individuals to identify their own intentions at given points, and also speculate on the intentions of others. A mismatch score would indicate the degree of predictive accuracy. In simulations, it should be easier to capture predictive accuracy. Team members can also be asked to predict the actions about to be taken by others, and their predictions assessed for accuracy. Another approach to measurement is to collect data such as videotapes and messages during actual events or simulation exercises and present these to outsiders. The better a team is at signaling status and intentions, the greater the accuracy of an outsider in predicting their actions.

Directability   Team members also have to be able to direct each other. Measures include time and effort (for example, number of messages needed) in redirecting the attention and behavior of others. One important aspect of directability is to direct the attention of others. During simulations and exercises, unexpected events can be introduced in order to measure the time and effort needed for all relevant team members to notice the new cues.

Conclusion

This chapter identifies ways of studying six key macrocognitive functions: decision making, sensemaking, problem detection, planning, adapting, and coordinating. Key aspects for each of these functions are identified and types of measures are suggested. The set of aspects and measures are not intended to be exhaustive. Rather, they illustrate the kinds of measures that could be used for future research on and applications of macrocognitive models.

The chapter also identifies some of the most common paradigms for collecting data on macrocognitive functions. Again, the catalog of paradigms is not intended to be exhaustive. It is intended to help researchers realize that they are using the same methods even when they use different labels for their paradigms. Hopefully, by showing the similarity of paradigms, the chapter will strengthen common ground in the field of macrocognitive research, and will encourage comparison of findings.

For example, it appears that most experimental research on sensemaking falls into only seven categories. The use of different terms may obscure the commonality we observe, and make comparisons more difficult than necessary.

Researchers need to develop more, new, and better measures if we are to tap into the complexity of cognitive work and the complexity of macrocognitive phenomena. The collection and comparison of empirical evidence is central to scientific investigation. These data are not sufficient for explicating the phenomena (one also needs to work on refining theory), but they are critical for making progress in understanding macrocognition.

Acknowledgements

I would like to thank Robert Hoffman, Donald Cox, and Emily Patterson for their helpful comments and suggestions. The preparation of this chapter was through participation in the Advanced Decision Architectures Collaborative Technology Alliance, sponsored by the US Army Research Laboratory under Cooperative Agreement DAAD19-01-2-000.

References

Baxter, H. C., & Phillips, J. K. (2004). Evaluating a scenario-based training approach for enhancing situation awareness skills, *Interservice/Industry Trainig, Simulation, and Education Conference (I/ITSEC)*. Orlando, FL: Klein Associates.

Blickensderfer, E. (1998). *Shared expectation measure: Tennis doubles*. Orlando, FL: Naval Air Warfare Center Training Systems Division.

Bruner, J. S., & Potter, M. C. (1964). Interference in visual recognition. *Science, 144*(3617), 424-425.

Crandall, B., Klein, G., & Hoffman, R. R. (2006). *Working minds: A practitioner's guide to Cognitive Task Analysis*. Cambridge, MA: The MIT Press.

Clark, H. H., & Brennan, S. A. (1991). Grounding in communication. In L. B. Resnick, J. M. Levine, & S. D. Teasley (Eds.), *Perspectives on socially shared cognition*. Washington DC: APA Books.

de Groot, A. D. (1946/1978). *Thought and choice in chess*. New York, NY: Mouton.

Endsley, M. R. (1995). Measurement of situation awareness in dynamic systems. *Human Factors, 37*(1), 65-84.

Feltovich, P. J., Johnson, P. E., Moller, J. H., & Swanson, D. B. (1984). LCS: The role and development of medical knowledge in diagnostic expertise. In W. J. Clancey & E. H. Shortliffe (Eds.), *Readings in medical artificial intelligence: The first decade* (pp. 275-319). Reading: Addison-Wesley.

Kaempf, G. L., Klein, G., Thordsen, M. L., & Wolf, S. (1996). Decision making in complex command-and-control environments. *Human Factors, 38*(2), 220-231.

Klein, G. (1998). *Sources of power: How people make decisions*. Cambridge, MA: MIT Press.

Klein, G. (2004). *The power of intuition*. New York, NY: A Currency Book/ Doubleday.

Klein, G. (2006). The strengths and limitations of teams for detecting problems. *Cognition, Technology and Work, 8*(4), 227-236.

Klein, G. (2007). Flexecution as a paradigm for replanning, Part 1. *IEEE Intelligent Systems, 22*(5), 79-83.

Klein, G. (2009). Streetlights and Shadows: Searching for the keys to adaptive decision making. Cambridge, MA: MIT Press.

Klein, G., Calderwood, R., & Clinton-Cirocco, A. (1986). Rapid decision making on the fireground. In *Proceedings of the Human Factors and Ergonomics Society 30th Annual Meeting, 1*, 576-580.

Klein, G., Calderwood, R., & MacGregor, D. (1989). Critical decision method for eliciting knowledge. *IEEE Transactions on Systems, Man, and Cybernetics, 19*(3), 462-472.

Klein, G., & Hoffman, R. (2008). Macrocognition, mental models, and cognitive task analysis methodology. In J. M. Schraagen, M. L. G, T. Ormerod &

R. Lipshitz (Eds.), *Naturalistic decision making and macrocognition* (pp. 57-80). Aldershot, UK: Ashgate.

Klein, G., Lewis, W. R., & Klinger, D. W. (2003). Replanning the army unit of action command post, *Annual Project Report submitted under Contract DAAD19-01-S-0009 prepared for the Army Research Laboratory*. Fairborn, OH: Klein Associates Inc.

Klein, G., Long, W. G., Hutton, R. J. B., & Shafer, J. (2004a). Battlesense: An innovative sensemaking-centered design approach for combat systems, *Final report prepared under contract # N00178-04-C-3017 for Naval Surface Warfare Center, Dahlgren, VA*. Fairborn, OH: Klein Associates Inc.

Klein, G., and Peio, K. J. (1989). The use of a prediction paradigm to evaluate proficient decision making. *American Journal of Psychology, 102*(3), 321-331.

Klein, G., Phillips, J. K., Battaglia, D. A., Wiggins, S. L., & Ross, K. G. (2002). FOCUS: A model of sensemaking. Interim Report-Year 1 prepared under Contract 1435-01-01-CT-31161 [Department of the Interior] for the U.S. Army Research Institute for the Behavioral and Social Sciences, Alexandria, VA (I). Fairborn, OH: Klein Associates Inc.

Klein, G., Phillips, J. K., Rall, E., & Peluso, D. A. (2007). A data/frame theory of sensemaking. In R. R. Hoffman (Ed.), *Expertise out of context: Proceedings of the 6th International Conference on Naturalistic Decision Making*. Mahwah, NJ: Lawrence Erlbaum & Associates.

Klein, G., & Pierce, L. (2001). Adaptive teams. In *Proceedings of the 2001 Command and Control Research and Technology Symposium [CD-Rom]*. Monterey, CA: Naval Postgraduate School.

Klein, G., Pliske, R. M., Crandall, B., & Woods, D. (2005). Problem detection. *Cognition, Technology, and Work, 7*(1), 14-28.

Klein, G., Wiggins, S. L., & Dominguez, C. O. (in press). Team sensemaking. *Theoretical Issues in Ergonomic Science*.

Klein, G., Wolf, S., Militello, L., & Zsambok, C. (1995). Characteristics of skilled option generation in chess. *Organizational Behavior and Human Decision Processes, 62*(1), 63-69.

Klein, G., Woods, D. D., Bradshaw, J. M., Hoffman, R. R., & Feltovich, P. J. (2004b). Ten challenges for making automation a 'team player' in joint human-agent activity. *IEEE Intelligent Systems, 19*(6), 91-95.

Patterson, E. S., Roth, E. M., & Woods, D. D. (2001). Predicting vulnerabilities in computer-supported inferential analysis under data overload. *Cognition, Technology and Work, 3*(4), 224-237.

Pliske, R. M., Klinger, D., Hutton, R., Crandall, B., Knight, B., & Klein, G. (1997). Understanding skilled weather forecasting: Implications for training and the design of forecasting tools. Technical Report AL/HR-CR-1997-0003. Brooks AFB, TX: U.S. Air Force Armstrong Laboratory.

Pradhan, A. K., Hammel, K. R., DeRamus, R., Pollatsek, A., Noyce, D. A., & Fisher, D. L. (2005). Using eye movements to evaluate effects of driver age on risk perception in a driving simulator. *Human Factors, 47*(4), 840-852.

Rudolph, J. W. (2003). *Into the big muddy and out again.* Unpublished Doctoral Thesis, Boston College, Boston, MA.

Schraagen, J. M., Klein, G., & Hoffman, R. (2008). The macrocognitive framework of naturalistic decision making. In J. M. Schraagen, L. Militello, T. Ormerod & R. Lipshitz (Eds.), *Naturalistic decision making and macrocognition* (pp. 3-26). Aldershot, UK: Ashgate.

Shattuck, L. G., & Woods, D. D. (2000). Communication of intent in military command and control systems. In C. McCann & R. Pigeau (Eds.), *The human in command: Exploring the modern military experience* (pp. 279-291). New York, NY: Plenum Publishers.

Smith, P. J., Giffin, W. C., Rockwell, T. H., & Thomas, M. (1986). Modeling fault diagnosis as the activation and use of a frame system. *Human Factors, 28*(6), 703-716.

Snowden, D., Klein, G., Chew, L. P., & Teh, C. A. (2007). A sensemaking experiment: Techniques to achieve cognitive precision. In *Proceedings of the 12th International Command and Control Research and Technology Symposium.*

Vicente, K. J., & Wang, J. H. (1998). An ecological theory of expertise effects in memory recall. *Psychological Review, 106*(1), 33-5.

Chapter 5

Measuring Attributes of Rigor in Information Analysis

Daniel J. Zelik, Emily S. Patterson, and David D. Woods

Introduction

Information analysis describes the process of 'making inferences from available data' and determining 'the best explanation for uncertain, contradictory, and/or incomplete data' (Trent et al., 2007, p. 76). It is an accumulation and interpretation of evidence to support decision making (Heuer, 1999; Schum, 1987) and it represents an active, goal-directed, and often technologically mediated, sensemaking activity (Klein et al., 2007).

This macrocognitive framing of information analysis as 'sensemaking' (Klein, Chapter 4; Klein, et al., 2006a, 2006b) highlights that cognition employs narrative elements to recognize and communicate meaning, and that sensemaking—as with other macrocognitive functions such as (re)planning—is (1) joint activity distributed over time and space, (2) coordinated to meet complex, dynamic demands, (3) conducted in an uncertain, event-driven environment, (4) often performed with simultaneously complementary and conflicting goals with high consequences for failure, (5) made possible by effective expertise in roles, and (6) shaped by organizational constraints (see Preface; Klein et al., 2003; Schraagen et al., 2008).

Thus, information analysis in both professional domains such as intelligence analysis and more everyday contexts like online shopping is a deliberate process of collecting data, reflecting upon it, and aggregating those findings into knowledge, understanding, and the potential for action. However, even in contexts where there are relatively low consequences for inaccurate conclusions, there lurk ever-present vulnerabilities exacerbated by ever-increasing data availability and connectivity (Woods et al., 2002; Treverton & Gabbard, 2008; Betts, 2002; 1978). Of particular concern is the potential inherent in all analytical activity that an analysis process is prematurely concluded and is subsequently of inadequate depth relative to the demands of a situation—a vulnerability we characterize as the risk of shallow analysis.

This chapter examines rigor as a measure of sensemaking in information analysis activity that captures how well an analytical process reduces this risk. It describes eight attributes of analytical rigor that were synthesized from empirical studies with professional intelligence analysts from a variety of specialties and agencies (Patterson et al., 2001; Miller et al., 2006; Trent et al., 2007; Zelik et al., 2007; Grossman et al., 2007). Examples are then provided that illustrate how these attributes were calibrated for two different studies of analytical activity. Finally, potential future directions are discussed for using the attributes to assess, negotiate, and communicate the rigor of information analysis processes.

Analytical Rigor as a Measure of Process

Across domains, conventional perspectives for measuring analytical rigor focus on identifying gaps between what was actually done versus a prescribed or 'standard' method, with rigor variously defined as the 'scrupulous adherence to established standards' (Crippen et al., 2005, p. 188), the 'application of precise and exacting standards' (Military Operations Research Society, 2006, p. 4), 'methodological standards for qualitative inquiry' (Morse, 2004, p. 501), and the 'unspoken standard by which all research is measured' (Davies & Dodd, 2002, p. 280).

Unfortunately, such definitions suggest a conceptualization of analytical activity that is neither particularly likely to reflect rigorous analysis work as practiced (Dekker, 2005; Sandelowski, 1993; 1986), nor particularly useful in developing the concept of rigor as a measure of performance. The diverse nature of intelligence analysis in particular renders the identification of a single 'standard' process an intractable, if not impossible task (Berkowitz & Goodman, 1991; Krizan, 1999; Marrin, 2007). Consequently, rather than on a standards-based notion of rigor, our measurement approach focuses on how the risk of shallow analysis is reduced via analyst-initiated strategies that are opportunistically employed throughout the analysis process. These strategies are alternatively conceptualized as 'broadening' checks (Elm et al., 2005) insofar as they tend to slow the production of analytic product and make explicit the sacrifice of efficiency in pursuit of accuracy, a central tenet of the framework.

Stories of Shallow Analysis

In our research with professional information analysts, study participants were often asked to directly evaluate or critique the products and processes of other analysts (for example, Miller et al., 2006; Zelik et al., 2007). Across these studies, consistent patterns emerged as experienced analysts identified critical vulnerabilities in information analysis by way of 'stories' describing how analytical processes often can or did go wrong. We organize these stories into eight interrelated risks of shallow analysis.

Specifically, we found that shallow analysis:

1. Is structured centrally around an inaccurate, incomplete, or otherwise weak primary hypothesis, which analysts sometimes described as favoring a 'pet hypothesis' or as a 'fixation' on an initial explanation for available data.

2. Is based on an unrepresentative sample of source material, for example, due to a 'shallow search,' or completed with a poor understanding of how the sampled information relates to the larger scope of potentially relevant data, for example, described as a 'stab in the dark.'

3. Relies on inaccurate source material, as a result of 'poor vetting' for example, or treats information stemming from the same original source as if it stems from independent sources, labeled variously as 'circular reporting,' 'creeping validity,' or as the 'echo chamber' effect.

4. Relies heavily on sources that have only a partial or, in the extreme, an intentionally deceptive stance toward an issue or recommended action, often characterized by analysts in terms of 'biased,' 'slanted,' 'polarized,' or 'politicized' source material.

5. Depends critically on a small number of individual pieces of often highly uncertain supporting evidence proving accurate, identified by some individuals as an analysis heavily dependent upon 'hinge evidence' or, more generically, as a 'house of cards' analysis.

6. Contains portions that contradict or are otherwise incompatible with other portions, for example, via the inclusion of lists or excerpts directly 'cut and paste' from other documents or via an assessment that breaks an issue into parts without effectively reintegrating those parts.

7. Does not incorporate relevant specialized expertise, for example, an analyst who 'goes it alone,' or, in the other extreme, one who over relies on the perspectives of domain experts.

8. Contains weaknesses or logical fallacies in reasoning from data to conclusion, alternatively described as having a 'thin argument,' a 'poor logic chain,' or as involving 'cherry picking' of evidence.

In addressing each of these sources of risk, eight corresponding attributes of analytical rigor were identified (see Figure 5.1). Together these attributes represent a model of how experts critique analytical process, rather than a model of the analytical activities of those experts. Consequently, these attributes make salient the aspects of process that are attended to in making an expert judgment of analytical rigor.

Our approach to operationalize the model involves translating empirically-based descriptions of each attribute into differentiating categories of analytical behaviors. Each attribute is thus characterized in terms of low, moderate, and high indicators of rigorous process. Using this rubric, analytical processes can be rated or scored by comparing the methods employed during a given analysis to the categorical descriptions that appear in Figure 5.1. Scoring an analysis process across all attributes, in turn, reflects an overall assessment of analytical rigor.

Generally, if no broadening checks are observed in relation to a particular aspect of risk, we code it as 'low' since analysts, both casual and professional, typically show some awareness of the possible risks of being inaccurate in an

Attribute / Description	Indicators of...		
	LOW Rigor	**MODERATE Rigor**	**HIGH Rigor**
Hypothesis Exploration The construction and evaluation of potential explanations for collected data.	- Little or no consideration of alternatives to primary or initial hypotheses. - Interpretation of ambiguous or conflicting data such that they are compatible with existing beliefs. - Fixation or knowledge shielding behaviors.	- Some consideration of how data could support alternative hypotheses. - An unbalanced focus on a probable hypothesis or a lack of commitment to any particular hypothesis.	- Significant generation and consideration of alternative explanations via the direct evaluation of specific hypotheses. - Incorporation of "outside" perspectives in generating hypotheses. - Evolution and broadening of hypothesis set beyond an initial framing. - Ongoing revision of hypotheses as new data are collected.
Information Search The focused collection of data bearing upon the analysis problem.	- Failure to go beyond routine and readily available data sources. - Reliance on a single source type or on data that are far removed from original sources. - Dependence upon "pushed" information, rather than on actively collected information. - Use of stale or dated source data.	- Collection from multiple data types or reliance on proximal sources to support key findings. - Some active information seeking.	- Collection of data from multiple source types in addition to the use of proximal sources for all critical inferences. - Exhaustive and detailed exploration of data in the relevant sample space. - Active approach to information collection.
Information Validation The critical evaluation of data with respect to the degree of agreement among sources.	- General acceptance of information at face value, with little or no clear establishment of underlying veracity. - Lack of convergent evidence. - Poor tracking and citation of original sources of collected data.	- Use of heuristics to support judgements of source integrity, e.g. relying on sources that have previously proven to be consistently accurate. - A few "key" high-quality documents are relied on heavily. - Recognizes and highlights inconsistencies between sources.	- Systematic and explicit processes employed to verify information and to distinguish facts from judgements. - Seeks out multiple, independent sources of converging evidence. - Concerned both with consistency between sources and with validity and credibility within a given source.
Stance Analysis The evaluation of collected data to identify the relative positions of sources with respect to the broader contextual setting.	- Little consideration of the views and motivations of source data authors. - Recognition of only clearly biased sources or sources that reflect a well-defined position on an issue.	- Perspectives and motivations of authors are considered and assessed to some extent. - Incorporates basic strategies to compare perspectives of different sources, e.g. by dividing issues into "for" or "against" positions.	- Involves significant research into, or leverages a preexisting knowledge of, the backgrounds and views of key players and thought leaders. - May involve more formal assessments of data sources, e.g. via factions analysis, social network analysis, or deception analysis.
Sensitivity Analysis The evaluation of the strength of an analytical assessment given possible variations in source reliability and uncertainty.	- Explanations are appropriate and valid at a surface level. - Little consideration of critical "what if" questions, e.g., "What if a given data source turns out to be unreliable?" or "What if a key prediction does not transpire as anticipated?"	- Considers whether being wrong about some inferences would influence the overall best explanation for the data. - Identifies the boundaries of applicability for an analysis.	- Goes beyond simple identification to specify the strength of explanations and assessments in the event that individual supporting evidence or hypotheses were to prove invalid or unreliable. - Specifies limitations of the analysis, noting the most vulnerable explanations or predictions on which the analysis is at risk of erring.
Information Synthesis The extent to which an analyst goes beyond simply collecting and listing data in "putting things together" into a cohesive assessment.	- Little insight with regard to how the analysis relates to the broader analytical context or to more long-term concerns. - Lack of selectivity, with the inclusion of data or figures that are disconnected from the key arguments or central issues. - Extensive use of lists or the restatement of material copied directly from other sources with little reinterpretation.	- Explicit, though perhaps not systematic, efforts to develop the analysis within a broader framework of understanding. - Depiction of events in context and framing of key issues in terms of tradeoff dimensions and interactions. - Provides insight beyond what is available in the collected data.	- Extracted and integrated information in terms of relationships rather than components and with a thorough consideration of diverse interpretations of relevant data. - Re-conceptualization of the original task, employing cross-checks on abstractions. - Performed by individuals who are "reflexive" in that they are attentive to the ways in which their cognitive processes may have hindered effective synthesis.
Specialist Collaboration The extent to which substantive expertise is integrated into an analysis.	- Minimal direct collaboration with experts. - Little if any on-topic, "outside" expertise is accessed or sought out directly.	- Involves some direct interaction with experts, though usually via readily available specialists. - Expertise is drawn from within preexisting personal or organizational networks.	- Independent experts in key content areas are identified and consulted. - Efforts to go beyond a "core network" of contacts to seek out domain-relevant experts, with additional resources and "political capital" potentially expended to gain access to such specialist expertise.
Explanation Critiquing The critical evaluation of the analytical reasoning process as a whole, rather than in the specific details.	- Few if any instances of alternative or "outside" criticisms being considered. - Reliance on preexisting channels of critiquing, primarily those supervisory.	- Brings alternative perspectives to bear in critiquing the overall analytical process. - Leverages personal or organizational contacts to examine analytical reasoning, e.g. by way of peer analysts, proxy decision makers, etc.	- Familiar as well as independent perspectives have examined the chain of analytical reasoning, explicitly identifying which inferences are stronger and weaker. - Use of formal methods such as "red teams" or "devil's advocacy" to challenge and vet hypotheses and explanations. - Expenditure of capital, political or otherwise, in critiquing the analytical process.

Figure 5.1   Attributes of rigorous analysis organized with respect to indicators of low, moderate, and high rigor

assessment with respect to a given attribute. For 'moderate' rigor, encompassed are activities that display an active response to a particular type of risk, while strategies that are particularly effortful, systematic, or exhaustive are coded as 'high.' Finally, aspects of an analysis that are not applicable to a given context are generally not scored, for example, attributes involving collaboration on a task where an analyst was expected to work alone.

Attributes of Analytical Rigor

1. *Hypothesis exploration.* Hypotheses are among the most basic building blocks of analytical work, representing candidate explanations for available data (Elm et al., 2005) and one of the necessary components of inferential analysis (Schum, 1987). As noted by Kent (1966), 'what is desired in the way of hypotheses . . . is quality and quantity . . . a large number of possible interpretations of the data, [and] a large number of inferences, or concepts, which are broadly based and productive of still other concepts' (p. 174). Consistent with this perspective, rigorous analysis is identified by the depth and breadth of the generation and consideration of alternatives, by the incorporation of diverse perspectives in brainstorming hypotheses, by the evolution of thinking beyond an initial problem framing, and by the ongoing openness to the potential for revision.

   Hypothesis exploration also represents a check against 'premature narrowing' (Woods & Hollnagel, 2006; Cooper, 2005), given the concern 'that if analysts focus mainly on trying to confirm one hypothesis they think is probably true . . . [they] fail to recognize that most of this evidence is also consistent with other explanations or conclusions' (Heuer, 1999, p. 96). Accordingly, indicators of weak hypothesis exploration include a focus on confirming a single explanation, a failure to consider alternative perspectives on available data, an interpretation of ambiguous or conflicting data such that it is compatible with existing beliefs, and analyst behavior indicative of fixation (Pherson, 2005; De Keyser & Woods, 1990).

2. *Information search.* Similarly viewed as a fundamental component of analysis work, this attribute is alternatively described as the 'collection of data bearing upon the [analysis] problem' (Kent, 1966, p. 157) or as the focused 'extraction of an essential, representative, "on analysis" sample from available data' (Elm et al., 2005, p. 297). Thus, information search encompasses all activities performed to gather task-relevant evidence— including those to broaden as well as deepen, those that are active as well as passive, and those hypothesis-driven as well as data-driven. Note that this framing of information search reflects the diverse nature of analytical activity and emphasizes the fact that, for the professional analyst, supporting evidence comes in many forms, and not simply as raw data (Krizan, 1999).

   Information search is primarily concerned with where and how analysts look for supporting information. A strong information search process is characterized by the extensive exploration of relevant data, by the collection of data from multiple source types, and, most critically, by an active approach to information collection. In this context, active collection describes efforts to seek information beyond that which is readily available (Serra, 2008). A weak information search in contrast is identified by failure to go beyond routine and readily available data sources, by reliance on a single source type or on 'distant' data that is removed from original source material, and by passive dependence upon 'pushed' rather than actively collected data.

3. *Information validation*. This attribute is concerned with the 'critical evaluation of data' (Kent, 1966, p. 157) and with determining the level of conflict and corroboration—or agreement and disagreement—among sources (Elm et al., 2005). In rigorous analysis, analysts make an explicit effort to distinguish fact from judgment and are concerned with consistency and credibility among, as well as within, sources. Thus, a strong validation process involves assessing the reliability of sources, assessing the appropriateness of sources relative to the task question, and the use of proximate sources whenever possible (Krizan, 1999). It also involves an explicit effort to seek out multiple, independent sources of converging evidence for key findings.

   In contrast, weak information validation is reflected in the uncritical acceptance of data at face value, little or no clear effort to establish underlying veracity, and a failure to collect independent supporting evidence. Poor tracking and citation of original sources also identify such analyses. Between strong and weak characterizations, a moderate validation process involves the recognition of inconsistencies among sources and, often times, involves the use of heuristics to support judgments of source integrity—such as deference to sources that have previously proven highly reliable and avoidance of those that have not. On the aggregate, then, information validation can be described as an intense concern with issues of agreement, consistency, and reliability with respect to the set of collected data.

4. *Stance analysis*. 'Stance' refers to the perspective of a source on a given issue and is alternatively discussed in the language of viewpoint, outlook, or affiliation—though more often it is characterized in terms of slant, bias, or predisposition. Stance analysis refers to the evaluation of information with the goal of identifying the positions of sources with respect to a broader contextual understanding and in relation to alternative perspectives on an issue. In the extreme, this analysis serves as a guard against deception (Johnson et al., 2001; Moore & Reynolds, 2006; Heuer, 1987). More prosaically, however, it serves as a check against a reliance on unrepresentative data sets that reveal only partial perspectives on an issue. A process in which little attention is paid to issues of stance reflects weak analysis. In such instances, the analysis may identify heavily slanted sources or sources that support a well-defined position on an issue but yet reflect little in the way of a nuanced understanding. A somewhat improved stance analysis would incorporate basic strategies for considering the perspectives of different sources. For example, dividing evidence into camps that are 'for' or 'against' an issue represents a simplifying heuristic for organizing and making sense of various stances on that issue.

A significantly stronger stance analysis involves research into, or leverages a pre-existing knowledge of, the backgrounds and views of key individuals, groups, and thought leaders. Where appropriate, it may also include a more formal assessment that employs structured methods to identify critical relationships, to predict how the general worldview of a source is likely to influence his or her stance toward specific issues, or to detect the intentional manipulation of information. Social network analysis (Butts, 2008; Moon & Carley, 2007; Krebs, 2002), faction analysis (Feder, 1987; Smith et al., 2008), and analysis of competing hypotheses (Heuer, 1999; Folker, 2000), for example, are techniques that may be selectively employed to those ends.

5. *Sensitivity analysis.* The term 'sensitivity', as it is used here, has a meaning most similar to its usage in the statistical analysis of quantitative variables, wherein it describes the extent to which changes in input parameters affect the output solution of a model (Saltelli et al., 2000). However, rather than with the relationship between output and input variance, our concern is with the strength of an analytical assessment given the potential for low reliability and high uncertainty in supporting evidence and explanations. Phrased differently, sensitivity analysis describes the process of discovering the underlying assumptions, limitations, and scope of an analysis as a whole, rather than those of the supporting data in particular, as with the related attribute of information validation.

   Many in the intelligence community emphasize the importance of examining analytical assumptions. Schum (1987), for example, identifies assumptions as foundational to analytic work. Similarly, Davis (2002) notes that 'careful attention to selection and testing of key assumptions to deal with substantive uncertainty is now well established as the doctrinal standard for [analysis work]' (p. 5). As such, effective sensitivity analysis demands the identification and explication of the key beliefs that tie an analysis together and that link evidence with hypotheses.

   To that end, a strong sensitivity analysis goes beyond simple identification, meticulously considering the strength of explanations and assessments in the event that individual supporting evidence or hypotheses were to prove invalid. It also specifies the boundaries of applicability for the analysis. With weak sensitivity analysis, in contrast, explanations seem appropriate or valid at surface level, with little consideration of critical 'what if' questions—for example, 'What if a key data source misidentified a person of interest?' Likewise, the overall scope of a weak analysis process may be unclear or undefined.

6. *Information synthesis.* Often emphasized by experts more than casual analysts is that rigorous analytical work is as much about putting concepts together as it is about breaking an issue apart (Cooper, 2005; Mangio & Wilkinson, 2008). That is to say, rigorous analysis demands not only 'analytic' activity in the definitional sense, but 'synthetic' activity as

well. Thus, information synthesis is a reflection of the extent to which an analysis goes beyond simply collecting and listing data to provide insights not directly available in individual source data.

Weak information synthesis is reflected in analyses that succeed in compiling relevant and 'on topic' information, but that do little in the way of identifying changes from historical trends or providing guidance for broader or more long-term concerns. Indicators of weak synthesis include extensive use of lists, copying material from other sources with little reinterpretation, and a lack of selectivity in what is emphasized by the analysis. A stronger synthesis is reflected by explicit efforts to develop an analysis within a broader framework of understanding. The depiction of events in relation to historical or theoretical context and the framing of key issues in terms of tradeoff dimensions and interactions also identify such analysis.

Stronger still is synthesis that has integrated information in terms of relationships rather than components, with a thorough consideration of diverse interpretations of relevant data. In addition, such synthesis is performed by reflexive analysts who are attentive to ways in which their particular analytical processes may hinder effective synthesis and who are attuned to the many potential 'cognitive biases' that manifest in analytical work (Moore, 2007; Krizan, 1999; Heuer, 1999).

7. *Specialist collaboration.* Inevitably, analysts encounter topics on which they are not expert or that require multiple areas of expertise to fully make sense of (Kent, 1966). Even in instances where an analyst has expertise in pertinent topics, success for the modern analyst still demands the incorporation of multiple perspectives on an issue (Clark, 2007; Medina, 2002). Accordingly, analytical rigor is enhanced when substantive expertise is brought to bear on an issue (Tenet, 1998).

The level of effort expended to incorporate relevant expertise defines effective specialist collaboration. In a process with little collaboration, minimal outside expertise is sought out directly. A moderately collaborative analysis process involves some interaction with experts, though at this level such expertise is often drawn from existing personal or professional networks, rather than from organizationally external sources. In a high-rigor process, independent experts in key content areas are identified and consulted. Thus, a strong specialist collaboration process is defined by efforts to go beyond a 'core network' of contacts in seeking out domain-relevant expertise. In many cases, additional resources and 'political capital' are expended to gain access to such specialized knowledge.

8. *Explanation critiquing.* Specialist collaboration and explanation critiquing are related in that both are forms of collaborative analytical activity that reflect the influence of diverse perspectives (Hong & Page, 2004; Guerlain et al., 1999). However, whereas specialist collaboration primarily relates to the integration of perspectives relative to information search and validation,

explanation critiquing relates to the integration of perspectives relative to hypothesis exploration and information synthesis. More succinctly, explanation critiquing is concerned with the evaluation of the overall analytical reasoning process, rather than with the evaluation of content specifically.

Similar to specialist collaboration, however, this attribute is largely defined by the extent to which analysts reach beyond immediate contacts in collecting and integrating alternative critiques. A low-quality explanation critiquing process has limited instances of such integration, while a more moderate process leverages personal and professional contacts to examine analytical reasoning. In the latter case, it is often peer analysts, supervisors, or managers who serve as the primary source of these alternative critiques.

In a still stronger analysis process, independent as well as familiar reviewers have examined the chain of analytical reasoning and explicitly identified which inferences are stronger and which are weaker. In addition, the use of formal methods such as 'red teams' or 'devil's advocacy' (Defense Science Board Task Force, 2003; Pherson, 2005) as well as the expenditure of capital, political or otherwise, serve as hallmarks of strong explanation critiquing.

Examples of Calibrating the Attributes of Rigor to a Particular Task

The attribute-based framing of analytical rigor represents a model in the sense that it 'constrains what can be measured by describing what is essential performance'; it represents an operational framework in that 'the model parameters . . . become the basis for specifying the measurements' (Dekker & Hollnagel, 2004, p. 82). We now provide examples that illustrate how the attributes of rigor can be scored as low, moderate, and high for two very different analysis tasks.

The first example involves scoring the analytical rigor of twelve ad hoc, three-member teams that completed a logistics analysis and planning task set in a military context used in a prior study (Trent et al., 2007). The study task was for each team to develop a plan to move a predetermined number of troops and supplies to a desired location by the fastest, cheapest, and most secure routes possible, given a number of scenario-specific constraints. Half of the teams were collocated and worked together directly while the other half were distributed and limited to audio communication only.

For all cases, verbal transcripts of the study sessions were used to score the analysis and planning processes of the teams with respect to the attributes of rigor. Two independent coders rated each team as low, moderate, or high on each attribute as shown in Figure 5.2, with specialist collaboration not scored due to limitations imposed by the design of the study. Note that both coders were also provided with feedback as to how well the plan developed by each team met the stated scenario requirements.

Attribute	Team 1	2	3	4	5	6	7	8	9	10	11	12
Hypothesis Exploration	Low	Moderate	**High**	**Moderate Low**	Moderate	Moderate	**Moderate Low**	Moderate	**High**	Moderate	Moderate	Low
Information Search	**Moderate Low**	Moderate	**High**	**High**	Moderate	Moderate	Moderate	**High**	**Moderate High**	Moderate	**Moderate High**	Moderate
Information Validation	Low	Low	**Moderate High**	**Moderate High**	Moderate	Low	Low	Low	Low	Moderate	Low	Moderate
Stance Analysis	Low	Moderate	**Moderate High**	Low	Moderate	Moderate	Moderate	Moderate	Low	Moderate	Moderate	Low
Sensitivity Analysis	Low	Low	Moderate	Low	Low	Low	Low	Low	Low	Low	Low	Low
Information Synthesis	Low	Low	Moderate	Moderate	Low	Low	Low	**High**	Low	Low	Low	Low
Specialist Collaboration	–	–	–	–	–	–	–	–	–	–	–	–
Explanation Critiquing	Low	Moderate	**High**	Low	Low	Low	Moderate	Moderate	Moderate	Low	Moderate	Low
Solution Score (out of 60)	44/60	53/60	60/60	31/60	33/60	19/60	24/60	53/60	60/60	60/60	53/60	39/60

Figure 5.2 Scoring the rigor of the analysis processes of teams in the first example, with differences between raters highlighted

Of the 84 items scored (12 teams x 7 scored attributes), 76 were judged by both raters to fall into the same low, moderate, or high category, implying strong agreement between raters ($\kappa_w = 0.86$). Disagreements in coding were most common with respect to the attribute of information search, but also occurred in scoring hypothesis exploration, information validation, and stance analysis. Overall, however, there was general consistency in how the coders applied the framework to assess the analytical rigor of the processes employed by the teams.

The second example involves another study in a laboratory environment in which participants were asked to respond to a quick reaction task in the form of a recorded verbal briefing that answered the question: 'What were the causes and impacts of the failure of the maiden flight of the Ariane 501 rocket launch?' The task was conducted under conditions of data overload, time pressure, and low to no expertise in the area, and the scenario was based on the case of the Ariane 5 rocket, whose maiden launch on June 4, 1996, ended in a complete loss of the rocket booster and scientific payload when it exploded shortly after liftoff, an accident which was significant because it departed from historical launch failure patterns in its underlying causes.

Data originally collected by this study for other purposes were later reanalyzed with respect to the attributes of rigorous analysis for two junior intelligence analysts who had not yet received their security clearances (Miller & Patterson, 2008) and for eight senior intelligence analysts, all from the same intelligence agency (Patterson et al., 2001). In this case, a single individual coded the processes of each analyst (see Figure 5.3), however note that neither specialist collaboration nor explanation critiquing were scored, as the study design did not allow participants to access individuals with specialized expertise or to have their problem framing and briefing critiqued by other analysts. The results of this evaluation are represented graphically in Figure 5.4.

Note that this depiction of the attributes affords an integrated, pattern-based assessment of rigor that reveals an overall picture of the relationships among the attributes, as well as insight into the level of rigor achieved on individual attributes with respect to the demands of a given context (Bennett & Flach, 1992). In this example, interestingly, stance analysis—and to a lesser extent information synthesis and sensitivity analysis—were most diagnostic of expertise on the study task. In considering both examples presented here, it is our observation that there is flexibility in the rigor attribute framework that allows for its application in a variety of analytical contexts and to a variety of assessment tasks.

Discussion

We have presented a framework for and examples of applying an approach to codify the rigor of an information analysis process from a macrocognitive perspective that reflects the diversity of strategies employed by professional analysts to cope with data overload, to make sense of an uncertain and dynamic world, and to

Attribute	Novice		Expert							
	N1	N2	E1	E2	E3	E4	E5	E6	E7	E8
Hypothesis Exploration	**Moderate** No obvious initial bias.	**Moderate** No obvious initial bias.	**Moderate** Relied on high quality sources early on to formulate hypotheses.	**Low** Generally sloppy, rushed to get an acceptable answer.	**Moderate** Spent a lot of time working out the impacts, but less on the diagnosis of the failure.	**Moderate** Explicit strategy for sampling in time, bracketing the event, and playing forward to help break fixations and detect overturning updates.	**Moderate** Considered holistically if there is disagreement in what had been read. When there is disagreement, picked one to rely upon.	**Moderate** Framing evolves over time, even though participant has more subject-matter knowledge than other study participants.	**Moderate** Went quickly, but no obvious initial bias.	**Low** Seemed biased towards "mechanical failure" hypothesis and ended up there.
Information Search	**Moderate** Read 10 documents.	**Moderate** Read 9 documents (2 repeats).	**High** Time spent.	**Low** Time spent.	**High** Time spent.	**High** Time spent.	**Moderate** Time spent.	**High** Time spent.	**Moderate** Time spent, though it appeared more efficient than sloppy.	**Moderate** Time spent.
Information Validation	**Low** Wrong date for accident.	**Moderate** Corroborated some information.	**Moderate** Worried about cut and paste, past predictions, did not trust exact cost estimates.	**Low** Would rely on customer to clarify inconsistencies in data.	**Moderate** Noted discrepancies informally and kept track of when they came up again.	**Moderate** Copied and pasted notes to bring information together on the screen.	**Low** Since no explicit statement of disagreement at a high level in documents, believed details were right (some were not).	**Moderate** Clearly noticed and worked on judging how to resolve discrepancies (but without looking for new documents).	**Low** Seemed to trust everything that was copied and pasted in, but holistically judged no obvious disagreements.	**Moderate** Highlighted phrases that appeared more than once.
Stance Analysis	**Low** Malaysian MBA students treated the same as other sources.	**Low** No judgment of source stances.	**Moderate** Sensitive when judging documents contributions.	**Moderate** Showed sensitivity to potential for bias in what was read, but did not do anything when it was detected.	**Moderate** Impressively nuanced judgments of documents.	**Moderate** Sensitive to stances when reading.	**Moderate** Mostly looked for deception or disagreement and for reasons why, used time as a major cue.	**Moderate** Sensitive to stances when reading.	**Moderate** Sensitive to stances when reading.	**Moderate** Somewhat sensitive to stances when reading.
Sensitivity Analysis	**Low** No sensitivity to risk.	**Low** No sensitivity to risk.	**Moderate** Wanted to do timeline or ask customer to verify the right accident.	**Low** Little sensitivity to risk.	**Moderate** Particularly on impacts, circled the answers to see if similar from different sources.	**Moderate** Spent time and effort to see if new info was different.	**Low** No sensitivity to risk.	**Low** No sensitivity to risk.	**Low** No sensitivity to risk.	**Low** No sensitivity to risk.
Information Synthesis	**Low** Found document and forwarded as answer.	**Low** Personal interests guided what he read next, with no obvious strategy for pulling together information.	**Moderate** Provided insights that went beyond what was read, based on patterns and past experience.	**Low** Briefing at level of what was written and not put in context.	**High** Pulled it all together in a synthesized fashion, and also gave supporting details.	**Low** Gave detailed briefing at the level of what was read in documents; briefed a compiled list.	**Low** No time spent "getting a feel" or contextualizing what was found; briefed a compiled list.	**High** Clearly utilized knowledge to pull together ideas in coherent framework.	**High** Impressive coherent narrative showed a command in general of what he had read at the "gist" level.	**Moderate** Noticed payload not asked about, put things into his own words and framed by question, briefed more than a compiled list.
Specialist Collaboration	—	—	—	—	—	—	—	—	—	—
Explanation Critiquing	—	—	—	—	—	—	—	—	—	—

Figure 5.3   Scoring the rigor of the analysis processes of participants in the second example, with brief rationales for each rating

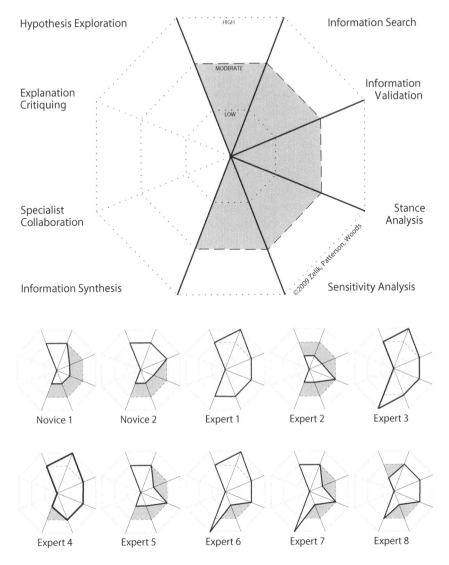

Figure 5.4   Graphical representation of the rigor attribute scores shown in Figure 5.3

reduce the risk of shallow analysis. In practice, however, scoring a given analytic process with respect to these eight attributes is not without challenges, and there are many opportunities for advancing this line of work.

Of principal concern is the fact that the rigor attribute model is grounded largely in the domain of intelligence analysis. Accordingly, the extent to which it proves relevant in other analytical domains remains an open issue, though preliminary research on activities such as information search by students working on educational projects, medical diagnosis using automated algorithms for x-ray interpretation, and accident investigation analyses (Patterson et al., 2008) suggest that the concepts may be fairly robust. Of course, the framework presented here does not purport to be the final word on rigor in information analysis either; rather, it represents a current iteration of an ongoing direction of exploration.

In contrast with the conventional, standards-based notions discussed earlier, our framework supports a view of analytical rigor that is better aligned with calls for guidance—as opposed to guidelines—in conducting analyses of complex and highly interconnected data (Patton, 2002). This is not to suggest that standard methods play an insignificant role in achieving rigor, but that such standards do not necessarily and completely define rigor for the practicing analyst. Rather it is the extent to which these methods and practices speak to the individual attributes that define the overall rigor of an analysis. Of course, in the intelligence community in particular, this is hardly a surprising conclusion (Mangio & Wilkinson, 2008; Marrin, 2007; Johnston, 2005).

A related implication of the rigor attribute model is that, although it elucidates the aspects of analytical process that are perceived as most diagnostic of rigor, it does not directly reveal what is required to be 'rigorous enough.' Again, this is not a surprising conclusion, though it is no less an important one.

In short, judging the sufficiency of rigor depends upon many contextual factors—analyst characteristics such as experience, specialization, and organizational affiliation; format and presentation characteristics such as writing style, presentation medium, and customer requirements; and production characteristics such as time pressure, resource availability, and priority, to name only a few (Krizan, 1999; Zelik et al., 2009). Moreover, just as there are situations where an analyst must rely on one or very few sources of information, so too are there situations in which exploring all available data is infeasible. Thus, contextual and situational demands largely dictate the best way in which to apply the rigor attribute model as a measure of process quality. Ultimately it is practitioners who negotiate and decide what is appropriate for a given set of circumstances (Gardiner, 1989; McLaughlin, 2008).

And so, in closing, we return to the risk of shallow analysis as a hallmark challenge of present-day information analysis. In the face of significant production pressures and rapidly proliferating data availability (Woods et al., 2002), it is increasingly easy for analysts and decision makers alike to fall into the trap of shallow analysis—believing an analysis to be sufficient in a context where it is not.

In responding to this challenge, the measurement of analytical rigor is advanced as a check against this risk. The model of analytical rigor presented here provides both direction for supporting this critical judgment task and a framework for a macrocognitive measure of performance that is based on how expert analysts assess rigor. To that end, the concept of analytical rigor in information analysis warrants continued exploration and diverse application as a macrocognitive measure of analytical sensemaking activity.

Acknowledgments

This research was supported by the Department of Defense (H98230-07-C-0406), the Air Force Research Laboratory (S110000012), and the Office of Naval Research (GRT00012190). The views expressed in this article are those of the authors and do not necessarily represent those of the Department of Defense, Air Force, or Navy. We thank Fernando Bernal for doing the analytic coding for the logistics analysis and planning task. We also thank David T. Moore, Richards J. Heuer, and Robert R. Hoffman for their insightful review comments.

References

Bennett, K. B., & Flach, J. M. (1992). Graphical Displays: Implications for divided attention, focused attention, and problem solving. *Human Factors, 34*(5), 513–533.

Berkowitz, B., & Goodman, A. (1991*). Strategic intelligence for American national security*. Princeton, NJ: Princeton University Press.

Betts, R. K. (1978). Analysis, war, and decision: Why intelligence failures are inevitable. *World Politics, 31*(2), 61–89.

Betts, R. K. (2002). Fixing intelligence. *Foreign Affairs, 81*(1), 43–59.

Butts, C. T. (2008). Social network analysis: A methodological introduction. *Asian Journal of Social Psychology, 11*(1), 13–41.

Clark, R. M. (2007). *Intelligence analysis: A target-centric approach*. Washington, DC: CQ Press.

Cooper, J. R. (2005, December). *Curing analytic pathologies: Pathways to improved intelligence analysis* (Center for the Study of Intelligence). Washington, DC: U.S. Government Printing Office.

Crippen, D. L., Daniel, C. C., Donahue, A. K., Helms, S. J., Livingstone, S. M., O'Leary, R., & Wegner, W. (2005). Annex A.2: Observations by Dr. Dan L. Crippen, Dr. Charles C. Daniel, Dr. Amy K. Donahue, Col. Susan J. Helms, Ms. Susan Morrisey Livingstone, Dr. Rosemary O'Leary, and Mr. William Wegner. In Return to Flight Task Group (Eds.), *Return to Flight Task Group final report* (pp. 188–207). Washington, DC: U.S. Government Printing Office.

Davies, D., & Dodd, J. (2002). Qualitative research and the question of rigor. *Qualitative Health Research, 12*, 279–289.

Davis, J. (2002). Improving CIA analytic performance: Strategic warning. *The Sherman Kent Center for Intelligence Analysis Occasional Papers, 1*(1). Retrieved October 27, 2008, from https://www.cia.gov/library/kent-center-occasional-papers/vol1no1.htm

De Keyser, V., & Woods, D. D. (1990). Fixation errors: Failures to revise situation assessment in dynamic and risky systems. In A. G. Colombo & A. Saiz de Bustamante (Eds.), *Systems reliability assessment* (pp. 231–251). Dordrecht, The Netherlands: Kluwer Academic.

Defense Science Board Task Force. (2003). *Final report of the Defense Science Board Task Force on the role and status of DoD red teaming activities.* Washington, DC: U.S. Government Printing Office.

Dekker, S. (2005). *Ten questions about human error: A new view of human factors and system safety.* Mahwah, NJ: Lawrence Erlbaum.

Dekker, S., & Hollnagel, E. (2004). Human factors and folk models. *Cognition, Technology & Work, 6*(2), 79–86.

Elm, W., Potter, S., Tittle, J., Woods, D., Grossman, J., & Patterson, E. (2005). Finding decision support requirements for effective intelligence analysis tools. In *Proceedings of the Human Factors and Ergonomics Society 49th Annual Meeting* (pp. 297–301). Santa Monica, CA: Human Factors and Ergonomics Society.

Feder, S. (1987). FACTIONS and Politicon: New ways to analyze politics. *Studies in Intelligence, 31*(1), 41–57.

Folker, R. D. (2000). *Intelligence analysis in theater joint intelligence centers: An experiment in applying structured methods* (Joint Military Intelligence College Occasional Paper Number Seven). Washington, DC: U.S. Government Printing Office.

Gardiner, L. K. (1989). Dealing with intelligence-policy disconnects. *Studies in Intelligence, 33*(2), 1–9.

Guerlain, S. A., Smith, P.J., Obradovich, J. H., Rudmann, S., Strohm, P., Smith, J. W., Svirbely, J., & Sachs, L. (1999). Interactive critiquing as a form of decision support: An empirical evaluation. *Human Factors, 41*(1), 72–89.

Grossman, J., Woods, D. D., & Patterson, E. S. (2007). Supporting the cognitive work of information analysis and synthesis: A study of the military intelligence domain. In *Proceedings of the Human Factors and Ergonomics Society 51st Annual Meeting* (pp. 348–352). Santa Monica, CA: Human Factors and Ergonomics Society.

Heuer, R. J. (1987). Nosenko: Five paths to judgment. *Studies in Intelligence, 31*(3), 71–101.

Heuer, R. J. (1999). *Psychology of intelligence analysis.* Washington, DC: U.S. Government Printing Office.

Hong, L., & Page, S. E. (2004). Groups of diverse problem solvers can outperform groups of high-ability problem solvers. *Proceedings of the National Academy of Sciences, 101*, 16385–16389.

Johnson, P. E., Grazioli, S., Jamal, K., & Berryman, R. G. (2001). Detecting deception: Adversarial problem solving in a low base-rate world. *Cognitive Science, 25*(3), 355–392.

Johnston, R. (2005). *Analytic culture in the US Intelligence Community: An ethnographic study.* Washington, DC: U.S. Government Printing Office.

Kent, S. (1966). *Strategic intelligence for American world policy.* Princeton, NJ: Princeton University Press.

Klein G. A. (2010). Macrocognitive measures for evaluating cognitive work. In E. S. Patterson & J. Miller (Eds.), *Macrocognition metrics and scenarios: Design and evaluation for real-world teams.* Aldershot, UK: Ashgate Publishing.

Klein, G., Moon, B., & Hoffman, R. R. (2006a). Making sense of sensemaking 1: Alternative perspectives. *IEEE Intelligent Systems, 21*(4), 70–73.

Klein, G., Moon, B., & Hoffman, R. R. (2006b). Making sense of sensemaking 2: A macrocognitive model. *IEEE Intelligent Systems, 21*(5), 88–92.

Klein, G., Phillips, J. K., Rall, E., & Peluso, D. A. (2007). A data/frame theory of sensemaking. In R. R. Hoffman (Ed.), *Expertise out of context: Proceedings of the 6th International Conference on Naturalistic Decision Making*. Mahwah, NJ: Erlbaum.

Klein, G., Ross, K. G., Moon, B., Klein, D. E., Hoffman, R. R., & Hollnagel, E. (2003). Macrocognition. *IEEE Intelligent Systems, 18*(3), 81–85.

Krebs, V. (2002, April 1). Uncloaking terrorist networks. *First Monday, 7*(4). Retrieved September 30, 2009, from http://firstmonday.org/htbin/cgiwrap/bin/ojs/index.php/fm/article/view/941/863

Krizan, L. (1999). *Intelligence essentials for everyone* (Joint Military Intelligence College Occasional Paper Number Six). Washington, DC: U.S. Government Printing Office.

Mangio, C. A., & Wilkinson, B. J. (2008, March). *Intelligence analysis: Once again*. Paper presented at the International Studies Association 2008 Annual Convention, San Francisco, CA.

Marrin, S. (2007). Intelligence analysis: Structured methods or intuition? *American Intelligence Journal, 25*(1), 7–16.

McLaughlin, J. (2008). Serving the national policymaker. In R. Z. George & J. B. Bruce (Eds.), *Analyzing intelligence: Origins, obstacles and innovations* (pp. 71–81). Washington, DC: Georgetown University Press.

Medina, C. A. (2002). The coming revolution in intelligence analysis: What to do when traditional models fail. *Studies in Intelligence, 46*(3). Retrieved February 10, 2007, from https://www.cia.gov/csi/studies/vol46no3/article03.html

Military Operations Research Society. (2006, October 3). *MORS special meeting terms of reference*. Paper presented at Bringing Analytical Rigor to Joint Warfighting Experimentation, Norfolk, VA. Retrieved December 11, 2008, from http://www.mors.org/meetings/bar/tor.htm

Miller, J. E., & Patterson, E. S. (2008). Playback technique using a temporally sequenced cognitive artifact for knowledge elicitation. In *Proceedings of the Human Factors and Ergonomics Society 52nd Annual Meeting* (pp. 523–527). Santa Monica, CA: Human Factors and Ergonomics Society.

Miller, J. E., Patterson, E. S., & Woods, D. D. (2006). Elicitation by critiquing as a cognitive task analysis methodology. *Cognition, Technology & Work, 8*(2), 90–102.

Moon, I., & Carley, K. (2007). Modeling and simulating terrorist networks in social and geospatial dimensions. *IEEE Intelligent Systems, 22*(5), 40–49.

Moore, D. T. (2007). *Critical thinking and intelligence analysis* (Joint Military Intelligence College Occasional Paper Number Fourteen). Washington, DC: U.S. Government Printing Office.

Moore, D. T., & Reynolds, W. N. (2006). So many ways to lie: The complexity of denial and deception. *Defense Intelligence Journal, 15*(2), 95–116.

Morse, J. M. (2004). Preparing and evaluating qualitative research proposals. In C. Seale, G. Gobo, J. F. Giubrium, & D. Silverman (Eds.), *Qualitative research practice* (pp. 493–503). London: SAGE.

Patterson, E. S., Roth, E. M., & Woods, D. D. (2001). Predicting vulnerabilities in computer-supported inferential analysis under data overload. *Cognition, Technology & Work, 3*, 224–237.

Patterson, E. S., Zelik, D., McNee, S., & Woods, D. D. (2008). Insights from applying rigor metric to healthcare incident investigations. In *Proceedings of the Human Factors and Ergonomics Society 52nd Annual Meeting* (pp. 1766–1770). Santa Monica, CA: Human Factors and Ergonomics Society.

Patton, M. Q. (2002). *Qualitative research and evaluation methods* (3rd edn). Thousand Oaks, CA: Sage.

Pherson, R. (2005, March). *Overcoming analytic mindsets: Five simple techniques.* Paper presented at the annual meeting on Emerging Issues in National and International Security, Washington, DC.

Saltelli, A., Chan, K., & Scott, E. M. (Eds.) (2000). *Sensitivity analysis.* New York: John Wiley & Sons.

Sandelowski, M. (1986). The problem of rigor in qualitative research. *Advances in Nursing Science, 8*(3), 27–37.

Sandelowski, M. (1993). Rigor or rigor mortis: The problem of rigor in qualitative research revisited. *Advances in Nursing Science, 16*(2), 1–8.

Schraagen, J. M., Klein, G., & Hoffman, R. R. (2008). The macrocognition framework of naturalistic decision making. In J. M. Schraagen, L. G. Militello, T. Ormerod, & R. Lipshitz (Eds.), *Naturalistic decision making and macrocognition* (pp. 3–26). Aldershot, UK: Ashgate.

Schum, D. A. (1987). *Evidence and inference for the intelligence analyst* (Vol. 1). Lanham, MD: University Press of America.

Serra, J. (2008). Proactive intelligence. *Futures, 40*(7), 664–673.

Smith, M. W., Branlat, M., Stephens, R. J., & Woods, D. D. (2008, April). *Collaboration support via analysis of factions.* Papers presented at the NATO Research and Technology Organisation Human Factors and Medicine Panel Symposium, Copenhagen, Denmark.

Tenet, G. (1998). *Unclassified recommendations of the Jeremiah report.* Retrieved December 26, 2008, from http://ftp.fas.org/irp/cia/product/jeremiah-decl.pdf

Trent, S. A., Patterson, E. S., & Woods, D. D. (2007). Challenges for cognition in intelligence analysis. *Journal of Cognitive Engineering and Decision Making, 1*(1), 75–97.

Treverton, G. F., & Gabbard, C. B., (2008). *Assessing the tradecraft of intelligence analysis.* (RAND National Security Research Division). Retrieved October 21, 2008 from http://www.rand.org/pubs/technical_reports/2008/RAND_TR293.pdf

Woods, D. D., & Hollnagel, E. (2006). *Joint cognitive systems: Patterns in cognitive systems engineering.* New York, NY: Taylor & Francis.

Woods, D. D., Patterson, E. S., & Roth, E. M. (2002). Can we ever escape from data overload? A cognitive systems diagnosis. *Cognition, Technology & Work, 4*(1), 22–36.

Zelik, D. J., Patterson, E. S., & Woods, D. D. (2007). Judging sufficiency: How professional intelligence analysts assess analytical rigor. In *Proceedings of the Human Factors and Ergonomics Society 51st Annual Meeting* (pp. 318–322). Santa Monica, CA: Human Factors and Ergonomics Society.

Zelik, D., Woods, D. D., & Patterson, E. S. (2009). *The supervisor's dilemma: Judging when analysis is sufficiently rigorous.* Paper presented at the CHI 2009 Sensemaking Workshop, Boston, MA.

Chapter 6

Assessing Expertise When Performance Exceeds Perfection

James Shanteau, Brian Friel, Rick P. Thomas, John Raacke, and
David J. Weiss

Introduction

In many domains involving expert judgment, the base rate of errors is quite low. In Air Traffic Control (ATC), for instance, operational errors for en route aircraft are quite rare. Even after approaching zero errors, however, controllers continue to improve with experience. In such cases, traditional measures of performance based on percent correct 'gold standard' answers may be insensitive to performance improvements.

The research question, therefore, is how to assess performance that keeps improving 'beyond perfection' (as defined by no errors)? To answer this question, various measures were compared for evaluating skilled performance that is 'better than perfect.'

Background

As part of a longitudinal study of the effect of training on the development of skills, two groups of student operators participated for eight weeks using a simulation of ATC developed by the Federal Aviation Administration (Bailey et al., 1999). In one condition, all operators quickly learned to control aircraft with no operational errors. By the traditional gold definition, they performed at maximum level, that is, 'perfection.'

Their overall performance, however, continued to show improvement. For instance, the time-through-sector (TTS) of aircraft decreased with continued experience. In addition, both a Discrimination index (the ability to distinguish one aircraft from another) and a measure of Consistency (doing the same thing with similar aircraft in similar situations) increased with experience.

These indices were compared to a recently developed measure combining discrimination and consistency. Specifically, a ratio of discrimination divided by Inconsistency (the obverse of consistency) was evaluated. This ratio, labeled CWS for Cochran-Weiss-Shanteau (Weiss & Shanteau, 2003), has been found effective in evaluating expertise across a wide variety of domains and situations (Shanteau et al., 2002; Weiss & Shanteau, 2003; Weiss et al., 2006).

CWS integrates two necessary conditions for expert skill. The first is consistency, as argued by Einhorn (1972, 1974). Experts should make reliable judgments of similar stimuli; unreliable judgments serve as evidence against expertise. The second necessary condition is discrimination (Hammond, 1996), that is, experts should be able to differentiate similar stimuli on the basis of differences on which non-experts may lack sensitivity. CWS integrates these two by taking the ratio of discrimination to inconsistency, such that higher CWS scores are indicative of better performance. That is, experts should be consistent in their discriminations of stimuli in a particular domain.

CWS is parallel to the suggestion made by Cochran (1943) to use the ratio of between-stimulus variance to within-stimulus variance for assessing quality of response instruments. Therefore, it is only appropriate that Cochran is included in the CWS acronym.

CWS has been successfully applied as a performance measure to several pre-existing datasets, three of these applications are described in Shanteau et al. (2002), drawing on data from: auditing (from Ettenson, 1984), personnel hiring (from Nagy, 1978), and livestock judging (from Phelps & Shanteau, 1978). Performance in each of these studies was asymptotic, that is, participants performed at steady state levels. Although they were very good, they did make some errors.

In contrast, there are situations in which expert behavior is near perfect, as reflected by number of errors. One such domain is en route ATC. The goal of this chapter is to evaluate the application of CWS to this real-world environment where expertise in sensemaking can take years to decades to develop competence.

Methodology

As part of a larger longitudinal study of the effect of training on the development of skills, two groups of student operators participated for two months using a simulation of ATC developed by the Federal Aviation Administration (Bailey et al., 1999). In one condition, all operators quickly learned to control aircraft with few operational errors. By the gold standard definition, therefore, they performed near maximum level, that is, 'perfection.'

In prior analyses using CWS, there were no gold standard answers available for comparison. In the present case, however, an absolute standard of performance does exist. Operational errors, in which two aircraft are allowed to get too close together, provide a widely-accepted gold standard.

Study participants, divided into two groups of six, worked for two months on Controller Teamwork and Assessment Methodology (C-TEAM), an ATC simulation. Participants were paid $12 per session, with each session lasting 60–90 minutes.

The C-TEAM scenarios had three levels of Aircraft Density: low (= 12 aircraft), medium (= 24 aircraft), high (= 36 aircraft). Additionally, half of the scenarios in each level of aircraft density included restricted airspace (RA). Any aircraft directed into this space was scored as a Crash. Each scenario involving different aircraft densities lasted roughly eight minutes. In the low condition, operators quickly reached near-perfect performance. The results for that condition form the basis for the present analysis.

Variables

The study included two manipulations of scenario complexity. The first involved three levels of aircraft density (Low, Medium, and High), whereas the second involved the presence or absence of RA.

Three dependent measures were collected: (1) the number of Separation Errors (aircraft within five scale miles of the sector boundary or another aircraft at the same altitude) made by participants, (2) the number of crashes, and (3) TTS (the amount of time between the aircraft's first appearance on the screen to when it reached its destination). The last measure was converted into CWS scores.

We expected that CWS scores would vary as a function of scenario complexity. That is, lower CWS scores should be obtained in the High Aircraft Density scenarios than in Low and Medium Aircraft Density scenarios. CWS scores were also anticipated to be lower in RA scenarios, as participants should have a more difficult time routing aircraft to their destinations with an obstacle present in the airspace.

Apparatus, Design, and Procedure

In a single-sector version of C-TEAM, scenarios were presented on 17-inch Sceptre monitors connected to NCR-3230 486 computers. The participants' task was to direct aircraft to exit gates or airports using Kensington Expert Mouse trackballs.

The study lasted eight weeks. Participants completed three repetitions (hereafter called replicates) of the same scenario in each session. Participants completed three sessions per week, yielding 24 sessions per participant. Two two-week blocks of scenarios were created, each presented twice. For all participants, the order of scenario presentation was the same.

Results

Two participants dropped out of the study, one after 13 sessions, the other after 23 sessions. The data that they provided were included up to those points, since their behavior was similar to that of the other participants.

CWS scores were calculated using TTS. For each participant and each session, discrimination was calculated using the variance between different aircraft, whereas inconsistency was assessed using the variance within the same aircraft over the three replicates. The discrimination-to-inconsistency ratio formed the CWS score.

Because there was a strong positive correlation ($r = +.84$) between the means and variances of CWS scores, the CWS scores were log transformed so they no longer violated the homoscedasticity assumption. Objective measures of performance (that is, crashes and separation errors) were also collected for comparison.

Scenario Complexity Manipulations

Mean log CWS scores for each scenario condition are presented in Figure 6.1. These data were submitted to a 3 (aircraft density) x 2 (RA) repeated measures ANOVA. The main effects of aircraft density, $F (2, 20) = 19.35, p < .05, h^2 = .66$, and RA, $F (1, 10) = 5.08, p < .05, h^2 = .34$, were both statistically significant. As aircraft density increased, log CWS scores decreased, consistent with the idea that more complex scenarios are more difficult for participants.

Oddly, scenarios with RA yielded higher CWS scores. It was expected that such scenarios with restrictions would lead to lower CWS scores, because participants had an obstacle to route aircraft around.

There are two possible explanations for this result. One is that RA scenarios were presented after corresponding scenarios without RA. Thus, experience with a particular level of aircraft density may account for the higher CWS scores in the RA-present conditions. A more likely explanation is that the presence of restrictions leaves participants with fewer response options to direct aircraft. Thus, inconsistency scores may have been lower because aircraft are less free to vary in terms of TTS, leading to increased CWS scores.

This is, in fact, what we found, as scenarios with RA yielded lower values of inconsistency (that is, better consistency) over each level of aircraft density (Low-RA: 27.95, Low: 70.75; Medium-RA: 50.03, Medium: 187.85; High-RA:

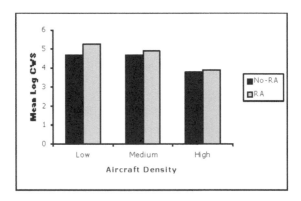

Figure 6.1   Mean log CWS values for three density levels for restricted/non-restricted airspace

238.31, High: 260.44). The interaction between aircraft density and RA was not significant, $F(2, 20) < 1$, $h^2 = .08$.

Overall Results

Five measures were used to assess performance: (1) mean number of errors per session; (2) mean TTS for aircraft in the airspace; (3) consistency of behaviors over repetitions of the same scenarios (Einhorn, 1974); (4) discrimination between different scenarios (as measured by mean squares); and (5) discrimination/ inconsistency (= CWS).

Three results illustrate the superiority of CWS over other measures. First, for low-scenario session with no errors (see below), CWS improved 500 percent from start to end of the study. Since there were no errors, the gold standard answer was at asymptote throughout. Thus, CWS was sensitive to gains in performance.

In comparison, for low-scenario session with no errors, raw TTS times showed a 5 percent improvement from session one to two; thereafter there was no change. Thus, raw TTS scores are not as revealing as CWS about the increase in performance.

Finally, separate analyses of discrimination and consistency revealed modest improvement—the gains were less than half that seen for CWS scores.

Low-Density Results

In the low-density condition, there were no errors by any operator for any session. That is, the performance was perfect from the earliest session to the last.

As shown in Figure 6.2, CWS scores reflected sizable improvement in performance with increased practice—even though no errors were made.

Three findings illustrate the superiority of CWS over other measures. First, CWS scores improved from start to end of the study. Since there were no errors, the gold standard was at asymptote throughout. Still, CWS revealed a sizable gain in performance.

In contrast, raw TTS times (as shown in Figure 6.3) showed a 5 percent improvement from session one to two; thereafter there was no change. Thus, raw TTS scores were less sensitive than CWS to improvements in performance.

Finally, separate analyses of discrimination and consistency revealed modest improvement. However, the gains were less than half that seen in CWS scores.

Discussion

CWS scores provide more information about performance than other measures. For instance, the TTS measurements tended to asymptote before CWS scores did, suggesting that raw measures of time are insufficient for assessing later performance improvements. Furthermore, in the present study, CWS scores

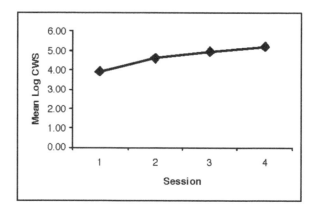

Figure 6.2   CWS scores for low-density condition for trials for which there were no errors

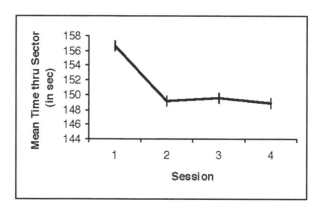

Figure 6.3   Mean Time-Through-Sector (TTS) values for low-density condition for trials for which there were no errors

suggest performance improvements that traditional gold standard measures failed to capture; specifically, CWS scores continued to increase over time in sessions that did not involve errors.

We acknowledge that in the absence of gold standard data, one can not be absolutely certain that CWS truly captures performance improvements. It is possible that increases in CWS scores reflect some measurement artifact or capture some other aspect of performance irrelevant to skill improvement. However, other research in similar contexts (Shanteau et al., 2005; Thomas et al., 2001) has shown CWS scores to be moderately negatively correlated with errors. Also, in a recent paper Weiss et al. (2009) showed that CWS was moderately correlated with performance in tasks, golf, and intuitive arithmetic, where absolute performance metrics exist.

We also have reason to assume that CWS score increases reflect strategy development and implementation. That is, an ATC operator who employs a strategy should be consistent in their treatment of aircraft across similar situations and discriminate among different aircraft, leading to high CWS scores. An operator who is in the process of developing a strategy may try different approaches, leading to less consistency across similar situations, which in turn would lead to lower CWS scores. Confirming evidence comes from Friel et al. (2002). In this study, half of the participants were provided with additional instructions that included ATC strategies for success in the C-TEAM task (for example, 'Line up aircraft like a "string of pearls"', 'Always slow the aircraft in trail first', and so on), whereas the other half were given basic instructions regarding the use of the C-TEAM interface. The group of participants who received extra tips on ATC strategies had higher CWS scores ($M = 3.46$) than those given only basic instructions ($M = 2.42$). Those participants who received the strategy tips also maintained their advantage in CWS scores over several sessions, but the gap narrowed toward the end of the study. Presumably, the participants who did not receive special instructions developed their own strategies (or discovered the ATC ones on their own).

The results demonstrate that the CWS index of expert performance can be applied to assessing skill development in dynamic environments. Participants' CWS scores increased with practice, suggesting that the index is sensitive to performance improvements. CWS scores were also sensitive to the scenario manipulations in C-TEAM, revealing that participants had more difficulty with more complex scenarios. Finally, the finding that CWS scores were negatively correlated with objective measures (that is, crashes and separation errors) demonstrates that CWS does capture elements of performance.

The results also suggest that CWS may be useful for training situations. The finding that CWS scores were sensitive to performance improvements in the present study suggests that the index could be used to assess skill development in trainees. In the present study, CWS scores also revealed aspects of development that error and raw TTS data were unable to capture, suggesting that it could be used to supplement objective measures in assessing training effectiveness. CWS score improvements could be used to determine whether a 'sink-or-swim' approach is preferable to instruction during training.

Future research with CWS will involve development of a real-time, continuous version and its application to comparisons of training methods. We believe that the power of CWS to capture performance in dynamic tasks has to do with strategy development and the consistent application of strategies.

Finally, we believe that a primary reason why CWS captures skill development is its sensitivity to the metacognitive ability of the operators. The ability to regulate and organize one's learning experience in a dynamic complex task likely aids the development of efficient tactics. Metacognitive regulation also allows operators to learn control strategies—when best to employ particular tactics or switch tactics due to task demands and/or cognitive limitations. Thus, metacognitive regulation could well be the basis of CWS's sensitivity to skill development in complex dynamic tasks. Another avenue of future research would be to formally explore the relation between individual differences in metacognitive skill, performance in dynamic complex task and CWS.

Acknowledgements

This research was supported in part by the Federal Aviation Administration, Department of Transportation (Grant 90-G-026). Correspondence concerning this research can be addressed to James Shanteau at Kansas State University, Department of Psychology, 492 Bluemont Hall, 1100 Mid-Campus Drive, Manhattan, KS 66506-5302.

References

Bailey, L. L., Broach, D. M., Thompson, R. C., & Enos, R. J. (1999). *Controller teamwork evaluation and assessment methodology: (CTEAM): A scenario calibration study.* (DOT/FAA/AAM-99/24). Washington, DC: Federal Aviation Administration Office of Aviation Medicine. Available from: National Technical Information Service, Springfield, VA 22161.

Cochran, W. G. (1943). The comparison of different scales of measurement for experimental results. *Annals of Mathematical Statistics, 14*(3), 205-216.

Einhorn, H. J. (1972). Expert measurement and mechanical combination. *Organizational Behavior and Human Performance, 7*(1), 86-106.

Einhorn, H. J. (1974). Expert judgment: Some necessary conditions and an example. *Journal of Applied Psychology, 59*(5), 562-571.

Ettenson, R. (1984). *A schematic approach to the examination of the search for and use of information in expert decision making.* Unpublished doctoral dissertation, Kansas State University.

Friel, B. M., Thomas, R. P., Raacke, J. D., & Shanteau, J. (2002, November). *Exploring strategy development with the CWS index of expert performance.* Poster presented at the 2002 annual meeting of the Society for Judgment and Decision Making. Kansas City, MO.

Hammond, K. R. (1996). *Human judgment and social policy.* New York, NY: Oxford University Press.

Nagy, R. H. (1977). *How are personnel selection decisions made? An analysis of decision strategies in a simulated personnel selection.* Unpublished doctoral dissertation, Kansas State University.

Phelps, R. H., & Shanteau, J. (1978). Livestock judges: How much information can an expert use? *Organizational Behavior and Human Performance, 21,* 209-219.

Shanteau, J., Friel, B. M., Thomas, R. P., & Raacke, J. (2005). Development of expertise in a dynamic decision-making environment. In Betsch, T., &

Haberstroh, S. (Eds.), *The routines of decision making* (pp. 251-270). Mahwah, NJ: Erlbaum.

Shanteau, J., Weiss, D. J., Thomas, R., & Pounds, J. (2002). Performance-based assessment of expertise: How to decide if someone is an expert or not. *European Journal of Operations Research, 136*(2), 253-263.

Thomas, R. P., Willems, B., Shanteau, J., Raacke, J., & Friel, B. M. (2001). CWS applied to controllers in a high-fidelity simulation of ATC. *Proceedings from the 11th Annual International Symposium on Aviation Psychology.* Columbus, OH.

Weiss, D. J., & Shanteau, J. (2003). Empirical assessment of expertise. *Human Factors, 45*(1), 104-116.

Weiss, D. J., Shanteau, J., & Harries, P. (2006). People who judge people. *Journal of Behavioral Decision Making, 19*(5), 441-454

Chapter 7

Demand Calibration in Multitask Environments: Interactions of Micro and Macrocognition

John D. Lee

Introduction

Technology continues to extend humans' span of control, forcing people to manage multiple processes. As a consequence, multitask situations represent an increasingly important challenge in many domains. Driving is an important example of such a domain. Driving is an inherently multitask environment and the migration of increasingly complex technology into the automobile raises the threat of distraction. Distraction-related crashes impose a substantial cost on society. By one estimate, cellphone-related crashes cost $43 billion each year in the US alone (Cohen & Graham, 2003). Cellphones are one of a large number of entertainment and information systems that range from MP3 players to navigation systems and Internet content (Lee & Kantowitz, 2005; Walker et al., 2001). Some of the costs associated with these In-Vehicle Information Systems (IVIS) are balanced by the value associated with the distracting activities. As an example, IVIS can reduce the monotony of driving and enable people to accomplish other tasks while driving. The challenges in managing distractions while driving reflect the challenges of managing attention to multiple, competing activities that confront operators in domains as diverse as healthcare and management of remotely piloted vehicles. Distraction, in which attention is diverted away from a safety-critical activity to some competing activity, represents an important problem in a broad range of domains.

In-vehicle technology adds additional demands to the already complex activity of driving. Drivers interacting with such technology must balance the demands associated with IVIS (for example, entering a destination in a route navigation system or selecting a playlist from an MP3 player) with those of the roadway. Distraction-related crashes reflect a failure to balance these demands. Much current research reflects a microcognitive perspective in which distraction-related mishaps occur as a consequence of a dual-task performance decrement. Microcognitive approaches often use an information-processing framework and multiple resource theory to describe driver distraction in terms of dual-task performance decrements (Wickens, 2002). Distraction is more complex and a more complete description includes concepts of macrocognition (Klein, et al., 2003), such as demand calibration. Demand calibration concerns the relationship between the combined demands of the safety-critical and competing activities and the perceived demands of these activities. Poor calibration of drivers to IVIS and roadway demands is central to distraction-related mishaps and suggests that demand miscalibration represents a general challenge to creating robust and resilient systems in many domains.

Considering distraction-related mishaps as breakdowns of a multilevel control process highlights the role of demand calibration. Poor demand calibration undermines feedback, feedforward, and adaptive control, which contributes to distraction-related mishaps. This perspective also shows how micro and macrocognition can interact through cascade effects to undermine performance in multitask environments. Cascade effects occur when the outcome at one level of control affects control at another level. Such cascade effects reflect an interaction between macro and microcognition that is often neglected when the information-processing framework is used to describe distraction. Conventional microcognitive approaches (for example, resource theory) tend to ignore the contextual nesting of the multiple constraints at each of these levels. To understand driving performance it will be critical to understand how aspects of higher levels (for example, expectations about traffic) interact with lower levels (for example, sampling of information, or headway maintenance).

Considering distraction in terms of micro and macrocognitive perspectives provides both theoretical and practical value for establishing metrics of individual and team performance. Each perspective implies a different cause of distraction and identifies different design and policy considerations to reduce distraction-related crashes. For example, considering distraction in terms of dual-task demands draws upon the theoretical constructs of multiple resource theory (Wickens, 2002), and involves experimental protocols used to assess dual-task interference, whereas considering distraction in terms of task planning would lead to a very different experimental approach and results. For example, simulator studies of driver performance show a systematic impairment associated with hands-free cellphone conversations (Caird, 2008; Horrey et al., 2006), whereas crash data do not (Young, 2009). One explanation is that drivers time their interactions to avoid high-demand situations and the surrounding traffic compensates for impaired performance. Likewise, a theory of multitask situations that considers only the individual driver neglects the traffic and societal contributions to distraction. Addressing the challenge of distraction requires a more comprehensive description of driver behavior than just an information-processing description of dual-task performance. This chapter describes micro and macrocognitive contributions to drivers' failure to manage IVIS and roadway demands. A central problem in demand management is that of calibration, where calibration is the appreciation for the magnitude of demands, the capacity to address those demands, and the consequence of neglecting those demands.

Defining Distraction

The definition of metrics of multitask management might benefit from the challenge associated with defining and measuring driver distraction. Distraction, and more generally inattention, has been a concern for road safety for many years (Brown et al., 1969; Sussman, et al., 1985; Treat, et al., 1979), but the recent controversy regarding cellphones has prompted a surge of research (Haigney & Westerman, 2001; Horrey & Wickens, 2006; McCartt et al., 2006). Many definitions of distraction and related phenomena have accompanied this increased interest in distraction. These definitions have important implications for assessing the magnitude and cause of the distraction problem. A recent naturalistic study found that nearly 80 percent of crashes and 65 percent of near crashes included inattention as a contributing cause (Klauer et al., 2006). Importantly, the authors defined inattention as including general inattention to the road, fatigue, and secondary task demand. Inattention represents a broad class of situations in which the driver fails to attend to the demands of driving, such as when sleep overcomes a fatigued driver. One way to distinguish between inattention and distraction is that distraction involves an explicit activity that competes for the driver's attention, as compared to a cognitive state that leads to diminished capacity to attend to the roadway: 'Driver distraction is a diversion of attention away from activities critical for safe driving towards a competing activity.' (Lee et al., 2008a).

This and other definitions of distraction face a challenge similar to that of definitions of human error (Rasmussen, 1990; Senders & Moray, 1991). There is a danger of post-hoc attribution of distraction as a cause of a crash. Only after the fact is it obvious what particular activities were critical to safe driving and how the driver should have distributed his or her attention. Because driving is varied and complex, it is not feasible to define a normative model of how drivers should attend to the roadway and therefore it is difficult to assess the degree to which there is a distraction-related inappropriate diversion of attention. The degree to which IVIS interactions pose a distraction depends on the combined demands of the roadway and the IVIS, relative to the available capacity of the driver and society's judgment of acceptable risk. Similar definitional challenges face other domains where people must distribute their attention across competing demands (Moray, 2003).

Distributions of Demand

Avoiding distraction can be considered as an exercise in minimizing the co-occurrence of high roadway and high IVIS demand situations. The overlap of high demand situations depends on separation of the means of these distributions and on the variability of the distributions (Lee et al., 2008b). The distribution of roadway demand reflects the road geometry—demand increases in traversing a curve. The traffic situation also influences roadway demand—demands increase with lead vehicles that brake periodically. The distribution of IVIS demands reflect the complexity of tasks drivers choose to perform and their experience performing the tasks. Distraction-related mishaps occur if the separation of the distributions diminishes or if the variance of the distributions increases. IVIS and driving demands vary at time horizons that range from seconds to days.

The time horizon affects the shape and rate of change of the distributions of roadway and IVIS demand. A short time horizon has narrow distributions—there is little uncertainty in the current demand, but the mean of these distributions could change quickly. At a long time horizon (hours or days) the distribution of demand is broad, but the mean of the distribution changes slowly, reflecting the slow changes in the typical demands of the roadway and IVIS interaction. The short timescale can lead to distraction-related mishaps when the driver is not able to respond to rapid changes in demands, whereas the long timescale can lead to mishaps because events from the tails of the distributions surprise drivers. Avoiding distraction-related mishaps depends on managing demand at short time horizons (to minimize situations in which demands change so quickly the driver cannot adjust) and at long time horizons to compensate for the differences between the typical and actual roadway demands. Distraction represents a failure to keep these distributions separated at a societally acceptable distance.

Considering distraction as a problem of managing distributions of demand emphasizes the factors that contribute to overlaps in these distributions that vary according to the time horizon. The overlap of long tails dominates at long time horizons and rapid changes in the separation of the distributions dominate at the short time horizons. People must be calibrated to the demand, the distribution of demand, and the rate of change of demand.

At a short time horizon, distraction depends on the competition for processing resources associated with the concurrent demands of the road and the IVIS (McCarley, et al., 2004; Strayer et al., 2003). At the long time horizon, distraction depends on a random walk process in which productivity pressures and indistinct safety boundaries lead drivers to adopt unsafe practices in which the demands are likely to exceed the capacity of the person (Rasmussen, 1997). Between these two extremes, distraction depends on the task timing and the ability of drivers to avoid overlapping demanding activities (Carbonell, 1966; Senders et al., 1967). Distraction-related mishaps reflect a breakdown in control at one or more of these time horizons and the interactions between them. These breakdowns often reflect a poor calibration of demand relative to the capacity to respond.

Levels of Control in Managing Demand

Anticipating a long drive, Sam decides to insert an MP3 player into his car's audio system and select a playlist as he begins to drive out of the city. After inserting the player, he glances down to view the catalog of playlists and begins to scroll through the list. Meanwhile a car abruptly merges in front to exit the

freeway, suddenly shrinking the gap that once existed with the car ahead. Sam
begins to slow to widen the gap and looks back to the playlist and continues to
scroll down the list. Overshooting the desired playlist, Sam continues to look at
the MP3 player as the car ahead brakes suddenly to accommodate other vehicles
entering the highway. Sam looks up to find himself crashing into the vehicle
ahead.

As with most safety-related mishaps, the distraction-related crash in this scenario
is a complex event with multiple causes. In describing this scenario one can pursue
an arbitrarily long and broad causal network (Rasmussen, 1990). Figure 7.1
shows some causes of distraction-related mishaps in terms of a multilevel control
process. Each level of this framework reflects control at a different time horizon.
In this framework, distraction-related mishaps result from a breakdown of control
at any one level or from cascade effects in which control problems grow as they
propagate across levels.

Figure 7.1 highlights several contributions to the distraction-related crash in
this scenario (Michon, 1985; Ranney, 1994). One contribution is a breakdown
of control at the operational level in which the visual demands of selecting the
playlist interfered with Sam's attention to the road. Another contribution was
a cascade effect in which Sam overshoots the playlist at the operational level,
leading to a longer than expected task duration, which then undermines control
at the tactical level. At the strategic level, the decision to interact with the MP3
player in a relatively demanding roadway environment also contributed to the
crash. Each of these causes represents equally valid explanations of the crash
and each demands a different set of theoretical considerations. This scenario also
points to the consequence of poor demand calibration at each level.

Types of Control

Three types of control are active in each level in Figure 7.1. Breakdowns in
any one of three types of control contribute to driver distraction. Each type of
control has important limits as shown by the columns on the right of Table 7.1.
Feedback control uses the difference between a goal state and the current state
to guide behavior. To be successful, feedback control depends on timely, precise
information regarding the difference between the current state of affairs and the
goal state. In driving, the variables concerning the state of the vehicle, such as lane
position and speed, are immediately available through the windshield. However,
the consequence of poor control of attention to a competing task (for example,
with a long glance away from the road) may not have any observable negative
effect. As a consequence, drivers may fail to modulate their attention appropriately
and not have any indication of this failure because driving is often forgiving and
only occasionally merciless. Furthermore, for feedback control to be effective the
time constant of drivers' response to the error signal must be fast enough so that

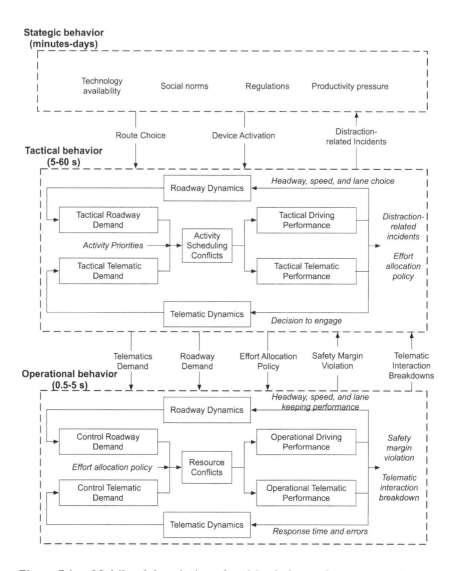

Figure 7.1   Multilevel description of multitask demand management

responses can be made before the system diverges by an unacceptable degree from the desired state.

Feedforward control uses the anticipated future state of the system to guide behavior. Feedforward control is critical for safe driving; it enables experienced drivers to detect hazards more reliably than inexperienced drivers (McKenna et al., 2006). Feedforward control can compensate for the limits of feedback control, but it suffers from other problems. Feedforward control requires an accurate internal model to anticipate the future state of the system and is vulnerable to unanticipated disturbances. The uncertainty associated with poor mental models of the roadway and the IVIS coupled with inherent variability limits the effectiveness of feedforward control to modulate attention between the road and the IVIS.

Adaptive control typically refers to adjusting the control laws to enhance performance (for example, changing scanning patterns to compensate for a change in traffic density), but can also refer to the process of reducing the difference between the goal state and the current state by redefining the goal state. As such, adaptive control represents a type of meta-control that is critical for adaptive and resilient systems. Adaptive control is particularly relevant for tactical behavior. Here drivers actively negotiate with other drivers on the road to widen their safety margins. Likewise, a driver might adjust the pace of a conversation or delay an interaction based on the road demands. In these situations drivers engage in interactional adaptation in which they adjust their interaction to the driving situation (Esbjornsson et al., 2007). Such adaptation depends on developing common ground between the driver, the roadway, and the IVIS device (Wiese & Lee, 2007).

With adaptive control, drivers adjust their response to task demands by changing their performance criteria and change control laws. The success of adaptive control depends on how these new criteria and control laws fit the dynamics of the situation. A challenge for adaptive control is the potentially poor calibration regarding the expected achievement of driving and IVIS. Drivers may not realize the consequences of adopting less ambitious goals for performing the driving task. Adaptive control can also lead to suboptimal control laws that fail change because driving often lacks the feedback needed to promote more safe behavior. Table 7.1 summarizes some of the reasons why each control type might fail at each time horizon.

At each level in Figure 7.1 calibration can be problematic. At the operational level, ambient vision undermines calibration because it enables good lateral control even when focal vision is directed toward the IVIS, which greatly compromises event detection (Horrey et al., 2006; Leibowitz & Owens, 1977). Drivers may overestimate their ability to accommodate roadway demands because their success at lateral control masks their impaired event detection. At the tactical level, drivers often lack awareness of roadway hazards (Pollatsek et al., 2006) and appreciation for how IVIS demands might diminish their awareness of these hazards (Lee, 1999). At the strategic level, drivers systematically overestimate their abilities (Waylen, et al., 2004a) and fail to adjust their behavior in the face of even dramatic feedback, such as surviving a fatal crash (Rajalin & Summala, 1997).

In summary, contrary to many views of distraction, drivers are not the passive recipients of IVIS and roadway demands. Instead they actively control the distraction they experience. This control occurs at three interacting time horizons (operational, tactical, and strategic) and is achieved by three types of control (feedback, feedforward, and adaptive control). The limits of control combine with problems of calibration at each time horizon to undermine demand management. Labeling distraction as simply a problem of information overload neglects the

Table 7.1   Types of control for each time horizon page (Lee & Strayer, 2004)

Control type	Operational—control attention to tasks (ms to s)	Tactical—control task timing (s to min)	Strategic—control exposure to tasks (min to days)
Feedback—reactive control based on past outcomes	Time constant of response slower than that of roadway demands	Feedback too delayed to be useful	Poor choices might not affect performance
Feedforward—proactive control based on anticipated outcomes	Task demands are unpredictable or unknown	Task timing is unpredictable or unknown	Potential demands are unpredictable or unknown
Adaptive—control by adjusting expectations, goal state, and task characteristics	Brittle tasks lack a graded effort/accuracy tradeoff	Biological and social imperatives that might not be calibrated to task importance	Poor calibration regarding driving and IVIS goals

failures that lead to temporally constrained roadway and IVIS demands that overwhelmed drivers' capacity to respond.

Cascade Effects that Couple Macro-micro Processes

Control can suffer from interactions across the three time horizons that can have particularly powerful consequences for driving safety. Cascade effects occur when the outcome at one level of control affects control at another level. Cascade effects occur when control breakdowns at one time horizon undermine control at the others—a small perturbation at one level could trigger instabilities at other levels. Such cascade effects are often non-linear in that small perturbations at one level lead to catastrophic effects at another. Poor choices at the strategic level make control at the tactical and operational levels more difficult. In the example above, the choice to start using the MP3 player in an urban environment imposed greater demands on task timing. Likewise, errors at the operational level can also propagate to tactical control, as in the overshooting when selecting a playlist leads to an unexpectedly long interaction.

Control at the longer time horizons constrains that of the shorter time horizons by specifying the goals and tasks at shorter time horizons. Control at the shorter time horizons creates disturbances that undermine control at the longer time horizons. Distraction-related incidents occur when roadway and IVIS demands combine to undermine control at any one time horizon or initiate cascade effects across the time horizons. Cascade effects imply that understanding diminished performance associated with macrocognitive behavior may depend on understanding the influence of microcognitive behavior.

Improving and Measuring Demand Calibration

Over time, feedback helps drivers develop skills to manage distractions. Driving generally provides poor feedback, leading drivers to grossly overestimate their capacity (Brehmer, 1980; Svenson, 1981). Drivers can drive in a dangerous manner for many years and not experience a negative outcome and so may not receive any feedback that they are driving dangerously. This may be particularly problematic for cognitive distraction, where drivers might not realize that they failed to notice a critical event (McCarley, et al., 2004; Richard, et al., 2002). Even when drivers experience a fatal crash, they tend not to adjust their driving behavior, except in situations that closely match the crash situation (Rajalin & Summala, 1997). Driving often accommodates a very poorly tuned control system, in that someone can drive dangerously for years without crashing. In adaptive control terms—the system falls into a local minima—adopting a strategy that is far from optimal. Driving often lacks the feedback that lead drivers to change control laws to become more and more skilled.

The feedback drivers naturally receive, particularly regarding distraction, is too intermittent and too difficult to generalize for it to moderate behavior effectively. Providing drivers with better feedback regarding the effect of distractions might be an effective way to reduce distraction-related crashes. Pairing immediate feedback and feedback over time might be an effective way to help drivers recognize the consequence of distraction and make them less willing to engage in distracting activities as they drive (Donmez et al., 2006; Engstrom & Hollnagle, 2007).

Most drivers are poorly calibrated regarding their driving ability—the vast majority believe they are above average. As an example, a survey of 1000 drivers found that 80 percent considered themselves to drive better than the average (Waylen et al., 2004b). The illusion of superiority refers to the persistent finding across nationality, age, and gender that drivers perceive themselves to drive better and more safely than their average peer or the average driver. Such a superiority bias exists regarding distraction—drivers may see themselves as less vulnerable to distraction than the average driver. Such a bias may lead drivers to engage in distracting activities, and eliminating this bias through enhanced feedback may diminish drivers' willingness to engage in distracting activities.

Waylen and colleagues (2004b) argue that a hallmark of expertise in many domains is self-knowledge. Specifically, experts can accurately judge their ability relative to that needed to succeed in a particular situation (Glaser, 1988). Enhancing feedback, particularly to support adaptive control, may be a powerful way to mitigate distraction by helping drivers become more expert by enhancing their self-knowledge, but it is currently unexplored. Enhanced feedback can calibrate drivers' estimation of their ability, particularly their ability to manage competing activities, and guide safe behavior. Such feedback might also enhance performance of operators in situations in which they must manage multiple demands with limited feedback concerning their success.

A recent study provides a technique and validated set of scales to quantify the illusion of superiority and provides a useful measure of demand calibration. Horswill and his colleagues (Horswill, et al., 2004) measured the illusion of superiority across 18 different driving skill components (for example, maintaining the appropriate distance to the car ahead). These driving skill components serve as a starting point for operationalizing distracted-related driving impairments and for relating subjective and objective distraction. Subjective ratings of these skill components and objective assessments of performance on these components reveals important failures of calibration.

Subjective ratings are reported on a scale where the midpoint has been defined as the population median, and endpoints are labeled as the 10 percent most and least skillful. A similar scale can be used for objective performance. Such objective performance might include aggregations of the duration and frequency of glances away from the road associated with extreme risk—glances away from the road longer than five seconds. Drivers can complete these scales for normal driving and driving while engaged in potentially distracting activities, such as tuning a radio, simple and complex conversations, and text messaging. The difference between the ratings of self, peers, and all other drivers indicate the degree of superiority bias. The differences in ratings between driving and driving while engaged in potentially distracting activities represent the degree of awareness of distraction as a problem, and the differences between ratings for self and others represent the degree of superiority bias for how well the driver thinks he or she can manage the activities. The 18 driving skill components also provide the basis for developing scenarios and for assessing the effect of distracting activities on driver performance in the simulator. For example, the driving skill associated with maintaining the appropriate distance to the car ahead can be measured in terms of the mean and variance of the time headway.

The two types of bias scores (subjective-objective performance and subjective self rating-subjective rating of other drivers) provide critical indicators regarding the effectiveness of the enhanced feedback (Murphy & Winkler, 1974). Beyond these bias scores one might include estimates of the strength of the relationship between subjective and objective distraction (percent of variance accounted for in a linear model relating subjective and objective distraction) and the amount the subjective distraction changes for every change in the objective distraction (slope parameter of a linear regression model) (Horrey & Wickens, 2006). This framework provides a basis for measuring the effect of feedback to make drivers more aware of the consequences of engaging in distracting activities.

Conclusion

Much research considers distraction purely as a dual-task performance decrement. This microcognitive explanation captures important contributions to distraction-related mishaps, but neglects many others. Considering multitask situations in terms of a multilevel control process suggests factors that undermine feedback, feedforward, and adaptive control at each level and contribute to distraction-related mishaps. Failures of calibration to IVIS and roadway demand at each level are central to these mishaps, suggesting that miscalibration represents a general challenge to creating robust systems in other domains. Cascade effects connect micro and macrocognition, making microcognitive processes and can undermine system safety. As in other engineering domains, the interaction of micro and macro processes contribute to failures (Farris et al., 1990).

In driving, as in other domains, the network of actors that comprise the driving community can adapt in ways contrary to safety (Reason, 1998). Societal attitudes regarding acceptable risk represent a powerful force in this adaptation. Societal response to traffic deaths illustrates this tendency. Recent high-profile catastrophes—the Oklahoma City bombing, shootings at Columbine High School, terrorist attacks on September 11, 2001, and Hurricane Katrina—led to less than 5,000 fatalities, but these events have had considerable influence on the American political, economic, and cultural landscapes. In contrast, the 42,636 lives lost in 2004 *alone* as a result of vehicle crashes on US roadways barely registered. The American public seems to consider the loss of an average of 116 lives *each day* in crashes as an acceptable risk of transportation. Calibrating the public to the human toll of distraction while driving through better measures and feedback mechanisms could lead to cultural changes that promote substantially fewer accidents (Moeckli & Lee, 2006).

References

Brehmer, B. (1980). In one word—Not from experience. *Acta Psychologica, 45*(1-3), 223-241.

Brown, I. D., Tickner, A. H., & Simmonds, D. C. V. (1969). Interference between concurrent tasks of driving and telephoning. *Journal of Applied Psychology, 53*(5), 419-424.

Caird, J. K., Willness, C.R., Steel, P., & Scialfa, C. (2008). A meta-analysis of the effects of cell phones on driver performance. *Accident Analysis and Prevention, 40*(4), 1282-1293.

Carbonell, J. R. (1966). A queueing model of many-instrument visual sampling. *IEEE Transactions on Human Factors in Electronics, HFE-7*(4), 157-164.

Cohen, J. T., & Graham, J. D. (2003). A revised economic analysis of restrictions on the use of cell phones while driving. *Risk Analysis, 23*(1), 1-14.

Donmez, B., Boyle, L. N., & Lee, J. D. (2006). The impact of driver distraction mitigation strategies on driving performance. *Human Factors, 48*(4), 785-804.

Engstrom, J., & Hollnagle, E. (2007). A general conceptual framework for modelling behavioral effects of driver support functions.

Esbjornsson, M., Juhlin, O., & Weilenmann, A. (2007). Drivers using mobile phones in traffic: An ethnographic study of interactional adaptation. *International Journal of Human-Computer Interaction, 22*(1-2), 37-58.

Farris, J., Lee, J., Harlow, D., & Delph, T. (1990). On the scatter in creep-rupture times. *Metallurgical Transactions A-Physical Metallurgy and Materials Science, 21*(2), 345-352.

Glaser, R., & Chi, M. T. H. (1988). Overview (the nature of expertise). In R. G. M. T. H. Chi, & M. J. Farr (Ed.), *The nature of expertise* (pp. xv–xxvi). Hillsdale, NJ: Erlbaum.

Haigney, D., & Westerman, S. J. (2001). Mobile (cellular) phone use and driving: a critical review of research methodology. *Ergonomics, 44*(2), 132-143.

Horrey, W. J., & Wickens, C. D. (2006). Examining the impact of cell phone conversations on driving using meta-analytic techniques. *Human Factors, 48*(1), 196-205.

Horrey, W. J., Wickens, C. D., & Consalus, K. P. (2006). Modeling drivers' visual attention allocation while interacting with in-vehicle technologies. *Journal of Experimental Psychology-Applied, 12*(2), 67-78.

Horswill, M. S., Waylen, A. E., & Tofield, M. I. (2004). Drivers' ratings of different components of their own driving skill: A greater illusion of superiority for skills that relate to accident involvement. *Journal of Applied Social Psychology, 34*(1), 177–195.

Klein, G., Ross, K. G., Moon, B. M., Klein, D. E., Hoffman, R. R., & Hollnagel, E. (2003). Macrocognition. *IEEE Intelligent Systems, 18*(3), 81-85.

Lee, J. D. (1999). Measuring driver adaptation to in-vehicle information systems: Disassociation of subjective and objective situation awareness measures. In *Proceedings of the Human Factors and Ergonomics Society* (Vol. 2, pp. 992-996). Santa Monica, CA: Human Factors and Ergonomics Society.

Lee, J. D., & Kantowitz, B. K. (2005). Network analysis of information flows to integrate in-vehicle information systems. *International Journal of Vehicle Information and Communication Systems, 1*(1/2), 24-43.

Lee, J. D., Regan, M. A., & Young, K. L. (2008a). Defining driver distraction. In M. A. Regan, J. D. Lee & K. L. Young (Eds.), *Driver Distraction: Theory, Effects, and Mitigation* (pp. 31-40). Boca Raton, FL: CRC Press.

Lee, J. D., Regan, M. A., & Young, K. L. (2008b). What drives distraction? Distraction as a breakdown of multi-level control. In M. A. Regan, J. D. Lee & K. L. Young (Eds.), *Driver distraction: Theory, effects, and mitigation* (pp. 41-56). Boca Raton, FL: CRC Press.

Lee, J. D., & Strayer, D. L. (2004). Preface to a special section on driver distraction. *Human Factors, 46*(4), 583-586.

Leibowitz, H. W., & Owens, D. A. (1977). Nighttime driving accidents and selective visual degradation. *Science, 197*(4302), 422-423.

McCarley, J. S., Vais, M., Pringle, H. L., Kramer, A. F., Irwin, D. E., & Strayer, D. L. (2004). Conversation disrupts visual scanning of traffic scenes. *Human Factors, 3*(3), 424-436.

McCartt, A. T., Hellinga, L. A., & Bratiman, K. A. (2006). Cell phones and driving: Review of research. *Traffic Injury Prevention, 7*(2), 89-106.

McKenna, F. P., Horswill, M. S., & Alexander, J. L. (2006). Does anticipation training affect drivers' risk taking? *Journal of Experimental Psychology-Applied, 12*(1), 1-10.

Michon, J. A. (1985). A critical view of driver behavior models: What do we know, what should we do? In L. Evans & R. C. Schwing (Eds.), *Human behavior and traffic safety* (pp. 485-520). New York, NY: Plenum Press.

Moeckli, J., & Lee, J. D. (2006). The making of driving cultures *Improving Traffic Safety Culture in the US: The Journey Forward*.

Moray, N. (2003). Monitoring, complacency, scepticism and eutactic behaviour. *International Journal of Industrial Ergonomics, 31*(3), 175-178.

Murphy, A. H., & Winkler, R. L. (1974). Subjective probability forecasting experiments in meteorology: Some preliminary results. *Bulletin of the American Meteorological Society, 55*(10), 1206-1216.

Pollatsek, A., Fisher, D. L., & Pradhan, A. (2006). Identifying and remedying failures of selective attention in younger drivers. *Current Directions in Psychological Science, 15*(5), 255-259.

Rajalin, S., & Summala, H. (1997). What surviving drivers learn from a fatal road accident. *Accident Analysis and Prevention, 29*(3), 277-283.

Ranney, T. A. (1994). Models of driving behavior – a review of their evolution. *Accident Analysis and Prevention, 26*(6), 733-750.

Rasmussen, J. (1990). Human error and the problem of causality in analysis of accidents. *Philosophical Transactions of the Royal Society Series B, 327*(1241), 449-462.

Rasmussen, J. (1997). Risk management in a dynamic society: A modelling problem. *Safety Science, 27*(2-3), 183-213.

Reason, J. (1998). Achieving a safe culture: theory and practice. *Work and Stress, 12*(3), 293-306.

Richard, C. M., Wright, R. D., Ee, C., Prime, S. L., Shimizu, Y., & Vavrik, J. (2002). Effect of a concurrent auditory task on visual search performance in a driving-related image-flicker task. *Human Factors, 44*(1), 108-119.

Senders, J. W., Kristofferson, A. B., Levison, W. H., Dietrich, C. W., & Ward, J. L. (1967). The attentional demand of automobile driving. *Highway Research Record, 195*, 15-33.

Senders, J. W., & Moray, N. P. (1991). *Human error: cause, prediction, and reduction*. Hillsdale, NJ: Lawrence Erlbaum Associates.

Strayer, D. L., Drews, F. A., & Johnston, W. A. (2003). Cell phone-induced failures of visual attention during simulated driving. *Journal of Experimental Psychology-Applied, 9*(1), 23-32.

Sussman, E. D., Bishop, H., Madnick, B., & Walter, R. (1985). Driver inattention and highway safety. *Transportation Research Record, 1047*, 40-48.

Svenson, O. (1981). Are we all less risky and more skillful than our fellow drivers? *Acta Psychologica, 47*(2), 143-148.

Treat, J. R., Tumbas, N. S., McDonald, S. T., Shinar, D., Hume, R. D., Mayer, R. E., et al. (1979). *Tri-level study of the causes of traffic accidents: executive summary*. Washington, DC: NHTSA, U.S. Department of Transportation.

Walker, G. H., Stanton, N. A., & Young, M. S. (2001). Where is computing driving cars? *International Journal of Human-Computer Interaction, 13*(2), 203-229.

Waylen, A. E., Horswill, M. S., Alexander, J. L., & McKenna, F. P. (2004a). Do expert drivers have a reduced illusion of superiority? *Transportation Research Part F-Traffic Psychology and Behaviour, 7*(4-5), 323-331.

Waylen, A. E., Horswill, M. S., Alexander, J. L., & McKenna, F. P. (2004b). Do expert drivers have a reduced illusion of superiority? *Transportation Research Part F: Psychology and Behavior, 7*, 323-331.

Wickens, C. D. (2002). Multiple resources and performance prediction. *Theoretical Issues in Ergonomics Science, 3*(2), 159-177.

Wiese, E. E., & Lee, J. D. (2007). Attention grounding: a new approach to IVIS implementation. *Theoretical Issues in Ergonomics Science, 8*(3), 255-276.

Young, R. A., & Schreiner, C. (2009). Review of 'Real-world personal conversations using a hands-free embedded wireless device while driving: Effect on airbag deployment crash rates.'

Chapter 8

Assessment of Intent in Macrocognitive Systems

Lawrence G. Shattuck

Introduction

Example One

An office manager begins to type a memo using a word-processing software program. The memo is to have five paragraphs numbered 1 through 5. The first paragraph will include three subparagraphs labeled a, b, and c. The manager begins the memo by typing '1.' After typing the first paragraph, he pushes the 'Enter' key and prepares to type subparagraph a. However, the software program responds by moving the cursor to the left margin of the next line; a '2' appears on the computer screen. The manager has to correct the actions of the software by pressing the 'Backspace' key. Why did this happen? The software developers attempted to design the program to infer the intent of the user. This feature saves a keystroke or two when the computer's actions match the intent of the user but is an annoyance when its actions are not what the user intended.

Example Two

A weary businesswoman walks into the lobby of a very respectable hotel chain after an arduous 18 hours of cancelled and delayed flights. At this point all she wants is to get a few hours of sleep before her board of directors meeting in the morning. She gives the desk clerk her name and searches through her wallet for her credit card. Before she can find it, the clerk informs her that there are no vacancies. 'How can that be? I have a confirmation number,' she says. She shows the clerk her travel itinerary with the confirmation number. The clerk checks his computer and verifies that the confirmation number is correct. He then explains that there was a serious maintenance problem an hour ago in another room in the hotel. Since she had not checked in by that time, the family occupying the damaged room had been moved to her room—the last vacant room in the hotel. 'Now what are you going to do?' she asks the clerk. The clerk considers the hotel chain's policies, his training and experience, and the situation. This corporate intent determines the range of options available to him in accommodating the weary traveler.

Example Three

In his autobiography, General Norman Schwarzkopf reflected on the days that led up to the 1991 ground war of Desert Storm. He wrote, 'When the time came for crucial decisions to be made on the battlefield, I wasn't going to be there. I was absolutely dependent on the individual skills, temperaments, and judgments of my generals. But I could establish a clear *framework* and convey my *intentions* and the spirit in which I wanted the campaign carried out' [emphasis added] (Schwarzkopf & Petre, 1992, p. 435).

Near the end of the ground war, Chairman of the Joint Chiefs of Staff General Colin Powell directed Schwarzkopf to set up a site for ceasefire talks. After considering several possibilities, it was decided to conduct the talks at Safwan airfield, an Iraqi landing strip just north of the Kuwaiti border. Schwarzkopf had ordered the VII Corps Army formation to take the airfield the previous morning because it was along the route that Iraqi forces would have to travel as they retreated from Kuwait. Reports from VII Corps indicated that the mission had been accomplished. However, when General John Yeosock, the Commander of all US Army Forces, contacted VII Corps to verify the mission had been carried out, he was told that VII Corps had no forces on the ground. Schwarzkopf felt as if he had been lied to by VII Corps and immediately ordered them to get combat troops on the ground at the airfield. He also said, 'I want to know in writing why my order was violated and why this mission was reported as carried out when it wasn't' (Schwarzkopf & Petre, 1992, p. 475).

In his recollection of the incident, VII Corps Commander General Fred Franks recalled the order from Yeosock to get to the crossroad near Safwan. Franks said, 'I had interpreted [Yeosock's] order as one to stop movement through the road junction. The tactics were up to me...' (Clancy & Franks, 1997, p. 440). Franks also stated:

> 'We had interdicted the road junction but not seized it. We had done so with attack helicopters of the 1st [Infantry Division] . . .. There had been no intent to disobey orders. I had selected the tactics to accomplish what I had interpreted to be the intent of the order. I had gotten those verbal orders from [Yeosock] at about 0330 on 28 February. The written order came later that morning. My interpretation of [Yeosock's] verbal orders to me was for us to stop Iraqi movement through that road junction. My selection of tactics was interdiction with air, and the assumption that the 1st [Infantry Division] attack would probably get there by 0800 if they continued on their attack axis. [VII Corps G-3] Stan [Cherrie] gave the 1st [Infantry Division] the mission to interdict the road junction, which they did. I and everyone else missed the 'seize' in the written order. That was my fault ...' (Clancey & Franks, 1997, pp. 445-456).

The first example is a relatively harmless example of what happens when technological systems are designed to 'understand' the intent of other agents.

Correcting the software's action is relatively easy, although perhaps somewhat frustrating for the manager. The second example illustrated the role that intent plays in the corporate world. Decisions made by personnel at lower levels of the organization often are guided by the policies established by those at the top of the organization. Typically, these policies are broad in nature and must be interpreted and applied to the local context. The hotel chain's policy may speak of 'putting the customer first' but when events conspire in a manner that pits the needs and comfort of one customer against another, the clerk is left to use his or her judgment to resolve the conflict. The possible courses of action available to the clerk are bounded by the guidance contained in the corporate policy. The third example illustrates that even in a complex organization in which the communication of intent is explicit, instantiated in doctrine, and practiced daily, misunderstandings, miscommunications, and misinterpretations can still happen. These miscues can often result in more than just the inconvenience of a few people. They can have dire consequences.

The Role of Intent in Macrocognitive Systems

What is Intent?

The Merriam-Webster's Collegiate Dictionary (2006, p. 651) defines intention as 'a determination to act in a certain way.' This definition, however, is inadequate for describing the concept of intent in macrocognitive systems. The definition suggests that intention (or intent) is the result of the cognitive processes of a single individual that then drives the actions of that individual. Pigeau & McCann (2000, p. 165) define intent in terms of military command and control (C^2) as 'an aim or purpose along with all of its associated connotation.' They further state that there are two elements that comprise intent: *explicit* (or public) intent; and, *implicit* (or personal) intent.

An order given to a subordinate is an example of explicit intent. According to Pigeau & McCann (2000, p. 167), implicit intent 'refers to all of the connotations latent within a specific (that is, explicit) aim. An individual's implicit intent is a combination of habits, experiences, beliefs, and values that reflect personal, military, cultural, and national expectations. Implicit intent is also consistent with the concept of tacit knowledge.' The definition developed by Pigeau & McCann implies that intent is something that is communicated from one agent to another for the purpose of guiding the actions of the other agent. The explicit intent indicates what must be done while the implicit intent guides the way in which the intent is interpreted and implemented.

US Army military doctrine provides the following definition of commander's intent. It is 'a clear, concise statement of what the force must do and the conditions the force must meet to succeed with respect to the enemy, terrain, and the desired end state' (FM 6-0, 2003, p. G-4). This definition focuses on the written or spoken aspect of intent, what Pigeau & McCann would refer to as explicit intent. The Army definition also indicates that a commander's statement of intent should describe the expected outcome of a mission in terms of three specific characteristics—enemy, terrain, and end state.

Intent in the Team Macrocognitive System Context

As stated in the preface, there are several characteristics that make the team macrocognitive system setting unique. Those characteristics most relevant to the present discussion of intent are the following:

- An emphasis on the manner in which team members interact.
- The role of artifacts in supporting cognitive processing.
- The environment is event-driven rather than self-paced.
- Feedback and interaction with the world are highly important.
- Multiple agents interact with one another.
- Tradeoffs will have to be made due to conflicting goals and conditions of uncertainty.
- Joint activity is distributed over both time and space.
- Coordination is necessary to meet complex and dynamic demands.
- Systems are composed of human and/or machine agents at the blunt end (for example, supervisors) and sharp end (for example, operators).

Intent is superfluous in a system that functions perfectly, that never deviates from the canonical path prescribed by it designers and supervisors. Unfortunately, there are no perfect systems. Intent, therefore, plays a vital role in virtually any complex system. To understand the role of intent, consider Figure 8.1. A remote supervisor directs a local agent or operator to move from the start state to the end state in this event-driven system. The remote 'supervisor' could be an actual human, a technological system, a system designer, or even a prescribed organizational policy or standard operating procedure. The supervisor may initiate a plan and may also communicate explicit intent. The local agent is responsible for moving the system from the start state to the end state.

In Figure 8.1, the remote supervisor has initiated a plan that will direct the local agent to move along canonical path *a* through *e* in order to reach the end state. The local agent must accomplish this plan in the context of the agent's local environment. The remote supervisor may or may not have knowledge of the local agent's environment when the plan is crafted. Even if the local agent's environment is known by the remote supervisor, that environment is likely to have changed by the time the local agent begins the plan. If things proceed as planned, the local agent will move the system along the canonical path to the end state. But, as any good military strategist knows, a plan rarely survives contact with enemy. Deviation from predetermined plans and procedures is the norm

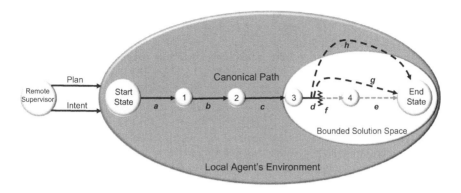

Figure 8.1   The role of intent in a macrocognitive system

(Roth et al., 1987). A resilient system, however, is one that is able 'to recognize and adapt to handle unanticipated perturbations...' (Woods, 2006, p. 22).

In Figure 8.1, the local agent moves the system successfully from the start state through intermediate states **1**, **2**, and **3** along canonical path *a*, *b*, and *c*. Just beyond intermediate state **3** an unexpected event (*f*) occurs. It is not possible to travel along path *d* to intermediate state **4**. The plan specified by the remote supervisor is no longer valid but the local agent is still expected to move the system to the end state. Consequently, the local agent must adapt (Woods et al., 1987; Suchman, 1987). The intent communicated to the local agent while the system was in the start state provides the framework for the adaptation. In Figure 8.1, this framework is represented by the bounded solution space. There are two paths (*g* and *h*) that lead from unexpected event *f* to the end state. Path *g* stays in the confines of the bounded solution space while path *h* does not. In a military context, path *h* may represent a violation of the rules of engagement; in an aircraft, path *h* may represent a case in which the design parameters of the system are exceeded.

The number of acceptable paths available to the local agent when confronted with an unexpected event is dictated by the remote supervisor. The remote supervisor's explicit and implicit intent establishes the size of the bounded solution space. A very narrow solution space provides little flexibility to the local agent. A very broad solution space may lead to unintended consequences such as collateral damage. The remote supervisor's intent must provide enough freedom of action for the local agent to respond to the variety of events or states that could emerge in a complex system. (See Ashby (1956) for a discussion of the Law of Requisite Variety.) It is no wonder that, even in organizations like the US military that routinely communicate intent explicitly, local agents often violate the boundaries of the solution space envisioned by the remote supervisor (Shattuck & Woods, 2000).

According to military doctrine, intent is developed by a senior commander and is given to subordinate commanders. Intent is embedded in the operation order for each mission and communicated via printed format, electronically, or orally. If the mission does not proceed as planned, the subordinate commanders will have to rely on the senior commander's intent to bound or guide their decision making and replanning. Note that the point at which the subordinate commanders must apply the intent is separated by both time and space from the point at which it was developed. The intent may have been created a few hours prior to the mission or perhaps several weeks earlier. The battlefield situation envisioned by the senior commander when the intent was crafted may be exactly what the subordinate commanders are experiencing or it may be vastly different. For intent to be implemented successfully, not only must the *explicit* intent be correct but the *implicit* intent must be accurate as well. Implicit intent is built up over time and requires senior and subordinate commanders to understand one another, to have an accurate and shared mental model.

As challenging as the intent process is in military organizations, it can be even more challenging in other types of complex systems. In military settings, intent is shared between humans embedded within a clear organizational hierarchy. In other complex systems, technological agents attempt to infer the intent of humans. This is often accomplished by having the technology monitor human actions and use the algorithms built into the system to infer the goals of the operator. That inference guides the 'behavior' of the technology thereafter. If the inference is correct, the technological agent appears to be a team player; if not, the human is left frustrated, or worse. In other complex systems, it may be necessary for humans to infer the intent of technology. Consider any situation in which humans monitor technology (for example, monitoring a nuclear power plant, flying a commercial aircraft, and so on). It is not enough for the human to understand the current state of the system. The human must accurately discern the goals of the system. Rarely are systems designed in a manner that makes their intent explicit to the human operators or supervisors. Finally, technological systems can be the communication channel for humans, sometimes unintentionally hampering intent communications. For example, electronic prescribing systems may use structured templates to increase ordering efficiency for physicians, unintentionally making it difficult to communicate the intent behind an order. When intent is misinterpreted, pharmacists could unintentionally make inappropriate substitutions of generic forms of medications that are not as effective. Regardless of whether intent is shared between two humans directly or mediated by technology, or between humans and technological systems, the explicit intent process is made up of five components.

Five Components of Explicit Intent

Intent is a process rather than a singular activity or event. As such, the intent process is comprised of five steps or components that typically are carried out in the following sequence: *formulation, communication, verification, interpretation,* and *implementation*. The first two components are normally executed by a remote supervisor, distant agent, or senior commander. The last two components are carried out by the local agent or subordinate commander. The third component (verification) is performed jointly by the remote supervisor and local agent.

Formulation. The remote supervisor considers the plan to be given to the local agent. The supervisor envisions how the plan may unfold, taking into consideration the context of the environment that will confront the local agent when the plan is implemented. If the plan and intent are to be executed a few hours after they have been formulated, predicting the future state of the system and the local agent's environment may be relatively easy. If the plan and intent will not be executed for weeks or months, then predicting the future state can be extremely challenging. In some cases (for example, the plan is a contingency in the event of a system failure), the remote supervisor may have no idea if or when the plan will be implemented.

The remote supervisor considers knowledge of the plan, the goal or end state, the local agent, the resources available to the agent, and a prediction of the future environment to develop a statement of intent. The intent establishes the boundary of the solution space—those options that are available to the local agent (should the plan fail) that do not violate system parameters. The system parameters may be physical, legal, or even moral in nature. A well-formulated intent establishes a clear boundary and provides sufficient discretion to the local agent to adapt the plan based on the local conditions of the system and the prevailing environment.

Communication. After the remote supervisor has formulated the explicit intent, it must be communicated effectively to local agents. There are a variety of ways in which intent can be communicated. CEOs of large corporations may publish vision, goal, or policy statements; they may share their intent verbally at board of director meetings or at annual addresses to employees and stockholders. Programmers may embed their intent for a technological system in the software code they develop well before the system is put into use. In other organizations, communicating intent is a formal process and dictated by doctrine. In the US Army, for example, doctrine dictates that intent must be stated explicitly in every operation order immediately following the mission statement. The doctrine also outlines the content of the written statement. (As cited previously, see FM 6-0, 2003, p. G-4.)

Verification. An important—but often neglected—component of the intent process is verification. After remote supervisors have communicated their intent they must check to see that the local agents have understood the intent *before* the intent is employed. In the case of a large corporation, this verification may be accomplished by senior managers speaking with a representative group of employees throughout the organization. When the remote supervisor is a programmer or designer and the local agent is a technological system, verification can be accomplished by running the system through a variety of scenarios and observing how the system responds. If time and circumstances permit, US Army senior commanders check the understanding of their subordinate commanders during face-to-face mission briefings and mission rehearsals. In these mission rehearsals, senior commanders often ask a series of 'what if' questions. What if the enemy does this instead of that? What if the unit on your left flank fails to accomplish its objective before you are supposed to launch your attack? What if you don't have close air support? These 'what if' scenarios are by no means exhaustive, but they do serve to calibrate the mental models (to some extent) of the senior and subordinate commanders.

Interpretation. Occasionally, local agents carry out the plan without incident. In these instances, there is no need to interpret or implement the remote supervisor's explicit intent. More often than not, an unanticipated event (or series of events) occurs and the task, order, or mission cannot be accomplished as specified. When this happens, it is the responsibility of the local agent to respond to the local conditions in a manner that continues to move the entire system toward its goal. The local agent relies on the remote supervisor's intent to guide the adaptation.

Interpreting the explicit intent of remote supervisors can be a difficult task. When remote supervisors formulate their intent (perhaps days or months before it must be interpreted) they must envision the environmental conditions and system characteristics with which the local agents will be confronted. The more complex the environment and the system, the more difficult it will be to envision all possible future states. If the remote supervisors do not envision the future correctly, their explicit intent will not be relevant. The future envisioned by them will be very different than the reality with which the local agents are confronted. An explicit intent that is overly specified could prove to be irrelevant to local agents because it creates a highly constrained solution space and substantially constrains the local agents' ability to adapt to the conditions with which they are confronted. An intent that is underspecified is also difficult to interpret because it does not provide local agents with any useful information on how to adapt to the local conditions.

Implementation. The last stage in the explicit intent process is for the local agent to implement the intent within the context of the current situation. As is often the case when intent must be implemented, the remote supervisor may be unavailable for consultation or clarification. This may be due to a breakdown in communications or because the remote supervisor is actually a designer, programmer, or some other agent far removed from the immediate circumstance. The decisions made by the local agent must keep the system within the boundaries of the solution space.

Many complex systems require the actions of local agents to be coordinated in order for the system to reach the end state. Such coordination is difficult to achieve because the complex system activities often are decentralized. Consider a military operation in which there are main and supporting attacks, or a ground attack that is preceded by an air attack or artillery fire. While decentralization affords local agents flexibility in responding to local situations, it also challenges these agents to respond in a manner that is both coordinated and synchronized. The more decentralized and flexible a complex system, the more challenging it will be to coordinate and synchronize the activities of the local agents in the system.

Assessing Explicit Intent in Team Macrocognitive System Settings

Assessing intent is challenging at best in most team macrocognitive system settings. However, the five steps in the process of explicit intent described above provide the basis for assessment. This section will consider how best to assess each of the five steps. While few empirical studies have been conducted in the explicit intent process, those that have been performed will be discussed. The following section will conclude the chapter with a description of how process tracing (Woods, 1993) can be used as an overarching strategy to assess the manner in which the explicit intent process is carried out.

Assessing the formulation of explicit intent. A well-formulated explicit intent should conform to both *structure* and *content* criteria. A consistent structure within a domain of practice provides practitioners at all levels with a reliable framework. In the 1990s, US Army doctrine dictated an intent structure and content that included the following: the *purpose* of the operation; the *method* to be used to accomplish the operation; and, the *end state* of the operation (Department of the Army, 1993). Klein (1993) analyzed 35 intent statements that were extracted from US Army tactical operation orders. He hypothesized a more detailed intent structure and content that contained seven distinct categories or slots. The slots are listed below. Klein's analysis of the 35 intent statements revealed seven slots. The total number of entries in the seven slots and their percentages listed below:

1. Purpose of the mission (higher goals)   9   (2.3%)
2. Mission objective (image of the desired outcome)   72   (18.6%)
3. Plan sequence   151   (38.9%)
4. Rationale for the plan   102   (26.3%)
5. Key decisions   22   (5.7%)
6. Anti-goals   6   (1.5%)
7. Constraints and considerations   26   (6.7%)
                                                         388

Entries in the 'plan sequence' slot made up nearly 40 percent of the 388 distinct entries identified in the 35 intent statements. These findings suggest that while the intent statements analyzed by Klein conformed to the general structure he hypothesized, they may have violated content guidelines because they were overly specified. This finding is consistent with the results of a study conducted by Shattuck and Woods (2000) in which the method sections of intent statements were found to be very detailed. When remote supervisors formulate intent statements that provide a detailed plan sequence, they increase the likelihood that their statement will be discarded by local agents because these agents may confront anomalies unanticipated by the remote supervisors that prevent that plan sequence from being carried out.

Another method to assess the formulation step is to consider the bounded solution space created by the explicit intent. The solution space should be large enough to cope with virtually all of the uncertainty with which the local agent could be confronted. The boundary of the solution space should be as discrete as possible. That is, it should be obvious to remote supervisors and local agents alike which potential solutions are inside and which are outside the solution space. The best way to assess the size of the solution space and the distinctness of its boundary is to test it against a variety of typical and atypical scenarios. For example, if a proposed solution would achieve the objective of capturing a military objective (or increasing the profit margin of a company) but, in the process, would result in significant collateral damage (or the firing of numerous valued employees), it clearly lies outside the boundary of the solution space. If the intent has been formulated correctly, it will not be difficult to determine on which side of the boundary a potential solution will fall.

Assessing the communication of explicit intent. Assessing the communication of intent is simply a matter of considering the method and the context. The methods for communicating explicit intent include the following: in person (face to face); in writing; graphically; remotely via voice only; and, video teleconferencing (VTC). Contextual issues include the domain of practice, the prevailing environmental conditions, the characteristics of the remote supervisor and the local agent, and the degree to which the supervisor and local agent share a common frame of reference. In a tactical military setting, if the stakes are high, the mission is complex, the subordinate commander is inexperienced and has a strong preference for visual learning, and the senior and subordinate commanders have not worked together previously, communicating intent in a purely written form is likely to be inadequate. In this case, a better match between method and context would be to communicate the intent face to face with graphical aids. The next step in the explicit intent process—verification—is critical to ensuring that the senior and subordinate commanders are calibrated.

Assessing the verification of explicit intent. Verifying explicit intent goes well beyond asking local agents whether they understood and agreed with what was communicated. The verification process is successful if the solution space envisioned by the remote supervisor and the local agent are identical in terms of both the size and shape of the space and the clarity of the boundary. The most effective approach for assessing this step in the explicit intent process is to generate a variety of possible future scenarios. If the context permitted, and remote supervisors were to develop ten scenarios to assess the extent to which verification has been completed successfully, one or two situations should reside well within the solution space while one or two other others should lie well outside the solution space. The vast majority of scenarios, however, should be located just inside or just outside the boundary of the solution space. These scenarios are then posed by remote supervisors to local agents. The local agents describe *what* they would do in each situation and *why*. Both the *what* and the *why* are important. Figure 8.2 depicts the four possible outcomes.

Any outcome other than the right decision made for the right reason indicates that the remote supervisor and the local agent are not calibrated. As discussed in Chapters 1, 2, and 3, measures need to separate processes and outcomes. A right decision made for the wrong reason suggests a lack of understanding of the system's goals and that the desired outcome would have been achieved by luck or chance. The wrong decision made for the right reason may indicate that the local agent understood and agreed with the higher-order goals of the system, but

Decision

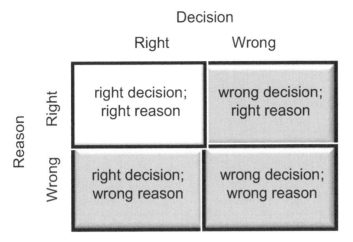

Figure 8.2   **Possible outcomes when local agents respond to the possible future scenarios posed by the remote supervisors**

chose a solution that violated a system parameter or the rules of engagement. The verification process should continue until the remote supervisors are confident that they are calibrated with the local agents. When completed, the system is set in motion and begins to move toward its goals. When the system is blocked from achieving its goal by an unanticipated event, local agents will have to interpret the explicit intent. The same is true when the system has been unexpectedly successful in achieving its goal. If the goal has been achieved faster than expected with fewer resources than predicted, the local agent will have to rely on the explicit intent to determine how to proceed.

Assessing the interpretation of explicit intent. Accurate interpretation of explicit intent begins with recognizing that the canonical path of the system has been blocked. The local agent must then properly assess the current state of the system with respect to the system's goals and the local conditions. Given the system's current state and the immediate context, local agents must recall the remote supervisor's intent and determine if it is still relevant. The explicit intent remains viable if it can apply to the system's current state and the local conditions. If still relevant, local agents use the explicit intent to guide them through the process of selecting a course of action that lies within the solution space.

Assessing the interpretation of intent is best accomplished by employing methods such as cued retrospective interviews (Militello & Hoffman, 2008), critical incident technique (Flanagan, 1954), and critical decision method (Klein et al., 1989). These and other similar methods will guide researchers and practitioners in answering the following questions:

- When, how, and why was the system not able to follow the planned path?
- What is the goal of the system and can it still be achieved?

- What are the local conditions?
- Is the remote supervisor's intent still relevant?
- What options are available to the local agent?
- What are other local agents doing or planning to do?
- Is/are the course(s) of action under consideration within the solution space boundary?
- Would the remote supervisor agree with the local agents' assessment? If not, why not?

Assessing the implementation of explicit intent. The final step in the explicit intent process is selecting a course of action and implementing it. The ultimate metric for this stage is whether the system has achieved its goal and whether it stayed within the bounded solution space. However, in loosely coupled systems, making the correct decision does not necessarily mean that the system will achieve its goal. Therefore, the preferred evaluation method is to have the remote supervisor determine if the local agents have complied with the explicit intent. This evaluation should go beyond the decisions of the local agents to include the strategies, tactics, techniques, and procedures employed to implement the intent. The decision selected by the local agents will provide insight into both the extent to which the remote supervisor and the local agents are calibrated and the solution space envisioned by the local agents. The strategies, tactics, techniques, and procedures chosen are an indication of the local agents' understanding of the system's structure and function.

Concluding Remarks

This chapter has discussed the five steps in the explicit intent process within complex cognitive systems. Each of the five steps was described and methods to assess them were introduced and explained. Just as important as evaluating each step is assessing the entire explicit intent process. Process tracing is a method that provides an end-to-end evaluation strategy. By employing a process tracing methodology, it is possible to determine where and how breakdowns occur. Woods (1993) describes process tracing as a technique that allows researchers to map the activities of a system as events unfold by considering the cues available to decision makers and by noting those cues of which the decision makers were actually aware. Process tracing uses data from a variety of sources to externalize internal thought processes and construct a sequence of information flow and knowledge activation. Once this protocol is complete, the point(s) at which the intent process broke down becomes obvious to practitioner and researcher alike. (See Shattuck and Woods (2000) for an example of a study that employed a process tracing methodology to evaluate the communication of intent processes from surrogate supervisors to four US Army combat units in a low-fidelity simulation.)

Assessing performance in a team macrocognitive system setting can be a daunting task. The system's complexity, plethora of artifacts, number of agents, and distributed nature are among the tasks that make this assessment so difficult. Given these attributes, no one human or machine agent is capable of operating such cognitively demanding systems. Remote supervisors must rely on local agents to carry out various functions. Inevitably, systems will experience unanticipated events that will require local agents to respond, often without access to or specific direction from a remote supervisor. It is at these times that the explicit intent process plays a vital role in overcoming the breakdown and moving the system to its ultimate goal or end state. The methods described in this chapter for measuring and assessing explicit intent are essential tools for the cognitive engineering researcher. What still remains to be accomplished is a similar explication of analytical tools and techniques to assess implicit intent.

References

Ashby, W.R. (1956). *An introduction to cybernetics*. New York, NY: Wiley.

Clancy, T., & Franks, F. (1997). *Into the storm: A study in command.* New York, NY: G. P. Putnam's Sons.

Flanagan, J. C. (1954). The critical incident technique, *Psychological Bulletin. 51(4)*, 327-358.

Klein, G. A. (1993). Characteristics of commander's intent statements. In *Proceedings of the 1993 Symposium on Command and Control Research* (pp. 62-69). McLean, VA: Science Applications International.

Klein, G. A., Calderwood, R., & MacGregor, D. (1989). Critical decision method for eliciting knowledge. *Systems, Man and Cybernetics, IEEE Transactions on Systems, Man, and Cybernetics, Vol 19 (3)*, 462-472.

Merriam-Webster's Collegiate Dictionary, 11th Edition. (2006). Springfield, MA: Merriam-Webster, Incorporated.

Militello, L. G. & Hoffman, R. R. (2008). The forgotten history of cognitive task analysis. *Proceedings of the Human Factors and Ergonomics Society 52nd Annual Meeting*, pp. 383–387.

Pigeau, R. & McCann, C. (2000). Redefining command and control. In R. Pigeau & C. McCann (Eds.), *The human in command: Exploring the modern military experience.* New York, NY: Plenum Publishers.

Roth, E. M., Bennett, K. B. & Woods, D. D. (1987). Human interaction with an 'intelligent' machine. *International Journal of Machine Studies*, *27(5/6)*, 479-525.

Schwarzkopf, H. N., & Petre, P. (1992). *It doesn't take a hero.* New York, NY: Linda Grey Bantam Books.

Shattuck, L. G. & Woods, D. D. (2000) Communication of intent in military command and control systems. In R. Pigeau & C. McCann (Eds.), *The human in command: Exploring the modern military experience.* New York, NY: Plenum Publishers.

Suchman, L. (1987). *Plans and situated actions: The problem of human – machine communication.* Cambridge, UK: Cambridge University Press.

U.S. Army Field Manual (FM) 6-0 (August 2003). *Mission Command: Command and Control of Army Forces.* Washington, DC: U.S. Government Printing Office.

U.S. Army Field Manual (FM) 100-5 (1993). *Operations.* Washington, DC: U.S. Government Printing Office.

Woods, D. D. (1993). Process- tracing methods for the study of cognition outside of the experimental psychological laboratory. In G. Klein, J. Orasanu, R. Calderwood, & C. Zsambok (Eds.), *Decision making in action: Models and methods.* Norwood, NJ: Ablex Publishing Co.

Woods, D. D. (2006). Essential characteristics of resilience. In E. Hollnagel, D. Woods, & N. Leveson (Eds.), *Resilience engineering.* Burlington, VT: Ashgate Publishing Company.

Woods, D. D., O'Brien, J. F., & Hanes, L. F. (1987). Human facors challenges in process control of nuclear power plants. In G. Salvendy (Ed.), *Handbook of human factors.* New York, NY:Miley.

Chapter 9

Survey of Healthcare Teamwork Rating Tools: Reliability, Validity, Ease of Use, and Diagnostic Efficacy

Barbara Künzle, Yan Xiao, Anne M. Miller, and Colin Mackenzie

Introduction

Hospital care is a uniquely collaborative endeavor. Exemplified by patient resuscitation and surgery, the success of care activities depends on skills to carry out highly technical procedures and to coordinate activities from team members. Analogous to other emergency, time-critical domains, numerous studies on human errors in healthcare reveal the adverse consequences of the lack of teamwork on patient safety (for example, Gawande et al., 2003; Greenberg et al., 2007; Reader et al., 2007; Zingg et al., 2008). These studies have prompted development and implementation of interventions to improve teamwork (Nielsen et al., 2007; Risser et al., 1999). However, researchers and training program designers are facing the challenge of measuring the impact of teamwork training, as well as the challenge of how to provide feedback on teamwork in real or simulated patient care (Guise et al., 2008; Manser 2008; Rosen et al., 2008).

An important method to collect data on team performance is through observation, either in real or simulated environments. The advantage of observational measures over others such as self-reporting or questionnaire-based surveys is the possibility to collect detailed data on teamwork under varying conditions (Carthey, 2003). Self-reporting, for example, tends to focus on information about things that went wrong (Fletcher et al., 2002). Observations can also provide information on how errors are influenced by teamwork skills that may be used for feedback for training purposes. In this chapter, we will review studies using observational methodologies to understand desirable characteristics of rating tools, especially those relevant to teams often found in healthcare.

Task Teams

In high-risk domains, team members tend to be highly skilled and specialized. They are organized to fulfill specific missions and work interdependently. Mission success requires deep knowledge about a domain, highly trained skills, use of rules and checklists, and strong commitment by team members. Additionally, teams in healthcare often work face-to-face. Their activity flow is poly-directional, flexible, and very intensive as teams repeatedly face novel situations. Teams are often formed by roles as opposed to by certain individuals, as individuals may change over shifts or workload constraints (Faraj & Xiao, 2006). In healthcare, individuals function in a team for the care of a specific patient. Teams often have fluid team membership with multiple disciplinary and professional backgrounds. Consequently, these teams have to learn how to rapidly integrate different professional cultures, such as nursing and medicine (Manser et al., 2008). Coordination of specialized expertise needs to occur in a time-compressed manner and in less familiar and high-stakes environments. Teams rely more on implicit modes of coordination to ensure rapid contribution from all team members who are highly interdependent (Blickensderfer et al. 1998). However, implicit coordination is only effective if the group members have shared and accurate mental models of the task and the team interaction. If one of these two requirements is not properly satisfied, implicit coordination can be very risky (Kolbe et al., 2009).

These characteristics of teams and team membership pose serious challenges to the measurement of teamwork. Those measures and methodologies developed for production, service, and project teams have limited applications for task teams found in healthcare. Survey measures that may be very appropriate for studies of groups or non-task oriented teams are unlikely to uncover critical teamwork skills and provide useful feedback information for training purposes. In particular, it is important to differentiate teams from groups where members are more homogeneous with less assigned roles or functions (Cannon-Bowers et al., 1993). In research settings, teams are frequently contrived to be devoid of elements of a social system and thus doubts have been raised about the validity of the findings of such teams (Driskell & Salas, 1992; Forsyth, 1999).

Rating Teamwork by Observation

Teamwork measures have been proposed in a framework that separates taskwork of individual activities from teamwork due to interactions among team members (Salas et al., 1992). A number of empirical teamwork measures have been proposed to assess the internal processes of teams and correlate them with external measures of performance (Cannon-Bowers & Salas, 1997). Baker & Salas (1992) emphasized observations of critical team behaviors or dimensions such as adaptability, coordination, and communication. Much research has focused on the development of team performance measurement for military teams (see Baker & Salas, 1992).

The potential for measuring teamwork by observation has several important benefits. During mission debriefing and after training sessions, teams may be provided with structured feedback through a set of measures of teamwork skills. Teams and individuals may be assessed by teamwork measures. Such assessment may form a basis for evaluating training programs. Below, we present examples of observational measurement and discuss their advantages and disadvantages regarding inter-observer reliability, validity, their ease of use, and diagnostic efficacy (for an overview see Table 9.1).

To assess team performance in surgery, two studies (Healey et al., 2004; Undre et al., 2006) applied Observational Teamwork Assessment for Surgery (OTAS), a tool that measures team rather than individual performance in surgery based on a generalized input-process-output model of teamwork (see also Yule et al., 2006). The performance assessment covers both tasks and behaviors. While a surgical observer completed a task checklist centered on the patient, equipment, and communication behavior, a psychologist rated teamwork on five dimensions such as cooperation or leadership by observing three phases of surgery (pre-, intra-, and post-operative). As reported by Healey et al. (2004) the rating of team behavior was feasible and an important complement to simply recording task completions. The authors intend to develop a short version suitable for use in training and simulation (Undre et al., 2006) which might also improve the ease of use. So far, no formal analyses of the reliability and validity have been published. However, Undre et al. (2006) suggest further assessing reliability and validity and refer to ongoing studies.

NOn-TECHnical Skills (NOTECHS) is another rating scale for surgery which was originally developed for assessment of nontechnical skills in aviation (Van Avermate & Kruijsen, 1998). It was modified for the assessment of nontechnical skills of surgeons (see Moorthy et al., 2005, 2006) and was applied during a surgical crisis simulation (Moorthy et al., 2006). The scale consists of four categories in order to assess nontechnical skills: 1) communication and interaction, 2) vigilance/situation awareness, 3) team skills, and 4) leadership and management skills. An additional category assesses decision making/crisis handling. Each of these categories consists of three to five elements which are rated on a six-point scale. The score for each category and the total score are expressed as a percentage. The assessment was performed by a human factors researcher and the surgical fellow who was trained by the researcher. The reliability of NOTECHS is appealing: the internal consistency between the five categories of the nontechnical skills assessment score was 0.87. Furthermore, high levels of inter-rater reliability were achieved.

Based on a behavioral marker system from aviation (see, for example, Klampfer et al., 2001), Helmreich et al. (1995) developed the Operating Room Checklist (ORCL) consisting of a list of behavior categories with rating scales that could be used to assess the so-called non-technical performance of operating teams. The term 'NOTECHS' is often used in healthcare domains to denote teamwork skills. It was adopted from European aviation research (Flin et al., 2003) and is defined as the cognitive and social skills of team members (for example, leadership, decision making, team coordination, situation awareness). It was primarily designed to measure team behaviors rather than to rate the NOTECHS of individual surgeons.

Table 9.1   Overview of observational rating tools for teamwork

Author/Year	Rating Tool	Constructs measured	Targeted Domain	Main Categories	Evaluation/ Training	Rating Agreement/ Instrument Accuracy
Healey et al. (2004) Undre et al. (2006)	OTAS	Team performance	Operating theatre	Task checklist: 1: Patient 2: Equipment 3: Task communications Teamwork behaviors: 1: Shared-monitoring 2: Communication 3: Cooperation 4: Coordination 5: Shared leadership Plus: Effective/ ineffective behaviors	Training	"Interobserver reliability is currently being explored and will be reported in future studies. Preliminary data suggest a good level of agreement, with exact agreement on ratings on a seven-point scale on over 75% of occasions for all five behavioral Dimensions." [P.1778]
Van Avermate (1998) Moorthy et al. (2005; 2006)	NOTECHS	Nontechnical Skills	Aviation / Surgery	1: Communication and interaction 2: Vigilance/situation awareness 3: Team skills 4: Leadership and management skills Plus: Decision making/ crisis-handling	Training	α: 0.87 [1] High interrater reliability
Helmreich et al. (1995)	ORCL	Non-technical team performance	Operating teams		Evaluation/ Training	

Table 9.1 *Continued*

Author/Year	Rating Tool	Constructs measured	Targeted Domain	Main Categories	Evaluation/ Training	Rating Agreement/ Instrument Accuracy
Thomas et al. (2004)	UTBMNR	Non-technical team performance	Neonatal Resuscitation teams	Team behaviours: 1: Information sharing 2: Inquiry 3: Assertion 4: Intentions verbalised 5: Teaching 6: Evaluation of plans 7: Workload management 8: Vigilance 9: Leadership 10: Teamwork overall	Evaluation/ Training	
Fletcher et al. (2003)	ANTS	Nontechnical Skills	Anesthesia	1: Task management 2: Team working 3: Situation awareness 4: Decision making	Evaluation/ Training	
Grote et al. (2003) Zala-Mezö et al. (2009)		Coordination behaviors	Anesthesia teams	1: Implicit/explicit coordination 2: Coordination via Leadership 3: Coordination via heedful interrelating		κ^2: 84%

Table 9.1   Concluded

Author/Year	Rating Tool	Constructs measured	Targeted Domain	Main Categories	Evaluation/ Training	Rating Agreement/ Instrument Accuracy
Kolbe et al. (2009)		Coordination behaviors	Anesthesia teams	1a: Implicit/explicit coordination 1b: Coordination of information exchange/ of actions 2: Heedful interrelating 3: Other behaviors		κ: 0.61 – 0.99 κ (implicit coordination/heedful interrelating): 0.29 - 0.49
Manser et al. (2007; 2008)		Coordination behaviors Adaptive coordination Clinical activities	Anesthesia teams	1: Information management 2: Task management 3: Coordination via work environment 4: Metacoordination Plus: Clinical activities		κ: Good to excellent feasible and sensitive for adaptive coordination
Gaba et al. (1998)		Skilled Performance in managing crisis situation	Anesthesiologists	1: Technical performance 2: Behavioral performance		
	EMCRM	Skilled Performance in managing crisis situation	Flight deck crews	1: Technical performance 2: Behavioral performance		κ(technical performance): Good. κ(behavioral performance): Somewhat low

1   α : Internal consistency; statistically referred to as alpha

2   κ: Interrater reliability; statistically referred to as kappa

There do not appear to be any published reports on its use as a teaching or research tool (Yule et al., 2006).

Similarly, Thomas et al. (2006) translated existing aviation team behaviors, named Line Operations Safety Audit (LOSA), to the setting of neonatal providers, resulting in the development of the University of Texas Behavioral Marker of Neonatal Resuscitation (UTBMNR) form. The behavioral marker audit form for neonatal resuscitation lists and defines ten team behaviors: information sharing, inquiry, assertion, intentions verbalized, teaching, evaluation of plans, workload management, vigilance, leadership, and teamwork overall. As the authors noted, the behavioral markers describe specific, observable behaviors but validity and reliability needs to be further investigated.

In the domain of anesthesia, several measures were developed. For example, Anaesthetists' Non-Technical Skills (ANTS) is a behavioral marker system for key NOTECHS required in anesthesia. The system was developed at the University of Aberdeen for NOTECHS assessment in the operating theatre, in the simulator, or from video recordings (Fletcher et al., 2003). It comprises four skill categories (task management, team working, situation awareness, and decision making) divided into 15 elements, each with examples for good and poor behaviors. NOTECHS are scored along with technical skills. Compared to the descriptive approach of behavior observation systems, ANTS provides a way to score teamwork in a structured manner by occurrences of desirable or undesirable behavior. The internal consistency and usability of this system was tested by asking 50 consultant anaesthetists to rate the behavior of an anaesthetist during eight simulator scenarios. The findings indicated that ANTS has a satisfactory level of validity, reliability, and usability and can be developed, alongside careful guidelines, within anaesthetic training curricula (Jeffcott & Mackenzie, 2008). Furthermore, the ANTS system appeared to be complete and the skills are observable, and could be rated with acceptable levels of agreement and accuracy (Fletcher, et al., 2003). It is the first tool for NOTECHS training in anaesthesia and supports the need for objective measures of teamwork to appropriately inform real and simulated training initiatives (Jeffcott & Mackenzie, 2008).

Descriptive observation systems were also developed in order to capture coordination behaviors within anesthesia teams. For example, the coding system developed by Grote et al. (2003) and Zala-Mezö et al. (2009) distinguishing between implicit and explicit coordination, coordination via leadership, and heedful interrelating as the individual effort to facilitate smooth coordination. The inter-rater reliability of coding was 84 percent (Zala-Mezö, et al., 2009). Despite the good inter-rater agreement, the system has its limitations. According to Zala-Mezö and colleagues (2009), the system allowed a detailed analysis of behavioral data but at the same time a less detailed categorization would have been easier to use allowing a more economical way of analyzing this kind of behavioral data. The category system has been further developed by Kolbe et al. (2009) consisting again of three main categories which are coordination, heedful interrelating, and other behavior. The coordination category distinguishes between two crucial coordination functions in anesthesia teams—information exchange and joint actions—which can be explicit and implicit. Inter-rater reliability was promising for most categories (for explicit coordination of information exchange $\kappa = 0.67$; for explicit coordination of actions $\kappa = 0.63$; for implicit coordination of actions $\kappa = 0.76$; for heedful interrelating $\kappa = 0.60$) except for implicit coordination of information exchange ($\kappa = 0.32$), indicating that the system is an appropriate tool to measure explicitness and heedfulness reliably while implicitness in sharing information remains difficult to observe. A similar approach to analyzing behavior of anesthesia was developed by Manser et al. (2008) which, in contrast to the systems discussed before, records not only coordination behavior such as information management, task management, coordination via the work environment, or meta-coordination but also clinical activities performed by the anesthesia teams. The coding system allowed for a good to excellent inter-rater reliability and proved to be feasible and sensitive for adaptive coordination. Gaba et al. (1998) developed an instrument measuring two separate aspects of skilled performance of anesthesiologists in managing crisis situations (technical and behavioral performance), later developed to EMCRM (Emergency Medicine Crisis Resource Management). An instrument used to rate flight deck crew behavior was adapted and included ratings for ten crisis management behaviors such as communication, leadership, or workload distribution. Behavior was rated at a five-point ordinal scale ranging from poor to outstanding performance. Technical performance was measured with a list of the appropriate medical and technical actions for recognition, diagnosis, and therapy (for example, ventilation and oxygenation, hyperthermia management). Raters recorded the presence or absence of each action, assigned point values for successful implementation of each action, and summed then the point values for all actions ranging from 0.0 to 1.0. Fourteen videorecorded teams were rated independently by five raters. There was good inter-rater agreement on technical performance, whereas the raters agreed less for behavior of team members. However, the study shows the feasibility of rating both technical and behavioral performance.

Wallin and colleagues (2007) used the behavioral categories developed by Gaba et al. (1998) in order to rate the quality of teamwork skills of medical emergency teams. Inter-rater reliability between the three raters ranged from good (0.60) for 'knowledge of the environment,' 'anticipation of and planning for potential problems,' and 'utilization of resources' to a fairly high rating (0.76) for 'communication with other team members' and 0.78 for 'recognition of limitations/call for help early enough.' Another attempt to rate trauma teams typically consisting of multiple professional groups, University of Maryland Team Observable Performance (UMTOP), was made by a multidisciplinary research team developing observable behaviors tool (Kuenzle et al., 2007). UMTOP was developed in order to measure team performance based on four main categories, tasks; clinical performance; leadership; and teamwork. The instrument underwent preliminary evaluation using two videorecorded trauma patient resuscitations rated by five surgeons and five nurses. Inter-rater agreement was dependent on the dimensions and ranged from slight to fair agreement. All reviewers rated the tool as simple to use. However, further analyses of the ease of use and/or diagnostic efficacy have yet to be completed.

Discussion

Progress in teamwork measures requires a systematic examination of important characteristics of such measures. Even after years of research on team performance, it is still elusive for researchers and practitioners alike to have comprehensive understanding of teamwork in high-risk, high-skill domains. Measures of teamwork based on observable behaviors can provide useful data on teams, even though they cannot reveal all relevant facets of team performance, such as team development and interpersonal dynamics. We reviewed recently developed observational methods to assess team performance in teams in general and more specifically teams in healthcare. Our review is based on a four-dimension matrix: reliability, validity, ease of use, and diagnostic efficacy. In summary, rating tools developed thus far show that they are useful and necessary in order to rate team performance. Most of the studies revealed the inter-rater agreements providing support for the reliability of the tools. However, only a few authors provide further statistical evidence to support a statement about the ease of use and diagnostic efficacy of the measurements. Validity is the most difficult dimension to confirm in teamwork measures, due to the many confounding variables and multiple conflicting influences found in teamwork.

It should be noted that observational rating tools have inherent shortcomings. For example, team cognitive processes have to be inferred, especially for teams with highly skilled team members under high workload, when teams will likely use implicit coordination modes (Zala-Mezö et al., 2009). A thorough assessment of team performance would ideally comprise a combination of data from multiple sources, addressing all disciplines and their intra-and interrelations, supported by both quantitative and qualitative data. However, comprehensiveness of measures have to be balanced against practical constraints in deploying measurement. Both simulated and real environment observations have limitations. In live settings, collecting data can be difficult because the events of interest do not always occur. Furthermore, it may difficult to observe the whole environment. The presence of a researcher may affect normal behavior and there might also be problems of consent and confidentiality (Fletcher et al., 2002).

Complexity in teamwork measures is likely to match complexity of team activities. Rating tools based on structured observation seem to be easier to develop for tasks that have a clear beginning and end point and well-defined team tasks, such as elective surgery. In emergency departments and intensive care units, team tasks are less well defined, with a higher degree of unpredictability. Consequently, teamwork measures may have to be limited to certain types of team activities, such as hand-offs or responses to emergencies (for example, cardiac arrest). A recent comprehensive analysis of team processes started to address this limitation by Miller and colleagues (Miller, 2009), who studied intensive care unit care coordination as a linear sequence over a five-day period.

In addition, tasks that involve verbal communication or a potentially high frequency of omission and commission errors may be suitable for observational ratings (see Carthey, 2003). Healey et al. (2004) define three steps in developing observational measures of team performance in surgery. First, the beginning and the end point of a task needs to be specified. Secondly, the objectives of each stage need to be defined. Third, a list of subsidiary goals and tasks within each category are defined. According to these authors, specifying tasks is the basis for assessing team performance.

Finally, we would recommend keeping observation systems simple and easy to use. Although systematic analysis of the ease of use and diagnostic efficacy of observation systems reviewed above are rare, we believe that they are essential for an economical way of recording observations. Furthermore, more analysis of the presented observation tools are needed, be they in-depth examinations of inter-rater reliability or systematic assessment of their usability.

References

Baker, D. P., & Salas, E. (1992). Principles for measuring teamwork skills. *Human Factors, 34*(4), 469-475.

Blickensderfer, E., Cannon-Bowers, J. A., & Salas, E. (1998). Cross training and team performance. In J. A. Cannon-Bowers & E. Salas (Eds.), *Making decisions under stress: Implications for individual and team training* (pp. 299–311). Washington, DC: American Psychological Association.

Cannon-Bowers, J. A., & Salas, E. (1997). A framework for developing team performance measures in training. In M. T. Brannick, E. Salas & C. Prince (Eds.), *Team performance assessment and measurement. Theory, methods, and implications* (pp. 45-62). Mahwah, NJ: Lawrence Erlbaum Associates.

Cannon-Bowers, J. A., Salas, E., & Converse, S. A. (1993). Shared mental models in expert team decision-making. In N. J. Castellan, Jr. (Ed.), *Current issues in individual and group decision making* (pp. 221-246). Hillsdale, NJ: Lawrence Erlbaum.

Carthey, J. (2003). The role of structured observational research in health care. *Quality & Safety in Health Care, 12*(Suppl II), ii13-ii16.

Driskell, J. E., & Salas, E. (1992). Can you study real teams in contrived settings? The value of small group research to understanding teams. In R. W. Swezey & E. Salas (Eds.), *Teams: Their training and performance* (pp. 101-122). Norwood, NJ: Ablex.

Faraj, S., & Xiao, X. (2006). Coordination in fast-response organization. *Management Science, 52*(8), 1155-1169.

Fletcher, G., Flin, R., McGeorge, P., Glavin, R., Maran, N., & Patey, R. (2003). Anaesthetists' non-technical skills (ANTS): Evaluation of a behavioural marker system. *British Journal of Anaesthesia, 90*(5), 580-588.

Fletcher, G. C. L., McGeorge, P., Flin, R. H., Glavin, R. J., & Maran, N. J. (2002). The role of non-technical skills in anaesthesia: a review of current literature. *British Journal of Anaesthesia, 88*(3), 418-429.

Flin, R., Fletcher, G., McGeorge, P., Sutherland, A., & Patey, R. (2003). Anaesthetists' attitudes to teamwork and safety. *Anaesthesia, 58*(3), 233-242.

Forsyth, D. R. (1999). *Group dynamics* (3rd edn). Belmont, CA: Wadsworth.

Gaba, D. M., Howard, S. K., Flanagan, B., Smith, B. E., Fish, K. J., & Botney, R. (1998). Assessment of clinical performance during simulated crisis using both technical and behavioral ratings. *Anesthesiology, 89*(1), 8-18.

Gawande, A. A., Zinner, M. J., Studdert, D. M., & Brennan, T. A. (2003). Analysis of errors reported by surgeons at three teaching hospitals. *Surgery, 133*(6), 614-621.

Greenberg, C. C., Regenbogen, S. E., Studdert, D. M., Lipsitz, S. R., Rogers, S. O., Zinner, M. J., & Gawande, A. A. (2007). Patterns of communication breakdowns resulting in injury to surgical patients. *Journal of the American College of Surgeons, 204*(4), 533-540.

Grote, G., Zala-Mezö, E., & Grommes, P. (2003). Effects of standardization on coordination and communication in high workload situations. In R. Dietrich (Ed.), *Communication in high risk environments* (pp. 127-154). Hamburg: Helmut Buske.

Guise, J. M., Deering, S. H., Kanki, B. G., Osterweil, P., Li, H., Mori, M., & Lowe, N. K. (2008). Validation of a tool to measure and promote clinical teamwork. *Simulation in Healthcare, 3*(4), 217-223.

Healey, A. N., Undre, S., & Vincent, C. A. (2004). Developing observational measures of performance in surgical teams. *Quality and Safety in Health Care, 13*(suppl 1), i33-40.

Helmreich, R., Schaefer, H., & Sexton, J. (1995). The operating room checklist (*Technical Report No. 95-10*). Austin, TX.

Jeffcott, S. A., & Mackenzie, C. F. (2008). Measuring team performance in healthcare: review of research and implications for patient safety. *Journal of Critical Care, 23*(2), 188-196.

Klampfer, B., Flin, R., Helmreich, L., Hausler, R., Sexton, B., Fletcher, G., Field, P., Staender, S., Lauche, K., Dieckmann, P., & Amacher, A. (2001). *Group interaction in high risk environments: enhancing performance in high risk environments, recommendations for the use of behavioural markers*. Berlin: GIHRE.

Kolbe, M., Künzle, B., Zala-Mezö, E., Wacker, J., & Grote, G. (2009). Measuring coordination behaviour in anaesthesia teams during induction of general anaesthetics. In R. Flin & L. Mitchell (Eds.), *Safer surgery. Analysing behaviour in the operating theatre*. Aldershot, UK: Ashgate.

Kuenzle, B., Xiao, Y., Mackenzie, C. F., Seagull, F. J., Grissom, T., Sisley, A., & Dutton, R. (2007). Development of an instrument for assessing trauma team performance. In *Proceedings of the Human Factors and Ergonomic Society Annual Conference.* 51, 678-682.

Manser, T. (2008). Teamwork and patient safety in dynamic domains of healthcare: A review of the literature. *Acta Anaesthesiologica Scandinavica, 53*(2), 143-151.

Manser, T., Howard, S. K., & Gaba, D. M. (2008). Adaptive coordination in cardiac anaesthesia: A study of situational changes in coordination patterns using a new observation system. *Ergonomics, 51*(8), 1153-1178.

Manser, T., Howard, S. K., & Gaba, D. M. (2009). Identifying characteristics of effective teamwork in complex medical work environments: Adaptive crew coordination in anaesthesia. In R. Flin, & L. Mitchell (Eds.), *Safer surgery: Analysing behaviour in the operating theatre* (pp. 223-239). Aldershot: Ashgate.

Miller, A., Scheinkestel, C., & Joseph, M. (2009). *Coordination and continuity of intensive care unit patient care.* Human Factors, *51*(3), 354-367.

Moorthy, K., Munz, Y., Adams, S., Pandey, V., & Darzi, A. (2005). A human factors analysis of technical and team skills among surgical trainees during procedural simulations in a Simulated Operating Theatre (SOT). *Annals of Surgery, 242*(5), 631-639.

Moorthy, K., Munz, Y., Forrest, D., Pandey, V., Undre, S., Vincent, C. A., & Darzi, A. (2006). Surgical crisis management skills training and assessment. A simulation-based approach to enhancing operating room performance. *Annals of Surgery, 244*(1), 139-147.

Nielsen, P. E., Goldman, M. B., Mann, S., Shapiro, D. E., Marcus, R. G., Pratt, S. D., Greenberg, P., McNamee, P., Salisbury, M., Birnbach, D. J., Gluck, P. A., Pearlman, M. D., King, H., Tornberg, D. N., & Sachs, B. P. (2007). Effects of teamwork training on adverse outcomes and process of care in labor and delivery: a randomized controlled trial. *Obstetrics & Gynecology, 109*(1), 48-55.

Reader, R. W., Flin, R., & Cuthbertson, B. H. (2007). Communication skills and error in the intensive care unit. *Current Opinion in Critical Care, 13,* 732-736.

Risser, D. T., Rice, M. M., Salisbury, M. L., Simon, R., Jay, G. D., & Berns, S. D. (1999). The potential for improved teamwork to reduce medical errors in the emergency department. *Annals of Emergency Medicine, 34*(3), 373-383.

Rosen, M. A., Salas, E., Wilson, K. A., King, H. B., Salisbury, M., Augenstein, J. S., Robinson, D. W., & Birnbach, D. J. (2008). Measuring team performance in simulation-based training: adopting best practices for healthcare. *Simulation in Healthcare, 3*(1), 33-41.

Salas, E., Dickinson, T. L., Converse, S. A., & Tannenbaum, S. I. (1992). Toward an understanding of team performance and training. In R. Swezey & E. Salas

(Eds.), *Teams: Their training and performance* (pp. 3-29). Norwood, NJ: Ablex.

Thomas, E. J., Sexton, J. B., & Helmreich, R. L. (2004). Translating teamwork behaviours from aviation to healthcare: development of behavioural markers for neonatal resuscitation. *Qual Saf Health Care 13*: i57-i64.

Thomas, E. J., Sexton, J. B., Lasky, R. E., Helmreich, R. L., Crandell, D. S., & Tyson, J. (2006). Teamwork and quality during neonatal care in the delivery room. *Journal of Perinatology, 26*(3), 163-169.

Undre, S., Healey, A. N., Darzi, A., & Vincent, C. A. (2006). Observational assessment of surgical teamwork: A feasibility study. *World Journal of Surgery, 30*(10), 1774-1783.

Van Avermate, J., & Kruijsen, E. (1998). *NOTECHS: the evaluation of non-technical skills of multi-pilot air crew in relation to the JAR-FCL requirements.* European Commission.

Wallin, C.-J., Meurling, L., Hedman, L., Hedegård, J., & Felländer-Tsai, L. (2007). Target-focused medical emergency team training using a human patient simulator: effects on behaviour and attitude. *Medical Education, 41*(2), 173-180.

Yule, S., Flin, R., Paterson-Brown, S., & Maran, N. (2006). Non-technical skills for surgeons in the operating room: A review of the literature. *Surgery, 139*(2), 140-149.

Zala-Mezö, E., Wacker, J., Künzle, B., Brüesch, M., & Grote, G. (2009). The influence of standardisation and taskload on team coordination patterns during anaesthesia inductions. *Quality and Safety in Health Care, 18*(2), 127-130.

Zingg, U., Zala-Mezoe, E., Kuenzle, B., Licht, A., Metzger, U., Grote, G., & Platz, A. (2008). Evaluation of critical incidents in general surgery. *British Journal of Surgery, 95*, 1420-1425.

Chapter 10

Measurement Approaches for Transfers of Work During Handoffs

Emily S. Patterson and Robert L. Wears

Introduction

There is a general consensus that transfers, including daily and weekday-to-weekend shift changes, changes of leadership personnel—particularly in highly knowledge-intensive roles like intelligence analysis—as well as larger-scale group personnel rotations like the end of an 18-month deployment in the military, are recognized points of vulnerability. In most domains, training, procedures, and support artifacts like standardized paperwork are embedded within orientation training, whether it be formal orientation or 'on the job' learning about 'how we do things around here.' Similarly, most Cognitive Task Analyses and other research studies focusing on macrocognitive phenomena have included transfers as a critical activity, both because it is a vulnerable point where practitioners proactively adapt to avoid high-consequence failures, and also because it is a naturally occurring opportunity to have a window into what otherwise is primarily internalized cognitive processes. By observing the verbal interactions and by viewing updates to paperwork in preparation for the transition, researchers can gain tremendous insight into the domain challenges in general.

Although transfers of responsibility and authority are often not researched as a topic worthy of independent consideration in most macrocognition studies outside of the healthcare domain (although see Lardner, 1996), there is a recurring interest in improving transfer processes, particularly through the use of introducing new software systems, procedures, and associated paperwork. From a systems engineering theoretical perspective, a transfer is a time when there is more than one 'controller,' which introduces the possibility of desynchronization, conflict, redundancy effort, and increases the communication burden to coordinate the handoff, which could be alleviated with technological innovation. For this reason, there is an interest in clearly defining and measuring transfers of authority and responsibility in complex, sociotechnical domains to see if innovations objectively improve the processes.

In healthcare, there is a growing awareness that high-quality handoff processes are critical to providing safe and effective patient care (Petersen et al., 1994; Singh et al., 2007; Borowitz et al., 2008; Christian et al., 2006; Kitch et al., 2008; Ye et al., 2007; Simpson, 2005; Wilwerding et al., 2004). Impacts of less-than-ideal handoffs likely include adverse events (Risser et al., 1999), delays in medical diagnosis and treatment, redundant communications, redundant activities such as additional procedures and tests, lower provider and patient satisfaction, higher costs, longer hospital stays, more hospital admissions, and less effective training for healthcare providers (Lawrence et al., 2008). A central tenet of quality improvement is that any process that is not systematically designed, measured, monitored, and provided objective feedback is likely to be ad hoc and highly variable. Several observational and survey studies have confirmed this to be the baseline condition for patient handoff processes (Wears et al., 2004; Kowalsky et al., 2004; Solet et al., 2004; Patterson et al., 2005; Beach et al., 2003; Bomba & Prakash., 2005; Nemeth et al., 2008).

The healthcare domain has unique lessons for measuring transfers of responsibility and authority based on the growing literature on the measurement of 'patient handoffs.' Over the last ten or so years, there has been a recent and impressive transition from relatively little emphasis of any kind on patient transfer methods to numerous quality improvement efforts and research studies across nearly every inpatient hospital setting. This transition has been largely motivated by new US hospital and training program accreditation requirements. Specifically, there is now a requirement (NPSG.02.05.01) to implement a standardized approach to handoffs by the Joint Commission on the Accreditation of Healthcare Organizations (JCAHO) and a requirement by the Accreditation Council for Graduate Medical Education (ACGME) to limit resident physician working hours, which has the effect of increasing physician handoffs for patients (Philibert & Leach, 2005). Since a central strategy of quality improvement projects is to compare baseline measures against post-implementation measures, these efforts have generated a substantial literature on attempts to measure the quality of patient handoffs. In this chapter, we synthesize this literature by classifying the work into six conceptual framings for the function of handoffs as well as four groupings of handoff quality measures: 1) outcomes, 2) content of interactions, 3) interaction processes, and 4) learning. We conclude with a discussion of the purposes of measurement approaches in general.

Conceptual Framings for Patient Handoffs

In Table 10.1, we provide six alternative conceptual framings and their associated functions, risks, and how the framing tends to point to different types of quality improvement interventions (note that Table 10.1 expands upon initial thoughts about four conceptual framings published in Table 2 of Cheung et al., 2009). These framings are not mutually exclusive, but rather can be viewed as levels of activity that occur simultaneously in parallel during a handoff. Others have made similar distinctions, including Cohen & Hilligoss (2009), who distinguished between correct transition of essential information, transfers of responsibility or control, error correction, and learning at the individual and organizational levels;

Table 10.1   Alternative conceptual framings for handoffs

Conceptual Frame	Primary Function	Risk	Intervention	Example
Information processing	Transfer data through a noisy communication channel	Missing, inaccurate data	Ensure accurate essential content is transmitted	Standardized handoff protocol (Arora & Johnson, 2006)
Stereotypical narratives	Label by stereotypical narrative and highlight deviations	Inappropriately applying default assumptions	Identify stereotypical narratives and support highlighting deviations	Daily goals for interdisciplinary teams (Provonost et al., 2003)
Resilience	Cross-check assumptions with a fresh perspective	Incorrect framing of problems/risks and solutions/ strategies	Create a supportive environment that encourages error detection and clarification questioning	Two-challenge rule for resident physicians questioning attending physicians (Pian-Smith et al., 2009)
Accountability	Transfer of responsibility and authority	Failing to clarify responsibility and authority for care-giving tasks	Ensure patients assigned to providers; Ensure providers accept responsibility for patients and tasks	Handover protocol explicitly assigned tasks to team members (Catchpole et al., 2007)
Social interaction	Co-construction of shared meaning	Failing to support sensemaking and anticipation	Encourage flexible, adaptive, tailored sharing of perspectives on data	Supporting interdisciplinary team communications during rounds (Uhlig et al., 2002)
Cultural norms	Negotiate and share group values	Failing to have a shared sense of values and social norms across an organizational unit	Training and socialization, particularly via apprenticeship mentoring about how to do handoffs	Simulation-based training on handoff processes (Slagle et al., 2008)

Kerr (2002), who distinguished between informational, social, organizational, and educational purposes; Smith et al. (2008), who distinguished between transfer of information, transfer of responsibility, and audit points; Philpin (2006), who distinguished between receiving necessary information, conveying essential meanings, and articulating group values; Lawrence et al. (2008), who distinguished between information exchange, socialization, averting or recovering from adverse events by using a fresh set of eyes, and team cohesion; and Lardner (1996, Table 1), who summarized ten theoretical framings for communication issues and associated recommendations.

The first frame, information processing, is the most prevalent in the healthcare literature. The information processing metaphor has dominated scientific thought for decades and is represented as a sequence of four mental operations: 1) Encoding, 2) Comparison, 3) Response Selection, and 4) Response Execution. These operations occur in-between inputs (stimuli) to an individual and outputs (responses) from an individual; these operations mediate the use of memory and attentional resources. This metaphor is oversimplified for complex, sociotechnical settings (Patterson & Miller, this volume) in that it does not model: 1) work that is coordinated with team members in a social context; 2) sophisticated computing capability that augments human memory and attentional resources; 3) parallel processing of simultaneous threads of prioritized activity; 4) management of conflicting goals; 5) activities that are influenced by time pressure, stress, data overload, uncertainty; 6) problems and tasks that are proactively avoided through strategic mechanisms; and 7) teams responding to unexpected events and impasses. Limiting the conceptualizing of a patient handoff to this frame will likely overly restrict an exploration of potentially useful quality improvement interventions (Patterson & Wears, 2009).

With this frame, the primary function of the handoff is to transfer data through a noisy communication channel (that gets noisier with background noise, interruptions, information overload, ambiguous language, and so on). The primary risk is that clinical judgments will be made with missing or inaccurate data. Defining and standardizing essential information (Arora & Johnson, 2006), perhaps coupled with readbacks to ensure that information was accurately received, is an intervention that is implied by this frame. The *Joint Commission Handbook* (2008, page 4) defines handoffs in a way that is sympathetic to this framing, whereby the rationale for their requirement to standardize handoff processes includes:

> The primary objective of a 'hand off' is to provide accurate information about a [patient's] care, treatment, and services, current condition and any recent or anticipated changes. The information communicated during a hand off must be accurate in order to meet [patient] safety goals.

The second frame, stereotypical narratives, takes advantage of the narrative structure to quickly and effectively communicate large amounts of information by associating the information with default narrative structures (Schank &

Abelson, 1977; Srull 1981; Behara et al., 2005; Orlikowski & Yates, 1998; Yates & Orlikowski, 1992). With this framing, the emphasis during the handoff is on highlighting deviations from typical narratives, such as a patient who is allergic to the preferred antibiotic for treating his diagnosed condition. Pronovost's (2003) daily goals intervention for interdisciplinary teams in the Intensive Care Unit uses typical patient narratives and deviations in a prominent fashion.

The third frame, resilience, involves taking advantage of the transparency of the thought processes revealed through the conversation in order to identify erroneous assumptions and actions (Patterson et al., 2007; Wears et al., 2003). Instituting check-out procedures that provide the opportunity for the oncoming provider to ask clarification and error detection questions (cf. Patterson & Woods, 2001), even when the handoff update is audio-taped or handwritten, is an example of an intervention for this frame. Although not specific to handoffs, the two-challenge rule for resident physicians questioning attending physicians encourages error detection questioning strategies (Pian-Smith et al., 2009).

The fourth frame, accountability, emphasizes the transfer of responsibility and authority that distinguishes a handoff update from an information update. Interventions to ensure that patients are assigned to providers, that providers are aware and have accepted the transfer, that others are aware of who is responsible for a patient, and reminders to complete tasks that have been handed off such as checking on lab results relate to this framing. The redesign of the handover instantiated in a protocol format by Catchpole et al. (2007) included explicitly assigning specific tasks to team members, such as ventilation to the anesthesiologist. One concern with this framing is whether handoff process redesigns are intended to shift power from front-end workers to more distant administrators and regulators (Murphy & Wears 2009).

The fifth frame, social interaction, emphasizes how the social interaction during a handoff allows a co-construction of essential meanings and co-orientation toward that essential meaning based on knowledge of the perspective of the participants in the exchange (Eisenberg et al., 2005). All communicative acts include both a content dimension and a relational dimension, which reinforces social relationships between parties. Handoffs from one discipline to another, such as anesthesiologist to Post- Anesthesia Care Unit nurse or emergency department physician to internal medicine physician, might be particularly important to consider how the different perspectives will likely impact what is meaningful for each perspective. The redesign of patient rounds to explicitly support interdisciplinary team communications and take into account their different perspectives, and in some cases including the patient and caregivers, is an example of this frame (Uhlig et al., 2002).

The final frame, cultural norms, relates to how group values (instantiated as social norms for acceptable behavior) in an organization or suborganization are negotiated and maintained over time. Socialization about 'how things are done here' can include practices for conducting handoffs. Training interventions can spread beyond the trained individuals through this framing (Slagle et al., 2008), and alternatively the 'hidden curriculum' can undermine policies, procedures, and training if what is documented and taught differs from what is learned during an apprenticeship period in the local work environment.

Methods: Literature Review and Classification Scheme

The literature review was conducted in electronic databases (PubMED, Google Scholar) using search terms, forward and backward citation searches from key articles, and by requesting papers from researchers. Approximately 400 relevant articles were identified. Synonyms for handoffs discovered during the search process include: handoffs, handovers, sign-outs, sign-overs, turnovers, intershift transfers, intershift handovers, shift change transfers, patient transfers, transitions of care, transfers of care, substitutions, bedside reports, shift reports, shift-to-shift communications, shift-to-shift reports, discharges, discharge communiqués, discharge summaries, discharge notes, post-operation updates, interdisciplinary transfers, multi-professional handovers, and admissions. Existing literature reviews that were discovered during the search process provided the majority of the referenced citations (Philibert, 2007; Cheung, 2009; Cohen & Hilligross, 2009; Wilson, 2007; Riesenberg et al., 2009; Djuricich & Logio, 2007; Lardner, 1996; ACSQH, 2005; Cohen & Hilligoss, 2009).

We propose the following classification scheme for organizing measures for patient handoff quality identified during the review: 1) outcomes, 2) content of interactions, 3) interaction processes, and 4) learning.

Outcome measures

Following a handoff, the incoming agent is ideally indistinguishable from the outgoing agent (Patterson & Woods, 2001; Singh et al., 2007) in:

1. performing technical work competently;
2. knowing the historical narrative (relevant patient history and chief complaint);
3. being aware of significant data or events;
4. knowing what data are important for monitoring changes and their associated levels of uncertainty;
5. managing impacts from previous events;
6. anticipating future events;
7. weighing trade-offs if diagnostic or therapeutic judgments need to be reconsidered;
8. planning patient care strategies; and
9. performing planned tasks.

In addition, other handoff outcomes have been identified:

10. alerting others to the completion of interdependent tasks;
11. supervising junior personnel and/or accessing senior personnel;
12. identifying warranted and avoiding unwarranted shifts in goals, priorities, plans, decisions, or stances toward key decisions (Patterson et al., 2005);
13. relationship building (Lally, 1999; Lamond, 2000; Miller, 1998; O'Connell & Penney, 2001) and other social functions (Kerr, 2002; Strange, 1996);
14. articulating and reinforcing group values (Philpin, 2006; Keenan et al., 2006);
15. protective functions, including support for grieving (Strange, 1996; Hopkinson, 2002);
16. increased patient involvement in decision making (Anderson & Mangino, 2006).

To illustrate how outcome measures relate to the first ten objectives, see Table 10.2. For most of these measures, comparisons could be made between outgoing and incoming personnel following an actual or simulated handoff. More often, studies have compared outcome variables between naturally occurring groupings. Examples of 'A vs. B' comparisons include: additional transfers to a cross-covering physician (Lofgren et al., 1990), coverage by a physician from another team (Petersen et al., 1994), short-call and cross-coverage (Philibert & Leach, 2005), information technology support vs. no information technology support for handoffs (Van Eaton et al., 2005). Note that few studies have specifically addressed the impact of hindsight knowledge bias on outcome measures (Fischhoff, 1975; Henriksen & Kaplan, 2003).

Measures of Interaction Content

There have been many calls to standardize the information content of handoffs (McCann et al., 2007; Brown-Lazzarra, 2004). Standardization reduces the cost of communication (Patterson, 2008) because: 1) the 'rules' for interaction do not need to be negotiated (including the function, process, content, timing, and who is directly or indirectly included in the conversation); 2) no information on a topic (usually) implies that there is nothing worthy of mention on that topic; and 3) information can be conveyed more efficiently and with higher reliability.

For measures, one approach would be to compare what is actually provided during a transfer via verbal or written mechanisms against some predefined essential content. For example, Raptis et al. (2009) found that there was a significantly higher number of completed data fields—specifically for the data fields of patient's demographic details, patient location, primary diagnosis, current problem, plan of action, and primary care team details—for electronically-supported handoffs (including automated data pulls) as compared to handwritten paperwork-supported

Table 10.2   Potential outcome measures for a handoff

Handoff objective	Measures
Perform technical work competently	Risk-adjusted mortality and morbidity (Young et al., 1998), cardiac arrests, preventable adverse events (Petersen et al., 1994; 1998; Wilwerding et al., 2004), length of stay (LOS), patient satisfaction, malpractice claims, timeliness of care, timeliness of disposition to hospital, boarding times in the emergency department, staff retention, staff satisfaction with quality of care (Meissner et al., 2007), patient satisfaction, unplanned transitions from acute care hospital wards to the Intensive Care Unit within 24 hours of a handoff (Horwitz et al., 2009), staff time (Footitt, 1997), technical errors (Catchpole et al., 2007)
Know historical narrative	Accuracy and completeness of retained information about relevant patient history and chief complaint (Talbot & Bleetman, 2007; Bhabra et al., 2007)
Aware of events	Awareness and accuracy of significant data or events; Redundant ordering of labs, tests, x-rays, procedures, etc.; Accurate billing; Legible, accurate, and up-to-date documentation (Fang, 2006)
Know what to monitor and data uncertainty	Accuracy and completeness of knowledge of leading indicators to monitor and deviations from typical levels of uncertainty for sensors
Manage impacts from previous events	Manage side effects of medications and other treatments and complications
Anticipate future events	Sensitivity and internal reliability of likelihood judgments about potential future events (Shanteau et al., 2002; Shanteau et al., in press, 2009)
Weigh trade-offs appropriately during diagnostic or therapeutic judgments	Accuracy of diagnoses, appropriateness of treatment plans; Percent compliance with recommended organizational practices (e.g., appropriate beta-blocker use, Smith, 2008)
Plan patient care strategies	Quality of care planning (Dowding, 2001), Know specific interventions and goals for the next shift (Priest & Holmberg, 2000)
Perform tasks	Number of 'dropped' patients (Van Eaton et al., 2005) or tasks (e.g., redundant x-ray orders)

handoffs during day to night handoffs from a team of specialist nurses, medical staff, and surgical staff. Extensions of this approach could include weighing some items as more important than others and/or separately reporting critical information and optional information. Similarly, one article suggests splitting information from Emergency Medical Service (EMS) to Emergency Department (ED) personnel into what is essential to stabilize and initially diagnose and treat the patient and information that is needed for longer-term care (Jenkin et al., 2007). Another extension of this approach is to rate whether information has been organized in a particular fashion (Reisenberg, 2009), such as based on Situation-Backaround-Assessment-Recommendation (SBAR) ordering (Haig et al., 2006). Alternative ordering schemes include physical locations in beds and on information technology systems, 'most important first' (Patterson, 2007), body systems, head-to-toe, patient problems, or modes (physiologic condition, self-concept, role function, interdependence) (McLaughlin et al., 2004).

These approaches can be further categorized into ones that are intended to apply across a wide range of settings (Table 10.3) and ones tailored to a particular setting, such as the operating room (Table 10.4). In discussing what is essential, there are challenges in striking a balance between completeness of information and directing attention toward particularly salient issues. Additional tradeoffs include minimizing

Table 10.3   Standardized handoff content across all hospital settings

Identify-Situation-Background-Assessment-Recommendation (I-SBAR) (Veteran's Administration VISN 9, personal communication, 2006)	Situation-Background-Assessment-Recommendation (SBAR) (JCAHO International draft document, 2007)
Identify: Full name Social Security number or birthdate	
Situation: Your name and unit Patient name Reason for the handoff	Situation: Identify patient (two identifiers) Age, gender Primary diagnosis Relevant co-morbidities, allergies
Background: Diagnosis and admission information Pertinent medical history Noteworthy treatments	Background: Current treatment or reason for transfer Recent changes in condition or treatment
Assessment: Vital signs if appropriate Medication and/or oxygen requirements Mental/neurologic status Wound care needs Cardiac concerns Any risk issues such as fall, elopement, or behavioral concerns	Assessment: Current condition
Recommendation: Any information regarding the need for future testing, consultation, family conference, etc.	Recommendation: Procedure to be done or anticipated changes in treatment Potential changes in condition Critical monitoring parameters

Table 10.4　Standardized handoff content for four operating room settings

Anesthesiologist to Relief Anesthesiologist (Rudolph et al., 2006)	Anesthesiologist to Post-Anesthesia Nurse (Slagle et al. 2007)	Surgical Physician to Surgical Physician (Cheah et al., 2005)	Surgical Physician to Surgical Physician and Rounds (Van Eaton et al., 2005)
Big picture: Your brief summary of case Background: Patient name, age, weight, surgery Diagnosis and HPI Allergies: and reactions Baseline status: functional status, mentation, ability to cooperate PMH and PSH Pertinent ROS: e.g., GERD Medications: and pertinent past meds Pertinent labs and studies	Identify: Chart number Patient name Account number Date of birth Age Gender ASA OR location	Minimum dataset: Patient's name Patient's location (ward and bed Admission Diagnosis Procedure (with date) Complications and progress Management plan Resuscitation plan Consultant availability (and instructions if not available) Expected need for review Name of doctor completing transfer Date to confirm that information is current	Automatically generated: Service Team Status (ICU, Floor, Consult) Patient name Patient ID Unit Room Admit date Service Attending
Intraoperative: Team members: Introductions Current progress of surgery Airway assessment, management, difficulties? ventilator settings Anesthetic and concerns: GA, Regional, MAC Antibiotics: given? next due? Other meds: anesthestic, narcotic, pressor, relaxant, antiemetic, or meds due? Concentrations of prepared meds: standard? Current vitals and trends Ins and Outs: fluids given, urine, ebl, available products, ongoing plans Lines Current labs and labs to be drawn Special equipment: e.g. tourniquet, laser Concerns intraop?	Situation: Date of surgery ID band location PACU time in PACU time out Allergies Anesthesia attending Anesthesia provider Type of anesthesia Surgical procedure Presenting diagnosis Attending surgeon Surgical resident First assist		Resident entered information: Admit diagnosis Problem list Code status Allergies Diet DVT Prophylaxis Medications Drips Antibiotics Signout/plans Resident Next HD day LEVD Pathology Spine status Mobilization precautions Next OR day Lines and start dates Procedures and dates Culture results

Table 10.4   *Concluded*

Anesthesiologist to Relief Anesthesiologist (Rudolph et al., 2006)	Anesthesiologist to Post-Anesthesia Nurse (Slagle et al. 2007)	Surgical Physician to Surgical Physician (Cheah et al., 2005)	Surgical Physician to Surgical Physician and Rounds (Van Eaton et al., 2005)
Plans: Tasks to do or follow-up Plan to extubate? Anticipate need for monitoring or equipment during transport? Postop destination? bed called for? Postop concerns: e.g., analgesia plan, need for consults, urine output Questions from relieving anesthetist?	Background: Surgical information and issues Preop vital signs: BP, HR, RR, SPO2, temp Problem list/medical history Airway management Anes. agents Paralyzed Reversed Intraop vital signs BP, HR (range), RR, SPO2, Temp on FiO2, PACU temp, NonInv BP Fluids-Inputs Fluids-Outputs Intraop labs Last PCV Other labs Glucose/insulin		
	Assessment: Events/complications Issues/special precautions		
	Recommendation: Plans		

the work to tailor a list to a particular setting, coordinating with other hospitals on standards, meeting the needs of different disciplines (Patterson et al., 1995), allowing flexibility for contingencies, and taking items off lists to avoid having them become unwieldy to manage over time. Note that both Tables 10.3 and 10.4 are primarily based on the framing of information transfer, and so we do not recommend either of these approaches due to the limitations of that conceptual framing.

Practically speaking, obtaining measures of what is said and written as compared to the essential content is likely to be labor-intensive. An alternative approach is to employ a global judgment that a rater with minimal training can efficiently and reliably make while watching a transfer real-time (Slagle et al., 2008). A similar approach was employed by Fenton(2006) in judging whether handoffs were more structured as an overall judgment.

An emerging approach to measuring transfer content is to measure content omissions rather than transmitted information (Catchpole et al., 2007). Variations on this approach include asking about satisfaction with handoffs (Meissner et al., 2007), perceived quality (Ye et al., 2007), whether a transfer was suboptimal

(Arora et al., 2005) or poor (Lee et al., 1996), whether a problem (McCann et al., 2007) or critical incident could be attributed to an inadequate transfer (Sabir et al., 2006), whether one or more patients experienced harm from problematic handoffs in a physician's most recent rotation (Kitch et al., 2008), and whether there were any surprises during a shift (Borowitz et al., 2008; Flanagan et al., 2009; Matthews et al., 2002).

After essential content is defined, it is rarely the case that implementation is the end of the story, and no measures were found in this literature review to capture content changes after an intervention. For example, paperwork or an electronic or digital audio support system could be implemented rather than continue to provide all of the information verbally (cf., Wallum, 1995; Horwitz et al., 2009). Approaches to automatically 'pull' data from electronic medical records can potentially reduce the data input burden (for example, Van Eaton et al., 2005; Flanagan et al., 2009; Frank et al., 2005; Kannry, 1999), although the reliability of the retrieved data can be an issue, creating an additional task of verifying data quality or leaving incoming personnel vulnerable to relying upon inaccurate or outdated information. After verbal information is transferred to another medium, verbal exchanges are usually shortened and some eliminate verbal interactions altogether. In addition, systems that increase time spent on handoff paperwork, particularly when there previously was none, have been accompanied by decreased time spent doing documentation in the patient chart and other traditional long-term repositories (Gurses & Xiao, 2006), particularly given that approximately 85 percent of handoff information is redundant with chart information (Sexton et al., 2004). For some, incorporating handoff paperwork into long-term repositories is judged to increase liability risks and thereby might reduce the willingness to include sensitive information that is helpful in the short term in interacting with others.

Measures of Handoff Interaction Processes

Although there is scant literature on handoff interaction processes (although see Wears et al., 2003; Lawrence et al., 2008), once they are identified, they can be measured based on their frequency of occurrence (cf., Patterson et al., 2004; Always, Routine, Usually, Common, Often, Rare, and Not Observed). In addition, interactions with existing and prototyped artifacts can be analyzed for gaps between intended and actual functionality in a setting of use (cf., use issues analysis of artifacts to support handovers by Wilson et al., 2007, user-centered design by Wong et al., 2007).

We have elected to describe nine handoff interaction processes for illustration purposes, which are grouped by the phase of the handoff:

    A. Before verbal/written interaction
   • Outgoing person:
      1. Complete and document activities to avoid handing them over.

      2. Synthesize a 'big picture' narrative for each patient.

      3. Identify tasks to be done by the incoming person.

- Incoming person:

      4. Review assignment and available documentation to 'warm up' for the interaction.

B. During verbal/written interaction

      1. Invite interactive questioning, topic initiation, and collaborative cross-checking by the incoming via policies (two-challenge rule, Pian-Smith et al., 2009) and respectful body language (Leonard et al., 2004).

      2. Minimize interruptions (Solet et al., 2005) by having a quiet, dedicated space away from main traffic area (McCann et al., 2007), non-interruptive communication devices (Coiera et al., 2002; Dahl, 2006), dedicated channels for requests that can be postponed such as pagers, and making accurate, up-to-date contact information easily accessible (Hiltz & Teich, 1994).

      3. Reduce multiple concurrent tasks (Strople & Ottani, 2006) by allocating a dedicated time for handoffs (Sabir et al., 2006), having adequate staffing and/or supplemental staffing during the transfer such as with short-term coverage by nurse managers, overlapping shifts and avoiding short shifts with minimal staffing, having personnel personally handoff to the incoming person, and handing off on-call responsibilities like responding to codes prior to the transfer.

      4. Use closed-loop verification communication techniques for critical information (Arora & Johnson, 2006). A similar, and likely more tractable, approach is having the incoming person summarize what was learned from the transfer update to the outgoing person and/or to another person while the outgoing person listens and is encouraged to make clarifications and corrections.

C. After verbal/written interaction

      1. Clear, unambiguous, transparent signal that a transfer of responsibility and authority has occurred (for example, transfer pager, cellphone, change name on whiteboard, handover laminated cards or stickers).

Measures of Learning

Many have expressed interest in having explicit education on how to conduct handoffs (Theorem & Morrison, 2001; McCann et al., 2007; Arora et al., 2008). Unfortunately, there is scant literature on what content should be included. The few publications that are available range in content and method of dissemination. One educational approach plays deidentified audiotaped handoffs during resident orientation and encourages group discussion about how they could be improved (Borowitz et al., 2008). Another approach uses a traditional one-hour class setting to teach effective communication skills during verbal sign-out interactions to medical students and residents (Horwitz et al., 2007), with a statistically significant increase in perceived comfort with providing sign-out. Finally, a simulation-based team approach combined with an individual web-based multimedia course was used for training on Post-Anesthesia Care Unit (PACU) handoffs, with statistically significant differences found for the probability of providing an acceptable handoff in the actual work setting post-intervention (Slagle et al., 2008).

Another way to view learning is 'learning how to learn.' Measurement in this context analyzes the number and/or quality of improvements that were recommended and/or implemented within a time period (Broekhuis & Veldkemp, 2007; cf., Welch & Welch, 2005, p. 56, work-out technique where groups of 30–100 are facilitated without the presence of leadership representatives in coming up with recommendations, for which 75 percent are given an immediate yes/no decision and the remainder is resolved within 30 days). One example was the application of the reflexivity method, which resulted in suggestions for handoff process improvement (Broekhuis et al., 2007). A similar approach used the appreciative inquiry technique where the staff were asked to discuss and build on their most effective handoff experiences (Shendell-Falike et al., 2007). Similar efforts included Kellogg and colleagues (2006), who explored new definitions for professionalism, Miller (1998) who conducted regular reviews of transfer processes, and McKenna et al. (1997) who made identifiable audiotaped handoff updates public to facilitate transparency and discussion.

Discussion

In the last decade, numerous quality improvement projects on patient handoffs have been conducted. However, measures for handoff quality have been highly variable.

The diversity of measurement approaches for handoff quality indicate a lack of consensus about the primary purpose of handoffs and how best to improve handoff processes. Given the challenges in coming to consensus on a clear definition of a handoff and how to measure a high-quality handoff, it is likely that there are additional conceptual frames that we have not included. As we move forward, clarifying and elaborating conceptual framings will be central to making progress on developing useful measures that objectively assess handoff quality.

We believe that all six of the conceptual framings that we have identified for handoffs are potentially useful, compatible framings for all complex, sociotechnical settings. Nevertheless, we caution against narrowly defining a handoff event based solely on an information processing conceptual frame. In addition, we suggest avoiding interventions that are intended to shift power from front-end workers to more distant administrators and regulators. We believe that measures based upon oversimplified communication models, as well as measures that overly restrict flexibility of sharp end practitioners, could potentially be used to justify the implementation of policies that will negatively impact patient safety and quality.

In this exploration of measurement approaches to patient handoffs, we noticed similar trade-offs that were pointed out by other authors in this book struggling to operationally measure complex, macrocognitive phenomena. For example, Roth & Eggleston (Chapter 13) point out that any evaluation is usually costly enough in time, money, and effort that both a summative evaluation goal of demonstrating cleanly that an innovation has had a statistically significant effect as compared to a baseline or alternative condition and a formative evaluation goal of identifying opportunities to improve compared to the evaluated innovation need to be pursued in parallel. Given the current state of knowledge of how to dramatically improve transfer processes (weak), approaches to measurement are likely to face these tensions moving forward, with no obvious solutions.

Similarly, Kunzle and colleagues (Chapter 9) pointed that observationally-based methods have the potential to change behavior due to the awareness of being observed and that complex measurement approaches can be unwieldy both during data collection and data analysis. On the other hand, transfers are so critical to safe operations that any decisions to 'push' the burden of data collection toward the personnel involved in the transfers should be done with great caution. Finally, as with the teamwork rating tools they surveyed, inter-rater reliability assessments often reveal the biases of the observers in terms of what they view 'good teamwork' to be, and preliminary experiences with handoffs suggest that similar disagreements exist in definitions of a 'good' handoff, with anecdotally less diversity as to perceptions about what constitutes a 'poor' handoff.

Acknowledgments

Partial funding support was provided by the National Patient Safety Foundation and the Association for Healthcare Research and Quality (AHRQ 1 P20 HS11592).

References

Anderson, C.D., & Mangino, R. R. (2006). Nurse shift report: who says you can't talk in front of the patient? *Nurse Administration Quarterly, 30*(2):112-22.

Arora, V., & Johnson, J. (2006) A model for building a standardized hand-off protocol. *Joint Commission Journal on Quality and Patient Safety, 32*(11), 646-655.

Arora, V., Johnson, J., Lovinger, D., Humphrey, H. J., & Meltzer, D.O. (2005). Communication failures in patient sign-out and suggestions for improvement: a critical incident analysis. *Quality and Safety in Health Care, 14*(6), 401-407.

Arora, V. M., Johnson, J. K., Meltzer, D. O., & Humphrey, H. J. (2008). A theoretical framework and competency-based approach to improving handoffs. *Quality and Safety in Health Care, 17*(1), 11-14.

Australian Commission on Safety and Quality in Healthcare (ACSQH). (2005). *Clinical handover and patient safety: Literature review report.* http://www.safetyandquality.gov.au/ Accessed September 5, 2007.

Behara, R., Wears, R. L., Perry, S. J., Eisenberg, E., Murphy, A.G., & Vanderhoef, M. (2005). Conceptual framework for the safety of handovers. In: K. Henriksen (Ed.), *Advances in Patient Safety* (pp. 309-321). Rockville, MD: Agency for Healthcare Research and Quality/Department of Defense. http://www.ahrq.gov/downloads/pub/advances/vol2/behara.pdf.

Bhabra, G., Mackeith, S., Monteiro, P., & Pothier, D. D. (2007). An experimental comparison of handover methods. *Annals of The Royal College of Surgeons of England, 89*(3), 298-300.

Bomba, D. T., & Prakash, R. (2005). A description of handover processes in an Australian public hospital. *Australian Health Review: A Publication of the Australian Hospital Association, 29*(1), 68-79.

Borowitz, S. M., Waggoner-Fountain, L. A., Bass, E. J., & Sledd, R. M. (2008). Adequacy of information transferred at resident sign-out (inhospital handover of care): A prospective survey. *Quality and Safety in Health Care, 17*(1), 6-10.

Broekhuis, M., & Veldkamp, C. (2007). The usefulness and feasibility of a reflexivity method to improve clinical handover. *Journal of Evaluation in Clinical Practice, 13*(1), 109-115.

Brown-Lazzara, P. (2004). Make your better best with a reporting system. *Nursing Management, 35*(8), 48A-48D.

Catchpole, K. R., de Leval, M. R., McEwan, A., Pigott, N., Elliott, M. J., McQuillan, A., MacDonald, C., & Goldman, A. J. (2007). Patient handover from surgery to intensive care: using Formula 1 pit-stop and aviation models to improve safety and quality. *Paediatric Anaesthesia, 17*(5), 470-478.

Cheah, L. P., Amott, D. H., Pollard, J., & Watters, D. A. K. (2005). Electronic medical handover: Towards safer medical care. *Medical Journal of Australia, 183*, 369-372.

Cheung D. (2009). Handoffs bibliography 1.6.09. Personal communication.

Cheung, D. S., Kelly, J. J., Beach, C., Berkeley, R. P., Bitterman, R. A., Broida, R. I., Dalsey, W. C., Farley, H. L., Fuller, D. C., Garvey, D. J., Klauer, K. M., McCullough, L. B., Patterson, E. S., Pham, J. C., Phelan, M. P., Pines, J. M., Schenkel, S. M., Tomolo, A., Turbiak, T. W., Vozenilek, J. A., Wears, R. L., White, M. L., (2010). Improving handoffs in the emergency department. *Annals of Emergency Medicine, 55*(2), 171-180.

Christian, C. K., Gustafson, M. L., Roth, E. M., Sheridan, T. B., Gandhi, T. K., Dwyer, K., Zinner, M. J., & Dierks, M. M. (2006). A prospective study of patient safety in the operating room. *Surgery, 139*(2), 159-173.

Cohen, M. D., & Hilligoss, P. B. (2009) Handoffs in hospitals: a review of the literature on information exchange while transferring patient responsibility or control. http://deepblue.lib.umich.edu/handle/2027.42/61522. Draft V8, Accessed June 10, 2009.

Coiera, E. W., Jayasuriya, R. A., Hardy, J., Bannan, A., & Thorpe, M. E. (2002). Communication loads on clinical staff in the emergency department. *Medical Journal of Austrailia, 176*(9), 415-418.

Dahl, Y. (2006). 'You have a message here': enhancing interpersonal communication in a hospital ward with location-based virtual notes. *Methods of Information in Medicine, 45*(6), 602-609.

Djuricich, A. M., & Logio, L. (2007). Handoffs Bibliography for Effective Patient Handoffs: Improving Resident Changeover. *APDIM Spring Meeting,* April 2007. www.aiamc.org/public/AIAMC_Hand-offs_Resources.pdf, Accessed June 10, 2009.

Dowding, D. (2001). Examining the effects that manipulating information given in the change of shift report has on nurses' care planning ability. *Journal of Advanced Nursing, 33*(6), 836-846.

Eisenberg, E., Murphy, A., Sutcliffe, K., Wears, R., Schenkel, S., Perry, S., & Vanderhoef, M. (2005). *Communication Monographs, 72*(4), 390-413.

Fang, L., Ming, Y., & Yu, W. C. (2006). A project to improve the completeness of nursing shift reports in the surgical ward [Article in Chinese]. *Hu Li Za Zhi, 53*(3), 52-59.

Fenton, W. (2006). Developing a guide to improve the quality of nurses' handover. *Nursing Older People, 18*(11), 32-37.

Fischhoff, B. (1975). Hindsight does not equal foresight: The effect of outcome knowledge on judgment under uncertainty. *Journal of Experimental Psychology: Human Perception and Performance, 1*(3), 288-299.

Flanagan, M. E., Patterson, E. S., Frankel, R. M., & Doebbeling, B. N. (2009). Evaluation of a physician informatics tool to improve patient handoffs. *Journal of the American Medical Informatics Associatio, 16*(5), 509-515.

Footitt, B. (1997). Ready for report. *Nursing Standard, 11*(25), 22.

Frank, G., Lawless, S. T., & Steinberg, T. H. (2005). Improving physician communication through an automated, integrated sign-out system. *Journal of Healthcare Information Management, 19*(4), 68-74.

Gurses, A. P., & Xiao, Y. (2006). A systematic review of the literature on multidisciplinary rounds to design information technology. *Journal of the American Medical Informatics Association, 13*(3), 267-276. Epub 2006 Feb 24.

Haig, K. M., Sutton, S., & Whittington, J. (2006). SBAR: a shared mental model for improving communication between clinicians. *Joint Commission Journal on Quality and Patient Safety, 32*(3), 167-175.

Henriksen, K., & Kaplan, H. (2003). Hindsight bias, outcome knowledge and adaptive learning *Quality & Safety in Healthcare, 12*(Suppl 2), ii46-ii50.

Hiltz, F. L., & Teich, J. M. (1994). Coverage List: a provider-patient database supporting advanced hospital information services. In *Proceedings—The Annual Symposium on Computer Applications in Medical Care*, 809-813.

Hopkinson, J. B. (2002). The hidden benefit: the supportive function of the nursing handover for qualified nurses caring for dying people in hospital. *Journal of Clinical Nursing, 11*(2), 168-175.

Horwitz, L. I., Moin, T., & Green, M. L. (2007). Development and implementation of an oral sign-out skills curriculum. *Journal of General Internal Medicine. 22*(10), 1470-1474.

Horwitz, L. I., Parwani, V., Shah, N. R., Schuur, J. D., Meredith, T., Jenq, G. Y., & Kulkarni, R. G. (2009). Implementation and evaluation of an asynchronous physician sign-out for emergency department admissions. *Annals of Emergency Medicine.*

Jenkin, A., Abelson-Mitchell, N., & Cooper, S. (2007). Patient handover: time for a change? *Accident and Emergency Nursing, 15*(3), 141-147.

Joint Commission. (2008). *National Patient Safety Goals Handbook.* pp. 1-19. http://www.jointcommission.org/NR/rdonlyres/82B717D8-B16A4442-AD00-CE3188C2F00A/0/08_HAP_NPSGs_Master.pdf. Accessed May 16, 2008.

Kannry, J., & Moore, C. (1999). MediSign: using a web-based SignOut System to improve provider identification. *AMIA Annual Symposium Proceedings.* 550-554.

Keenan, G. M., Stocker, J. R., Geo-Thomas, A. T., Soparkar, N. R., Barkauskas, V. H., & Lee, J. L. (2002). The HANDS project: Studying and refining the automated collection of a cross-setting clinical data set. *Computers, Informatics, Nursing, 20*(3), 89-100.

Kellogg, K. C., Breen, E., Ferzoco, S. J., Zinner, M. J., & Ashley, S. W. (2006). Resistance to change in surgical residency: an ethnographic study of work hours reform. *Journal of the American College of Surgeons, 202*(4), 630-636.

Kerr, M. P. (2002). A qualitative study of shift handover practice and function from a sociotechnical perspective. *Journal of Advanced Nursing, 37*(2), 125-134.

Kitch, B. T., Cooper, J. B., Zapol, W. M., Marder, J. E., Karson, A., Hutter, M., & Campbell, E. G. Handoffs causing patient harm: a survey of medical and surgical house staff. *Joint Commission Journal on Quality and Patient Safety, 34*(10), 563-570.

Lally, S. (1999). An investigation into the functions of nurses' communication at the inter-shift handover. *Journal of Nursing Management, 7*(1), 26-36.

Lamond, D. (2000). The information content of the nurse change of shift report. *Journal of Advanced Nursing, 31*(4), 794-804.

Lardner, R. (1996). Effective shift handover—a literature review. *Offshore Technology Report—OTO 96 003. The Keil Centre.

Lawrence, R. H., Tomolo, A. M., Garlisi, A. P., & Aron, D. C. (2008). Conceptualizing handover strategies at change of shift in the emergency department: a grounded theory study. *BMC Health Services Research, 8,* 256.

Lee, L. H., Levine, J. A., & Schultz, H. J. (1996). Utility of a standardized sign-out card for new medical interns. *Journal of General Internal Medicine, 11*(12), 753-755.

Leonard, M., Graham, S., & Bonacum, D. (2004). The human factor: the critical importance of effective teamwork and communication in providing safe care. *Quality and Safety in Health Care, 13*(Suppl 1), i85-90.

Lofgren, R. P., Gottlieb, D., Williams, R. A., & Rich, E. C. (1990). Post-call transfer of resident responsibility: its effect on patient care. *Journal of General Internal Medicine, 5*(6), 501-505.

Matthews, A. L., Harvey, C. M., Schuster, R. J., & Durso, F. T. (2002). Emergency physician to admitting physician handovers: an exploratory study. *Paper presented at the Proceedings of the Human Factors and Ergonomics Society 46th Annual Meeting.*

McCann, L., McHardy, K., & Child, S. (2007). Passing the buck: clinical handovers at a tertiary hospital. *New Zealand Medical Journal, 120*(1264): U2778.

McKenna, L., & Walsh, K. (1997). Changing handover practices: One private hospital's experiences. *International Journal of Nursing Practice, 3*(2),128-132.

McLaughlin, E., Antonio, L., & Bryant, A. (2004). Hospital nursing: Get an A+ on end-of-shift report. *Nursing, 34*(6), 32hn8-32hn10.

Miller, C. (1998). Ensuring continuing care: Styles and efficiency of the handover process. *Australian Journal of Advanced Nursing, 16*(1), 23-27.

Murphy, A. G., & Wears, R. L. (2009). The medium is the message: Communication and power in sign-outs. *Annals of Emergency Medicine, 54*(3), 379-380.

Nemeth, C. P., Cook, R. I., Kowalsky, J., & Brandwijk, M. (2004). Understanding sign outs: Conversation analysis reveals ICU handoff content and form. *Critical Care Medicine, 32*(12), A29.

Nemeth, C. P., Kowalsky, J., Brandwijk M., Kahana, M., Klock, P. A., Cook, R. I. (2008). Between shifts: healthcare communication in the PICU. In: C. P. Nemeth *Improving Healthcare Team Communication* (pp. 135-153). Aldershot, UK: Ashgate Publishing.

O'Connell, B., & Penney, W. (2001). Challenging the handover ritual. Recommendations for research and practice. *Collegian: Journal of the Royal College of Nursing, Australia, 8*(3), 14-18.

Orlikowski, W., & Yates, J. (1998). Genre Systems: Structuring interaction through communicative norms. *CCS Working Paper 205*; Cambridge, MA: MIT Sloan School of Management Center for Coordination Science; http://ccs.mit.edu/papers/CCSWP205/ Accessed 5 May 2003.

Patterson, E. S. (2007). Editorial: Communication strategies from high-reliability organizations: Translation is hard work. *Annals of Surgery, 245*(2), 170-172.

Patterson, E. S. (2008). Editorial: Structuring flexibility: the potential good, bad, and ugly in standardisation of handovers. *Quality and Safety in Healthcare, 17*(1), 4-5.

Patterson, E. S., & Miller, J. (2010) Preface. In E. S. Patterson & J. Miller (Eds.), *Macrocognition Metrics and Scenarios: Design and Evaluation for Real-World Teams*. Farnham, UK: Ashgate Publishing.

Patterson, E. S., Roth, E. M., & Render, M. L. (2005). Handoffs during nursing shift changes in acute care. *Proceedings of the Human Factors and Ergonomics Society 49th Annual Meeting*. Santa Monica, CA: Human Factors and Ergonomics Society. [CD-ROM]

Patterson, E. S., Roth, E. M., Woods, D. D., Chow, R., & Gomes, J. O. (2004). Handoff strategies in settings with high consequences for failure: lessons for health care operations. *International Journal for Quality in Health Care, 16*(2), 125-132.

Patterson, E. S., & Wears, R. L. (2009). Editorial: Beyond 'communication failure'. *Annals of Emergency Medicine, 53*(6), 711-712.

Patterson, E. S., & Wears, R. L. (2010). Measurement Approaches for Transfers of Responsibility During Handoffs. In E. S. Patterson, & J. Miller, (Eds.), *Macrocognition Metrics and Scenarios: Design and Evaluation for Real-World Teams*. Farnham, UK: Ashgate Publishing.

Patterson, E. S., & Woods, D. D. (2001). Shift changes, updates, and the on-call model in space shuttle mission control. Computer Supported Cooperative Work: *The Journal of Collaborative Computing, 10*(3-4), 317-346.

Patterson, E. S, Woods, D. D., Cook, R. I., & Render, M. L. (2007). Collaborative cross-checking to enhance resilience. *Cognition, Technology and Work, 9*(2), 155-162.

Patterson, P. K., Blehm, R., Foster, J., Fuglee, K., & Moore, J. (1995). Nurse information needs for efficient care continuity across patient units. *Journal of Nursing Administration 25,* (10), 28-36.

Petersen, L. A., Brennan, T. A., O'Neil, A. C., Cook, E. F., & Lee, T. H. (1994). Does housestaff discontinuity of care increase the risk for preventable adverse events? *Annals of Internal Medicine, 121*(11), 866-872.

Petersen, L. A., Orav, E. J., Teich, J. M., O'Neil, A. C., & Brennan, T.A. (1998). Using a computerized sign-out program to improve continuity of inpatient care and prevent adverse events. *Joint Commission Journal on Quality Improvement, 24*(2), 77-87.

Philibert, I. (2007). Selected articles on the patient hand-off. Available from: http://www.acgme.org/acwebsite/dutyhours/dh_annotatedbibliography_patienthandoff_1207.pdf, Accessed June 10, 2009.

Philibert, I., & Leach, D. C. (2005). Re-framing continuity of care for this century. *Quality and Safety in Health Care, 14*(6), 394-396.

Philpin, S. (2006). 'Handing over': transmission of information between nurses in an intensive therapy unit. *Nursing in Critical Care, 11*(2), 86-93.

Pian-Smith, M. C. M., Simon, R., Minehart, R. D., Podraza, M., Rudolph, J., Walzer, T., & Raemer, D. (2009). Teaching residents the two-challenge rule: A simulation-based approach to improve education and patient safety. *Simulation in Healthcare, 4*(2), 84-91.

Priest, C. S., & Holmberg, S. K. (2000). A new model for the mental health nursing change of shift report. *Journal of Psychosocial Nursing & Mental Health Service, 38*(8), 36-43.

Pronovost, P., Berenholtz, S., Dorman, T., Lipsett, P., Simmonds, T., & Haraden, C. (2003). Improving communication in the ICU using daily goals. *Journal of Critical Care, 18*(2), 1-75.

Ram, R., & Block, B. (1992). Signing out patients for off-hours coverage: comparison of manual and computer-aided methods. In *Proceedings—The Annual Symposium on Computer Applications in Medical Care,* 114-118.

Raptis, D. A., Fernandes, C., Chua, W., & Boulos, P. B. (2009). Electronic software significantly improves quality of handover in a London teaching hospital. *Health Informatics Journal, 15*(3):191-198.

Riesenberg, L. A., Leitzsch, J., & Little, B. (2009). Systematic review of handoff mnemonics literature. *American Journal of Medical Quality, 24*(3), 196-204.

Risser, D. T., Rice, M. M., Salisbury, M. L., Simon, R., Jay, G. D., & Berns, S. D. (1999). The potential for improved teamwork to reduce medical errors in the emergency department. The MedTeams Research Consortium. *Annals of Emergency Medicine, 34*(3), 370-372.

Rudolph, J. W., Simon, R., Dufresne, R. L., & Raemer, D. B. (2006). There's no such thing as 'non-judgmental' debriefing: A theory and method for debriefing with good judgment. *Simulation in Healthcare, 1*(1), 49-55.

Sabir, N., Yentis, S. M., & Holdcroft, A. (2006). A national survey of obstetric anaesthetic handovers. *Anaesthesia, 61*(4), 376-380.

Schank, R. C., & Abelson, R. P. (1977). *Scripts, plans, goals, and understanding.* Hillsdale, NJ: Lawrence Erlbaum Associates, Inc.

Sexton, A., Chan, C., Elliott, M., Stuart, J., Jayasuriya, R., & Crookes, P. (2004). Nursing handovers: do we really need them? *Journal of Nursing Management 12*(1), 37-42.

Shanteau, J., Friel B., Thomas, R. P., Raacke, J., & Weiss D.J. (2010). Assessing expertise when performance exceeds perfection. In E. S. Patterson & J. Miller (Eds), *Metrics and Scenarios: Design and Evaluation for Real-World Teams.* Farnham, UK: Ashgate Publishing.

Shanteau, J., Weiss, D. J., Thomas, R., & Pounds, J. (2002). Performance-based assessment of expertise: How to decide if someone is an expert or not. *European Journal of Operations Research, 136*(2), 253-263.

Shendell-Falik, N., Feinson, M., & Mohr, B. J. (2007)/ Enhancing patient safety: improving the patient handoff process through appreciative inquiry. *Journal of Nursing Administration, 37*(2), 95-104.

Sherlock, C. (1995). The patient handover. *Nursing Standard, 9*(25), 33-36.

Simpson, K. (2005). Handling handoffs safely. *American Journal of Maternal Child Nursing, 30*(2), 152.

Singh, H., Thomas, E. J., Petersen, L. A., & Studdert, D. M. (2007). Medical errors involving trainees: a study of closed malpractice claims from 5 insurers. *Archives of Internal Medicine, 167*(19), 2030-2036.

Slagle, J. M., Kuntz, A., France, D., Speroff, T., Madbouly, A., & Weinger, M. B. (2008). Simulation training for rapid assessment and improved teamwork—lessons learned from a project evaluating clinical handoffs. *Human Factors and Ergonomics Society Annual Meeting Proceedings*, 5, 668-672.

Smith, D. H., Kramer, J. M., Perrin, N., Platt, R., Roblin, D. W., Lane, K., Goodman, M., Nelson, W. W., Yang, X., Soumerai, S. B. (2008). A randomized trial of direct-to-patient communication to enhance adherence to β-blocker therapy following myocardial Infarction. *Archives of Internal Medicine, 168*(5), 477-483.

Smith, A. F., Pope, C., Goodwin, D., & Mort, M. (2008). Interprofessional handover and patient safety in anaesthesia: observational study of handovers in the recovery room. *British Journal of Anaesthesia, 101*(3), 332-337.

Solet, D. J., Norvell, J. M., Rutan, G. H., & Frankel, R. M. (2004). Physician-to-physician communication: methods, practice and misgivings with patient handoffs. *Journal of General Internal Medicine, 19*(suppl 1), 108.

Solet, D. J., Norvell, J. M., Rutan, G. H., & Frankel, R. M. (2005). Lost in translation: challenges and opportunities in physician-to-physician communication during patient handoffs. *Academic Medicine,* 80(12), 1094-1099.

Srull, T. K. (1981). Person memory: Some tests of associative storage and retrieval models. *Journal of Experimental Psychology: Human Learning and Memory, 7*(6), 440-463.

Strange, F. (1996). Handover: an ethnographic study of ritual in nursing practice. *Intensive and Critical Care Nursing, 12*(2), 106-112.

Strople, B., & Ottani, P. (2006). Can technology improve intershift report? What the research reveals. *Journal of Professional Nursing, 22*(3), 197-204.

Talbot, R., & Bleetman, A. (2007). Retention of information by emergency department staff at ambulance handover: do standardized approaches work? *Emergency Medicine Journal, 24*(8), 539-542.

Theorem, S., & Morrison, W. (2001). A survey of the perceived quality of patient handover by ambulance staff in the resuscitation room. *Emergency Medicine Journal, 18*(4), 293-296.

Uhlig, P. N., Brown, J., Nason, A. K., Camelio, A., & Kendall, E. (2002). System innovation: Concord Hospital. *Joint Commission Journal on Quality Improvement, 28*(12), 666-672.

Van Eaton, E. G., Horvath, K. D., Lober, W. B., Rossini, A. J., & Pellegrini, C. A. (2005). A randomized, controlled trial evaluating the impact of a computerized rounding and sign-out system on continuity of care and resident work hours. *Journal of the American College of Surgeons, 200*(4), 538-545.

Wallum, R. (1995). Using care plans to replace the handover. *Nursing Standard, 9*(32), 24-26.

Wears, R. L., Perry, S. J., Eisenberg, E., Murphy, A., Shapiro, M., Beach, C., Croskerry, P., & Behara, R. (2004). Conceptual Framework for Studying Shift Changes and other Transitions in Care. In *Proceedings of the Human Factors and Ergonomics Society 48th Annual Meeting*. New Orleans, LO.

Wears, R. L., Perry, S. J., Shapiro, M., Beach, C., & Behara, R. (2003). Shift changes among emergency physicians: best of times, worst of times. In *Proceedings of the Human Factors and Ergonomics Society 47th Annual Meeting* (pp 1420-1423) Denver, CO: Human Factors and Ergonomics Society.

Welch, J., & Welch, S. (2005). *Winning: The answers*. New York, NY: Harper Collins.

Wilson, M. J. (2007). A template for safe and concise handovers. *MedSurg Nursing, 16*(3), 201-206.

Wilson, S., Galliers, J., & Fone, J. (2007). Cognitive artifacts in support of medical shift handover: An in use, in situ evaluation. *International Journal of Human-Computer Interaction, 22*(1 & 2), 59-80.

Wilwerding, J. M., White, A., Apostolakis, G., Barach, P., Fillipo, B. H., &Graham, L. M. (2004). Modeling Techniques and Patient Safety. *Paper presented at the 7th International Conference on Probabilistic Safety Assessment and Management*. Berlin, June 14-18.

Wong, M. C., Turner, P. & Yee, K. C. (2007). Socio-cultural issues and patient safety: A case study into the development of an electronic support tool for clinical handover. *Studies in Health Technology and Informatics, 130*, 279-89.

Yates, J. F., & Orlikowski, W. (1992). Genres of organizational communication: A structural approach to studying communication and media. *Academy of Management Review, 17*, 299-362.

Ye, K., Taylor, M., Knott, J., Dent, A., & MacBean, C. E. (2007). Handover in the emergency department: Deficiencies and adverse effects. *Emergency Medicine Australasia, 19*(5), 443-441.

Young, G. J., Charns, M. P., Desai, K., Khuri, S. F., Forbes, M. G., Henderson, W., & Daley, J. (1998). Patterns of coordination and clinical outcomes: a study of surgical services. *Health Service Research, 33*(5 Pt 1), 1211-1236.

Chapter 11

The Pragmatics of Communication-based Methods for Measuring Macrocognition

Nancy J. Cooke and Jamie C. Gorman

Introduction

Teams can consist of two people (dyads) working over hours or can consist of dozens of coordinating components (for example, modern industry) working over months. Communications measures are complicated by looking at large, and sometimes highly unstructured, teams compared to dyads, but perhaps there is a common ground. In this chapter, we provide a pragmatic look at communications measures especially suited for measuring macrocognition. Many sources have provided the theoretical motivation for such measures, have offered empirical support, and have described their costs and benefits. However, due to the experimental and sometimes premature nature of many of these measures, there is a gap between theory and practice when it comes to actually conducting the measurement. In this chapter, we attempt to fill this gap by focusing on practical considerations, pitfalls, and guidelines for data collection, analysis, and interpretation of communication data. Issues such as selection of methods, data formatting, real-time implementation, and interpretation are critical and will be covered.

The Pragmatics of Communication-based Methods for Measuring Macrocognition

Macrocognition in modern complex systems often spans the heads of multiple individuals. Macrocognition at the team level, or team cognition, can be observed in many of the cognitive activities performed by teams such as collaborative planning and replanning, team decision making, group design, or joint problem solving. There are two general theoretical perspectives on team cognition (Cooke et al., 2007a, 2009). One, which is aligned with the shared mental models literature, focuses on team cognition as the aggregate of individual knowledge. The other perspective which the authors of this chapter support is one which places more weight on the interactions of team members than on individual knowledge. According to this latter perspective, much of team cognition can be understood by studying team communication or, more generally, team interactions. This perspective also lends itself to the fortuitous situation in which team cognition is directly observable in team communication. Communication in this sense *is* team cognition.

Communication is not a novel topic of study (for example, Emmert & Barker 1989; Poole et al., 1993) and it has been a consistent focus of team research. There are a number of empirical results linking team communication parameters to team performance in cognitive tasks. For instance, high-performing teams have been found to communicate with higher overall frequency than low-performing teams (Foushee & Manos, 1981; Mosier & Chidester, 1991; Orasanu, 1990), but communication frequency is reduced during high workload periods (Kleinman & Serfaty, 1989; Oser et al., 1991). Studies of communication content have also linked communication to team cognition and performance. Achille et al. (1995) found that the use of military terms, acknowledgments, and identification statements increased with experience. Similarly Jentsch et al. (1995) found that faster teams made more leadership statements and more observations about the environment than slower teams.

Extending communication to more complex patterns that take into account multiple dimensions including content, frequency, sequence, and communication flow has been fruitful as well. For instance, Bowers et al. (1998) analyzed the sequence of content categories occurring in communication in a flight simulator task. They found that high team effectiveness was associated with consistent responding to uncertainty, planning, and fact statements with acknowledgments and responses, in comparison to lower-performing teams. Similarly, Bowers et al. (1994) found that a two-category sequence was superior to simple frequencies at predicting performance on an aerial reconnaissance task. On the basis of results like these, Salas et al. (1995) conclude, 'It is likely that additional pattern-based analyses will emerge in future literature as a means to understand the impact of communication on team performance' (p. 64).

Since that time, however, the analysis of communication has been fraught with pragmatic challenges and bottlenecks that result in shelves of video and audio tapes, and now digital media, that often go unanalyzed. The data are rich, but the analytic techniques have been impoverished. Voice recognition is improving, but despite this technology, manual transcription of communication records is often a first time-consuming and error prone step. This transcription step is generally followed by another labor-intensive process of segmentation and coding in which transcripts are divided into discrete utterances and then categorized and labeled with a code consistent with the aims of the research. The codes could be speech acts such as *acknowledgement, comment,* and *question* or they could be relevant to critical task events such as *locate target, request target information,* and *identify target.* Alternatively, the researcher may want to tag the utterances with information about speaker and listener. The coding process is time consuming and subject to judgment, requiring multiple judges and measures of inter-rater reliability. Emmert & Barker (1989) cite an example of a study requiring 28 hours of transcription and encoding for each hour of communication (p. 244).

However, given the richness of communication data and its promise for facilitating our understanding of macrocognition, we have dedicated our research program to developing methodologies to facilitate, and even automate, communication analysis (Cooke et al., 2008; Kiekel et al., 2001). These methods take advantage of the rich data as well as the fact that communication is typically an ongoing and natural byproduct of team interaction. Team communication is like individual think-aloud data but it occurs naturally 'in the wild.' Automated communication analysis methods, combined with the fact that the data are task embedded (and therefore can be collected unobtrusively), can contribute to our understanding of team performance and cognition and can also enable real-time monitoring and analysis of teams.

The communication analysis methods that we have developed have been applied to a number of different team contexts with promising results (Gorman et al., 2003, 2007a, 2007b; Kiekel et al., 2004; Weil et al., 2008a). We continue to modify methods and develop new ones and encourage others to apply them to new contexts. In this chapter, we provide practical guidance for the application of a variety of communication analysis methods.

The Generic Procedure

Data Collection

Typically when we think of communication, we think of people speaking to each other. However, communication can occur using various other media including whiteboards, post-it notes, letters, emails, maps, flags, facial expressions, gestures, and charts (Hutchins, 1995) just to name a few. Knowing what type of data to collect and how to collect it is therefore a fundamental concern when considering communication-based measures of macrocognition. We will describe four types of communication data that have been useful for measuring macrocognition: audio, chat, email, and logged communication events.

Audio data can be collected from verbal communications. Dimensions of audio data can include communication content (that is, what was said), communication timing (who was talking and for how long), and sequential flow (that is, who talks to whom or what communication event follows another). We have collected audio data in the context of simulated Unmanned Aerial Vehicle (UAV) operations and a collaborative planning synthetic task environment in order to measure macrocognition using the content, timing, and flow dimensions (particular measures are described below).

In the UAV task, audio data are generated by team members pressing Push-to-Talk (PTT) buttons in order to communicate with each other over headsets (Figure 11.1). Communication content is continuously recorded on both digital video and audio. These records can be assessed retrospectively, but lack communication flow and timing data (that is, the records would have to be post-processed for flow and

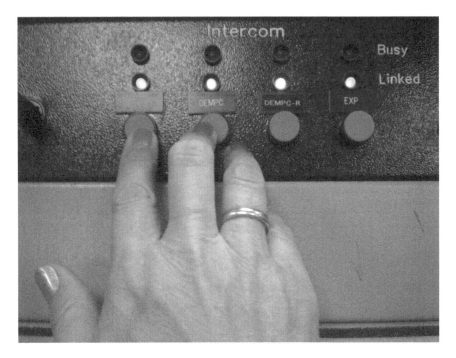

Figure 11.1   Push-to-talk communication system

timing). The PTT mechanism further allows us to record who is talking to whom, when, and for how long. This recording is called the ComLog, which is a matrix in which speaking (=1) versus not speaking (=0) for every team member-to-team member communication channel with a sampling frequency of 8 Hz.

In the collaborative planning task, we collect audio data generated by team members talking to each other in the presence of directional microphones, one positioned in front of each team member. As in the UAV task, communication content is recorded continuously during task performance using digital media. However, communication timing and flow data are collected using a voice key mechanism: if a team member's microphone picks up activity above a threshold then activity at that microphone is recorded at a sampling frequency of 10 Hz. In order to transform these data into a ComLog, a low-pass filter is applied to the data stream resulting in a record of speaking (=1) versus not speaking (=0) from which communication timing and flow can be collected (Figure 11.2). Thus far we have analyzed audio data post hoc after task performance.

As a medium for communication, text-based chat is increasingly prevalent in some organizational settings (for example, Air Operations Center), with verbal communications filling a more supplementary role. Chat communication consists of sequences of typed messages sent by team members. Two dimensions of chat data that need to be considered during data collection include communication content and message flow. We have collected chat communications data in the

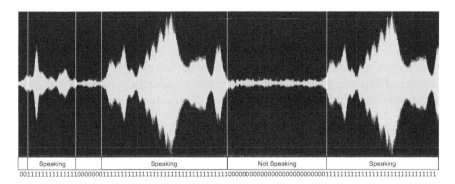

| Speaking | | Speaking | | Not Speaking | | Speaking |

001111111111111110000000111111111111111111111111111111111111000000000000000000000000000001111111111111111111111111111111111

Figure 11.2   Illustration of low-pass filter applied to sound pressure activity to generate a communication flow log

UAV task (Figure 11.3). Chat communication content includes the text within the message itself and also an aggregate measure of message length (for example, word count). Chat communication flow can be collected in terms of when the message was sent and by whom, and to whom the message was sent, and when it was opened.

As previously noted, communication can occur over a wide variety of media resulting in many possible data sources. Here we consider two other forms of communication data. Other forms of communication data that we have collected include email-based communication (Weil et al., 2008b) and logging of specific communication events. Data that can be collected in email-based communication

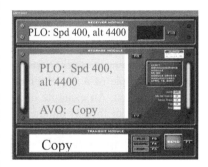

Figure 11.3   Chat interface

are similar to chat communication, in that it includes message content and message flow (who is sending, when, to whom, and when opened). We have also employed a technique in which trained observers monitor communication for specific events by specific team members and then time stamp the occurrence of these events. This approach to collecting communication data represents a combination of communication content and flow.

Post-Processing Communication Data

It is critical that the communications data that are collected contain the appropriate dimensions (content, timing, flow) in order to measure the macrocognitive property of interest. Even though the appropriate data are collected, communication analysis can still require a significant amount of post-processing. Content analysis will often require transcription, and possibly additional coding for speech acts, processes, or some other behavior. Ultimately, the coded content is what is analyzed. Importantly, recordings of communication can be post-processed for communication timing and flow, as well as content. Communication timing and flow do not necessarily require post-processing, however. This is because the timing and flow data do not require any syntactic or semantic parsers (for example, a transcriptionist). Of course, the ease of collection of timing and flow data trades off with the meaning inherent in the content data. Nonetheless, if collection of timing and flow data can be embedded in the task, as in the UAV and team collaboration tasks, this opens up new possibilities in terms of communication analysis and macrocognitive measurement. In combination with an efficient data collection format (for example, XML), communication timing and flow are amenable to real-time (online) analysis. We are currently developing techniques for real-time analysis of communication timing and flow.

Types of Communication Analysis Measures

Once data have been collected, measures can be selected. Classifying measures according to measurement types provides a framework for thinking about how different aspects of communication can be measured. The measurement classification system we have developed is shown in Table 11.1. Content, flow, and timing correspond to the data types described above. Static and dynamic refer to measurement types. Static measures collapse communication data over time. An example of a static content measure is the average number of words per UAV mission. An example of a static flow measure is the number of times one team member follows another per UAV mission (for example, 'dominance'; Cooke et al., 2008). An example of a static timing measure is the average time each communication follows another. Dynamic measures capitalize on temporal dependencies in the communication data. An example of a dynamic content measure is long-range correlation in semantic content (Gorman, 2005). An example of dynamic flow is the ChainMaster technique, which uses transition probabilities between speakers to identify the most probable sequence of speakers (Gorman et al., 2007b). Stability in the timing of how team members communicate (and coordinate) is an example of a dynamic timing measure.

Communication Analysis

Similar to the static/dynamic distinction in communication data types, analysis can be classified broadly as either a summary or pattern analysis. However, it is important to note that dynamic data can be submitted to a summary analysis and static data can be analyzed for patterns, given enough sequenced static snapshots. We have conducted both summary and pattern analysis on communication data using methods such as Latent Semantic Analysis for summary (Transcript Density) and patterns (Lag Coherence; Gorman et al., 2003), keyword indexing (Cooke et al., 2005), and ChainMaster (Gorman et al., 2007b).

Table 11.1   Classification of communication measures

	CONTENT	FLOW	TIMING
STATIC	Avg. # of words, Latent Semantic Analysis, Communication Density	Following behavior (Dominance)	Avg. time of following behavior
DYNAMIC	Semantic correlations, Latent Semantic Analysis Lag Coherence	ChainMaster, Procedural Networks (PRONET), transition analysis	Communication timing stability

A summary analysis does not explicitly incorporate a time or sequencing component, whereas a pattern analysis does. A summary analysis collapses communication across a relatively large interval of time (for example, tens of minutes to hours) for mean communication behavior and the variability of behavior. The assumption in a summary analysis is that sequence of communication events is essentially random, such that the frequency, mean, or variability is the best estimate of communication behavior. An example of a summary analysis is shown in Figure 11.4a. In the example, four types of communication behavior are summarized over a one-hour interval. The analysis reveals that communication behavior during that time can be fairly described as soliciting information.

In contrast to a summary analysis, pattern analysis examines how communication behavior varies over time. The assumption in the pattern analysis is that the timing or sequencing of communication events carries relevant information. It is therefore not appropriate to collapse across time in a pattern analysis as it is in a summary analysis. The communication content codes summarized in Figure 11.4a are shown across time in Figure 11.4b. If the temporal layout of communication events is random, then the pattern analysis will not yield any new information beyond the summary analysis. Alternatively, if the temporal layout is non-random, then a pattern analysis can add information that is not apparent in the summary analysis. Finally, there may also be some combination of summary and pattern analysis that yields important information about macrocognition. An example of this is a moving window analysis. In this type of analysis, a summary measure is computed over some window (sample) size and this measure is updated each time the window moves.

Interpretation of Analyzed Communication Data

The communication analysis itself is not the end goal. Rather, it is a means to the end goal of investigating macrocognitive phenomena. Beyond data collection, coding, post-processing, and other analysis considerations, interpretation of

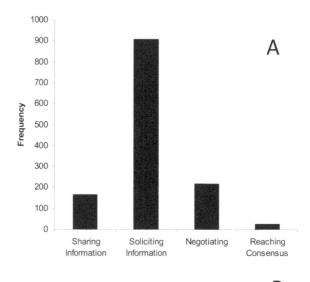

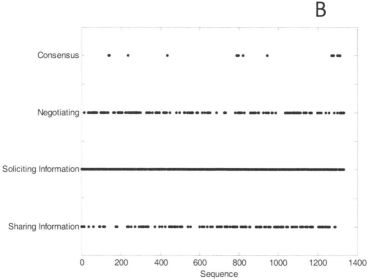

Figure 11.4 A. Summary of communication events over one hour; B. Temporal layout of the same communication events over one hour

communication data is the bridge that links communication analysis to broader questions of theoretical and applied interest (for example, communication of intent as described in Chapter 8). As with any method, interpretation is facilitated by having a model or criterion measure of macrocognition with which to compare the results of the communication analysis.

Interpreting communication relative to a model may consist of comparing the results of communication analysis to a set of predictions made by the model. To the extent that the sequencing prediction made by the model is supported by the data, the analysis may be interpreted as providing support for the model. Figure 11.5a shows a hypothesized model of team collaboration. This model hypothesized that team members loop between soliciting and sharing information, then iterate (that is, negotiate) on that information. The model further predicts that reaching consensus is a relatively uncommon event that separates the cycling behavior.

Figure 11.5b shows a model fit using the data from Figure 11.4. The model fit is based on content coding of events, and the lag-1 transition probabilities between events. The transitions above a threshold of *Prob.* > .10 are shown in Figure 11.5b.

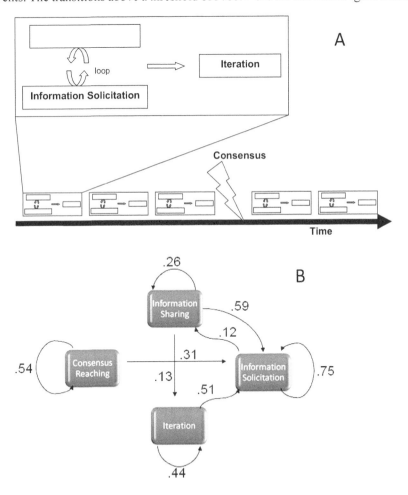

Figure 11.5   A.   Hypothesized   model   of   team   collaboration;   B.   Data interpreted in light of the hypothesized model

The data can be interpreted as showing the loop between sharing and solicitation as well as a transition from sharing to iteration. In addition, because there are no direct transitions to consensus, these data suggest that reaching a consensus is a behavior that is separate from the cycling behavior.

Interpreting communication relative to a criterion measure concerns how features of the communication analysis are related to other, specific aspects of macrocognition. For example, tying the results of communication analysis to team effectiveness, expert ratings of communication or individual-level factors such as team member knowledge may allow for the interpretation that some aspects of team communication are supportive of these constructs, whereas some are not.

Interpretation of analyzed communication data occurs against a backdrop of scientific or practical objectives. In our work, communication analysis has been useful in marking critical events in a team or organization (Gorman et al., 2007b; Kiekel et al., 2004), identifying signatures related to effective team performance and high workload conditions (Cooke et al., 2008; Cooke et al., 2005), distinguishing communication patterns in co-located from distributed teams (Cooke et al., 2005), and identifying recurring and nested patterns of macrocognitive processes (Gorman et al., 2009).

Pitfalls and Guidelines

Data Collection

At the data collection stage, decisions have to be made about what data to collect (the type of interactions—voice, chat, email, nonverbal, and so on), who to collect from (all individuals in a team or group, a subset, a single speaker), whether to record speaker identity, whether to record receiver identity, recording medium (audio, video, note-taking), availability of outcome metrics, and how to record timing or sequence or speaker/receiver identity.

Clearly, the type of communication data collected and from whom is largely driven by the objectives and context of the research project. If the objective is to study communications in the Air Operations Center where chat predominates, then it is important to collect and analyze chat data. However if the focus is on a bigger picture social network analysis stemming from interactions in an office context, then presumably all types of interactions would be logged with more of a focus on who is talking to whom than what is being said.

Other decisions about data collection will likely be guided by the facilities available for it. In the interest of capturing as much detail as possible in the interactions, it would be ideal to collect speaker and listener identities with all timings on audio or video tape and to ensure that at least one outcome measure is available. However, if the requisite facilities are not available, then accommodations must be made. In cases in which the data in the form of audio or written transcripts already exist, there may be little room for decision making about data collection and not only is the content constrained, but the format as well.

The biggest pitfall in the collection of communication data is the failure to collect enough data. That is, data are not collected at a sampling frequency that is small enough to be sensitive to differences in macrocognitive processes and data collected omit important aspects of communication (for example, flow, timing) or omit outcome measures for later validation. As a general rule of thumb, *collect all forms of communication data that you anticipate possibly using and collect it at the smallest sampling frequency possible.* Data collection in group settings tends to be costly and so it is wise to collect and store as much data as possible. The data are so rich that several different research questions can be addressed by different analyses applied to the same data set. At the same time, you can always collapse over several sampling windows to go from small sampling to large, but the reverse direction is not possible.

Post-processing and the Dangers of Other People's Data (OPD)

Post-processing is often a 'necessary evil' in communication analysis. Extensive transcription and coding of audio tapes is sometimes required. Transcription, coding, and other forms of post-processing can become even more complicated when analyzing Other People's Data (OPD). Aside from any post-processing the owner of the data may have performed, the data may require additional attention before it can analyzed. Are names used? It is likely that you have to get rid of these. Was the data obtained in an experiment? If so, what Institutional Review Board (IRB) approved the original data collection and analysis plan? What other restrictions are there on the data? How reliable are the data (for example, inter-rater reliability of codes)?

When deciding whether it is worth the time and effort involved in the additional post-processing introduced by OPD, it is important to establish some minimum data characteristics for the phenomenon under study. Table 11.2 shows an example of minimal and preferable data characteristics of team communication data that we used to determine appropriate OPD for studying team collaboration. We used these minimal data characteristics in deciding whether OPD sets were amenable to our analysis techniques for investigating team collaboration. Setting these criteria is important because you cannot repair bad data, and investing time and energy into attempting to do so can be a waste of both time and resources. For example, in a recent analysis, we were unable to examine a communication phenomenon of interest (following behavior) because the data were sampled at too large a grain (30-second snapshots). As mentioned previously, you cannot recover phenomena smaller than the minimum sampling rate, which is something else to consider when evaluating whether OPD are appropriate.

In general, it is important to anticipate data needs and screen data sets carefully before investing in analysis. Another guideline is to attempt to

Table 11.2	**Minimum and preferable communication data characteristics for collecting OPD for examining team collaboration**

Data Type	Minimum	Preferable
Flow/Timing	Begin time	Begin time
	Speaker	End time
		Speaker
		Receiver
Content	Utterance-level transcript	Utterance-level transcript
	Speaker	Speaker
		Receiver
		Content codes

streamline post-processing by well-planned data collection and formatting. General purpose data formats can be designed using Extensible Markup Language (XML). We have begun to streamline our analyses by taking in XML data. Therefore, any data collected in XML format is compatible with the analysis methods.

Communication Analysis Costs and Benefits

Communication analyses can be positioned in a cost-benefit tradeoff space which should help to guide the selection of specific measures and analytic techniques. At a general level, the more expensive the method in terms of analytic time and effort, the more information is gained (Figure 11.6). For example, the analysis of communication content requires transcription and content coding, is more difficult to automate, and can be quite costly. On the other hand, content analysis provides information that is much richer than who is talking to whom. This can be contrasted with speaker sequence information, or even word count measures which are easy and cheap to collect, but more difficult to meaningfully interpret.

One strategy, which we refer to as 'the wedding cake strategy,' involves a tiered process of moving from easy and cheap analyses to more expensive ones (Figure 11.7).

This movement is accompanied by a progressive narrowing of the data of interest. Easy and cheap methods are applied to all data and patterns that are interesting or anomalous are flagged for further processing. Instead of conducting a deep and expensive analysis of the larger data set, data are strategically and quickly culled for further analysis.

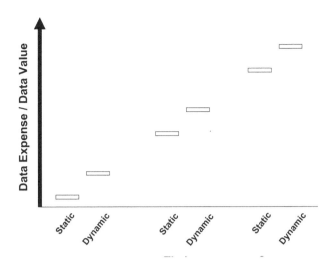

Figure 11.6   Tradeoff between data expense and data value for different types of data and measures

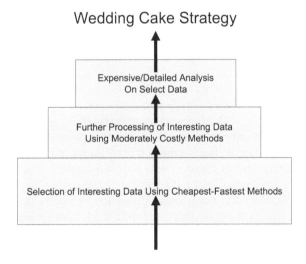

Figure 11.7   Strategy of progressive deepening of communication analysis

Real-time Data Collection and Analysis

If at all possible, it is preferable to collect communication data online, as the data are being generated. This saves time and money because the data do not have to be post-processed. In addition, online data collection can also lend itself to real-time communication analysis, given that the analysis methods can be performed at least as fast as the data become available.

Something that must first be considered when deciding to incorporate real-time data collection and analysis is the type of data collected and analysis to be performed. Content data, for example, are difficult to collect in real time, because in the case of audio data, they require speech to text conversion. Similarly, many content analysis methods may not amenable to real-time analysis because they require top-down interpretation of the data; for example, content coding. However advances have been made recently in this area (Foltz & Martin, 2009). Alternatively, flow and timing data types are optimal for online data collection and analysis. Dynamic analysis of flow patterns in real time is an example of an analysis technique that capitalizes on the temporal aspects of online data collection. Currently, we are developing real-time dynamical analysis of communication timing in order to detect non-routine events in team communication with the goal of intervening in the case of anomalous events.

Another issue that should be considered when conducting any analysis, but particularly with real-time analysis, is the amount of computational burden the technique introduces. As noted previously, real-time analysis only works if the technique can be applied at least as fast as the data come in. Computationally intense procedures may therefore not be appropriate for real-time analysis, or may have to be modified to handle the rate of data analysis. A good first test of any method is to assess the rate of computation on samples of existing data. If the method is fast enough for the communication behavior of interest, or can be appropriately streamlined, then it may be a candidate for use as a real-time method.

Interpretation Issues

Perhaps the greatest pitfall of communication analysis is the lack of sound criterion measures. In order to meaningfully interpret patterns of communication, it is necessary to juxtapose them against other measures of team performance or macrocognition. The interpretation is only as good as these criterion measures. Criterion measures should be valid and reliable and collected at a level of granularity that is compatible with the communication data. Often rich communication data are collected over a session lasting several hours and a single expert judgment is the sole criterion. Examples of criterion measures used in our lab include composite outcome measures for command-and-control teams (that is, team performance), acquisition and retention rates of team performance, observer ratings of team process at critical events, team situation awareness measures, and coordination metrics (Cooke & Gorman, in press; Cooke et al., 2007b). The guideline here is to pay as much or even more attention to criterion measures as you do to communication measures.

Conclusion

The analysis of communication opens the door to many exciting and rich measures of macrocognition and automated measurement can facilitate the process. Because communication is a natural byproduct of sociotechnical interaction, it is an unobtrusive source of data. New analytic techniques and improvements in speech recognition hold the promise of real-time analysis. There are useful approaches to data collection, post-processing, measurement, data analysis, and interpretation. However there are also common pitfalls as we have described and as are recapped in Table 11.3.

Table 11.3   Common pitfalls in the analysis of communications

- Failure to collect adequate data (timing, speaker ID, receiver ID).
- Failure to sample at fine enough level.
- Failure to screen OPD (Other Peoples' Data).
- Applying costly analyses to uninteresting data.
- Awkward application of real-time analysis.
- Failure to collect criterion measures.

Similarly there are guidelines for applying communication analysis and avoiding the pitfalls (Table 11.4).

Table 11.4   Guidelines for the effective analysis of team communications

- Let research objectives guide data collection.
- All else being equal collect as many dimensions of the communications as possible.
- Sample at frequencies higher than the phenomenon of interest (e.g., 8-10 Hz).
- Set minimum criteria for the acceptance of OPD (Other Peoples' Data).
- Adopt a 'wedding cake strategy' in which cheap, fast methods are applied to identify interesting data for progressively more expensive analyses.
- Identify data and measure types that lend themselves to real-time analysis.
- Pay special attention to criterion measure validity, reliability and sensitivity.

There has been significant progress on the analysis of communications over the last decade. Increased automation of techniques allows the researcher to fully exploit the richness of the communications data collected. Additional innovations in this area are critical prerequisites to interpreting analyzed communications data and applying the analysis to further our understanding macrocognition.

Acknowledgements

The work described in this chapter was partially supported by Office of Naval Research grants N00014-05-1-0625 and N00024-06-1-0446 (Mike Letsky, program manager). The authors also thank Nia Amazeen, Eric Hessler, Leah Rowe, and Steve Shope for their contributions toward this effort.

References

Achille, L. B., Schulze, K. G., & Schmidt-Nielsen, A. (1995). An analysis of communication and the use of military terms in Navy team training, *Military Psychology, 7*, 95-107.

Bowers, C. A., Braun, C. C., & Kline, P. B. (1994). Communication and team situational awareness. In R.D. Gilson, D. J. Garland, & J. M. Koonce (Eds.), *Situational awareness in complex systems* (pp. 305-311). Daytona Beach, FL: Embry Riddle Aeronautical University Press.

Bowers, C. A., Jentsch, F., Salas, E., & Braun, C. C. (1998). Analyzing communication sequences for team training needs assessment. *Human Factors, 40,* 672-679.

Cooke, N. J. & Gorman, J. C. (2009). Interaction-based measures of cognitive systems. *Journal of Cognitive Engineering and Decision Making, 3,* 27-46.

Cooke, N. J., Gorman, J. C., Duran, J. L., & Taylor, A. R. (2007b). Team cognition in experienced command-and-control teams. *Journal of Experimental Psychology: Applied, Special Issue on Capturing Expertise across Domains, 13*(3), 146-157.

Cooke, N. J., Gorman, J. C., & Kiekel, P. A. (2008). Communication as team-level ognitive processing. In M. Letsky, N. Warner, S. Fiore & CAP Smith, *Macrocognition in teams: Theories and methodologies* (pp. 51-64). Aldershot, UK: Ashgate Publishing Ltd.

Cooke, N. J., Gorman, J. C., Kiekel, P. A., Foltz, P., & Martin, M. (2005). Using team communication to understand team cognition in distributed vs. co-located mission environments. *Technical Report for ONR Grant no. N00014-03-1-0580.*

Cooke, N. J., Gorman, J. C., & Rowe, L. J. (2009). An ecological perspective on team cognition. In E. Salas, J. Goodwin, & C. S. Burke (Eds.), *Team effectiveness in complex organizations: Cross-disciplinary perspectives and approaches* (pp. 157-182). SIOP Organizational Frontiers Series. New York, NY: Routledge.

Cooke, N. J., Gorman, J. C., & Winner, J. L. (2007a). Team cognition. In F. Durso, R. Nickerson, S. Dumais, S. Lewandowsky, and T. Perfect, *Handbook of applied cognition, 2nd Edition* (pp. 239-268). Hoboken, NJ: Wiley.

Emmert, P., & Barker, L. L. (1989). *Measurement of communication behavior.* White Plains, NY: Longman, Inc.

Foltz, P. W., & Martin, M. J. (2009). Automated Communication Analysis of Teams. In E. Salas, J. Goodwin, & C. S. Burke (Eds.), *Team effectiveness in complex organizations: Cross-disciplinary perspectives and approaches* (pp. 411-431). SIOP Organizational Frontiers Series. New York, NY: Routledge.

Foushee, H. C., & Manos, K. (1981). Information transfer within the cockpit: Problems in intracockpit communications. In C. E. Billings & E. S. Cheaney (Eds.), Information transfer problems in the aviation system *(Report No. NASA TP-1875)*. Moffett Field, CA: NASA-Ames Research Center.

Gorman, J. C. (2005). The concept of long memory for assessing the global effects of augmented team cognition. Poster presented at Augmented Cognition International, HCII, July 22-27, Las Vegas, NV.

Gorman, J. C., Cooke, N. J., Amazeen, P., Hessler, E., & Rowe, L. (2009) Automatic tagging of macrocognitive collaborative processes through communication analysis. *Technical Report for ONR Grant N00014-05-1-0625.*

Gorman, J. C., Cooke, N. J., Warner, N. W., & Wroblewski, E. M. (2007a). Tagging macrocognitive processes using communication flow patterns. In *Proceedings of the Human Factors and Ergonomics Society 51st Annual Meeting*, (pp. 410-415). Baltimore, MD.

Gorman, J. C., Foltz, P. W., Kiekel, P. A., Martin, M. J., & Cooke, N. J. (2003). Evaluation of latent-semantic analysis-based measures of team communications. In *Proceedings of the Human Factors and Ergonomics Society 47th Annual Meeting* (pp. 424-428).

Gorman, J., Weil, S. A., Cooke, N., & Duran, J. (2007b). Automatic assessment of situation awareness from electronic mail communication: Analysis of the Enron dataset. In *Proceedings of the Human Factors and Ergonomics Society 51st Annual Meeting* (pp. 405-409). Baltimore, MD.

Hutchins, E. (1995). *Cognition in the wild.* Cambridge, MA: MIT Press.

Jentsch, F. G., Sellin-Wolters, S., Bowers, C.A., & Salas, E. (1995). Crew coordination behaviors as predictors of problem detection and decision making times. In *Proceedings of the Human Factors and Ergonomics Society 39th Annual Meeting* (pp. 1350-1353). Santa Monica, CA: Human Factors and Ergonomics Society.

Kiekel, P. A., Cooke, N. J., Foltz, P. W., & Shope, S. M. (2001). Automating measurement of team cognition through analysis of communication data. In M. J. Smith, G. Salvendy, D. Harris, & R. J. Koubek (Eds.), *Usability evaluation and interface design* (pp. 1382-1386). Mahwah, NJ: Lawrence Erlbaum Associates.

Kiekel, P. A., Gorman, J. C., & Cooke, N. J. (2004). Measuring speech flow of co-located and distributed command and control teams during a communication channel glitch. In *Proceedings of the Human Factors and Ergonomics Society 48th Annual Meeting*.

Kleinman, D. L., & Serfaty, D. (1989). Team performance assessment in distributed decision making. In *Proceedings of the Symposium on Interactive*

Networked Simulation for Training (pp. 22-27). Orlando, FL: University of Central Florida.

Mosier, K. L., & Chidester, T. R. (1991). Situation assessment and situation awareness in a team setting. In Y. Quéinnec & F. Daniellou (Eds.), *Designing for everyone: Proceedings of the 11th Congress of the International Ergonomics Association* (pp. 798-800). London: Taylor & Francis.

Orasanu, J. (1990). *Shared mental models and crew performance* (Report No. CSLTR-46). Princeton, NJ: Princeton University.

Oser, R. L., Prince, C., Morgan, B. B., Jr., & Simpson, S. (1991). An analysis of aircrew communication patterns and content, *NTSC Technical Report No. 90-009*. Orlando, FL: Naval Training Systems Center.

Poole, M. S., Holmes, M., Watson, R., & DeSanctis, G. (1993). Group decision support systems and group communication: a comparison of decision making in computer-supported and nonsupported groups. *Communication Research, 20*(2), 176-213.

Salas, E., Bowers, C. A., & Cannon-Bowers, J. A. (1995). Military team research: 10 years of progress. *Military Psychology, 7*(2), 55-75.

Weil, S. A., Duchon, A., Duran, J., Cooke, N. J., Gorman, J. C., & Winner, J. L. (2008a). Communications-based performance assessment for air and space operations centers: Preliminary Research. In *Proceedings of the Human Factors and Ergonomics Society 52nd Annual Meeting,* New York, NY.

Weil, S. A., Foster, P., Freeman, J., Carley, K., Diesner, J., Franz, T., Cooke, N., Shope, S., & Gorman, J. (2008b). Converging approaches to automated communications-based assessment of team situational awareness. In M. Letsky, N. Warner, S. Fiore & C. A. P. Smith, *Macrocognition in Teams: Theories and Methodologies* (pp. 277-303). Aldershot, UK: Ashgate Publishing Ltd.

Chapter 12

From Data, to Information, to Knowledge: Measuring Knowledge Building in the Context of Collaborative Cognition

Stephen M. Fiore, John Elias, Eduardo Salas, Norman W. Warner, and Michael P. Letsky

Introduction

Over the past two decades, cognitive psychology and cognitive engineering have examined team process and performance to gain a better understanding of how cognition contributes to effective team performance (for example, Cannon-Bowers et al., 1993; Hinsz et al., 1997; Hutchins, 1991; Larson & Christensen, 1993; Levine et al., 1993). Developing alongside these areas of inquiry, the concept of *macrocognition* arose primarily to describe how cognition emerges in naturalistic contexts, in 'the study of the role of cognition in realistic tasks, that is, in interacting with the environment' (Cacciabue & Hollnagel, 1995, p. 57). Here the goal was to understand how complex and emergent cognitive processes arise (that is, macrocognitive processes) as cognition unfolds in dynamic settings (for example, Hutton et al., 2003; Klein et al., 2003). More recently, the concept of *macrocognition in teams* was developed to describe cognition in collaborative settings and the internalized and externalized cognitive processes occurring during team problem solving (Letsky et al., 2008; Warner et al., 2005).

Following on recent work in *team cognition* (Salas & Fiore, 2004) and research on the application of cognitive concepts to teams, this chapter focuses on this notion of *macrocognition in teams* to consider teams in complex problem-solving environments. We follow recent thinking on team problem solving, which has differentiated between traditional team cognition approaches to understanding interaction and macrocognition in teams (Rosen et al., 2008). Specifically, team cognition tends to focus on coordinating actions between individuals, for example, understanding how team members are able to sequence their actions in service of their team tasks. But macrocognition in teams focuses more on the *knowledge work* done by a team (see Fiore et al., 2008; Letsky et al., 2007; Warner et al., 2005). As such, what is of interest is the coordination and integration of *informational inputs* in the process of building knowledge.

By adapting Rasmussen's (1983) skills, rules and knowledge classification of human performance, this approach more clearly explicates the relevance of rule and knowledge levels for understanding this distinction. In particular, rule-based performance can be said to relate more to behavioral coordination where teamwork requires executing previously learned task procedures in familiar environments. For example, traditional team cognition research seeks to understand, and thereby train for, situations such as an emergency response crew using shared mental models to coordinate behavior in time-stressed situations and use implicit and explicit processes to pursue a course of action. But, in macrocognition, the emphasis is on knowledge-based performance where situations are unfamiliar and, more often than not, lack pre-existing rules to guide action (Klein et al., 2003; Schraagen et al., 2008a, 2008b). Furthermore, fitting with what others have described as the conditions under which macrocognition functions and processes are performed (for example, Klein et al., 2003; Schraagen et al. 2008a), our focus is on teams operating in uncertain and time-stressed situations, with ill-defined goals, and high stakes, and working to solve one-of-a-kind problems. Thus, we suggest that this type of team interaction involves the generation or adaptation of rules to novel situations as opposed to executing predetermined rules. More specifically, macrocognition in teams involves individuals and teams generating new knowledge when faced with unusual problems (Fiore et al., 2008; Rosen et al., 2008).

In this chapter, we examine and emphasize the *knowledge-building* process and its intrinsic importance to how we are conceptualizing macrocognition in teams. Our goal is to explicate this process in order to take the steps necessary to measure what we argue is a foundational process of team macrocognition. We provide not operational definitions of this process; rather, we provide conceptual definitions of knowledge building as this is a critical precursor to measurement. Specifically, in order for us to truly understand knowledge building, we must have a more fine-grained explication of two intermingled issues—that of defining this process and that of measuring its occurrence. Scientifically, it is prudent to first provide a set of conceptual definitions from which we might build a better specified set of operational definitions that could drive robust methods of measuring this process. As such these conceptual definitions help us to more fully communicate something about the process of knowledge building and move the field closer to operational definitions.

Toward this end, first we describe some of the core elements of what is generally referred to as the Data-Information-Knowledge transformation process (DIK-T), and present representative definitions of data, information, and knowledge, and how these instantiate the knowledge-building process. Arising primarily out of knowledge management and information sciences, this hierarchy has been widely discussed in a number of articles[1] (see Ackoff, 2004; Cleveland, 1982; Zelany, 1987). Secondly, as any discussion of the nature of *knowledge* must pay some

---

1   The traditional hierarchy includes 'wisdom' at the end of this chain. Given the complexity inherent in operationalizing what can be construed of as wisdom in the context of collaborative problem solving, we have not included that in our discussion.

respect to the philosophical tradition, we draw on the well-established analysis of knowledge as Justified True Belief (also known as JTB), and explore how it may fruitfully map onto the DIK-T process and hierarchy and provide an avenue through which measurement can then be pursued. We then show how analogous approaches can be found in fields such as computer-supported collaborative learning, using the implementations from that discipline to illuminate processes of knowledge construction. By illustrating some of the similarities across these different disciplines we lay a broad foundation for our adaptation of these views to studies of macrocognition in teams. Further, this allows us to consider various distinctions that may be used to viably measure these processes of knowledge accrual and creation. We conclude with a definition of knowledge building as it occurs in macrocognition in teams, centering on the notion of actionability, and demonstrate how these can be fit together in support of measuring a core part of complex collaborative problem solving.

Foundational Concepts in Knowledge Building

To help ground our conceptualization of the knowledge-building process, following Fiore et al. (2008), we rely upon two intermingled concepts—that of context and integration. These two concepts, we argue, are crucial components of the problem-solving process, and are therefore foundational to developing a useful operationalization of the steps taken toward knowledge. To start with, it is perhaps informative to consider the etymological origins of the word *context*. Context has as its Latin root, *contextus*, 'a joining together,' which, in turn, was derived from *contexere*, 'to weave together,' with *com* meaning 'together' and *texure*, 'to weave.' This set of original meanings connotes an active, constructive quality, and from this we suggest that context enables us to actively weave together our understanding of events, in order, for example, to form a mental model of the world with which we are interacting at any given moment in time. Relatedly, we use the term *integration* to indicate the process by which information is organized and made meaningful and actionable in a particular context. What prevents this definition from being a mere tautology, however, what keeps it from saying, in effect, 'integration is the process by which information is *integrated*,' is precisely the distinction between *organization* and *integration*: we employ the latter to designate specifically the sort of organization that enables action, that snaps information together, as it were, into practical actionable knowledge. The literature on expertise sheds light on this view, in its account of how expert knowledge differs from novice knowledge, not just in content but in its organization (for example, Chi et al., 1981). Specifically, Glaser (1989) states that 'beginners' knowledge is spotty, consisting of *isolated* definitions and superficial understandings of central terms and concepts. With experience, these items of information become *structured*, are *integrated*... Thus, structuredness, coherence, and accessibility to *interrelated* chunks of knowledge. . . '

(p. 272, emphasis added) are the foundation of expert knowledge; which is arguably the highest form of understanding.

Yet these seemingly simple preliminary descriptions already begin to yield potentially instructive difficulties and ambiguities. And so to initiate a pattern of call and response that will turn up occasionally throughout this chapter, we will proceed, briefly, to complicate and qualify what we have just said, in the pursuit of clarifying an intrinsically complex process. For example, our etymological consideration of *context* would seem, again, to attribute an active character to it: but does our sense of context itself do the weaving, or does it, at least to some degree, arise from our weaving? This question of cause and effect introduces the theme of circularity that will come up again in our discussion. Whether context is the weaver weaving, or is to be woven itself, seems best answered by conceding that some sense of context must exist at the start, in order for understanding to even get a start. For it is difficult to make sense of the process of *making sense*, of generating and understanding meaning, without some basic contextual background. For instance, the elementary observation that an object in a room is a chair, the classification of such a thing as a *chair*, implies a network of categories and concepts (for example, that of *furniture* and different types therein). Furthermore, in viewing this chair, we bring to it a pragmatic sense of what can be done with it: our very perception of it is informed by its *affordances*, its manifold possible uses (Gibson, 1977), which in turn are bound up with our identifying and conceptualizing it as a particular kind of thing.

The above point has genuine implications for our elaboration of knowledge building, as to whether we should conceive it as a building up, or narrowing down, of context, which of course depends on how we conceive of context. If we view it broadly, in terms of a general semantic context, as a matrix of meanings within which our understanding necessarily operates, then progress toward pragmatically actionable knowledge, would involve funneling our vast set of generalized contextual information into a particular practical context, a specific situation within which we act and interact. However, if we limit what we mean by context to our understanding of some situation, then our advancement of knowledge would consist in the integration and enlargement of our sense of context, that is, the context of that particular situation, as we come to learn more about it. While this issue may seem abstract, our overarching conception of the process of knowledge building can differentially impact our analysis. Specifically, whether we think of context as being built up or winnowed down in the approach toward actionable knowledge may influence the extent to which we bring prior knowledge to a particular situation in order to then build upon it. On the one hand, emphasis on our starting with a basic contextual matrix, largely semantic and conceptual in nature, which is then filtered, specified, and applied to each newly encountered situation, would highlight previously attained and integrated information and knowledge. On the other hand, an approach that theorizes little or no context at the start would foreground newly acquired and organized information relevant to the current situation. As we lay out our conceptualizations, we suggest that knowledge building requires and utilizes both—prior knowledge brought into play with newly acquired information. We next more specifically describe some of the prevailing views on the process of knowledge building as it is found in various scholarly treatments.

Knowledge Building in Information Systems Theory: The DIK-T Hierarchy

With this consideration of context and integration as our stepping off point, we next review some definitions and descriptions of data, information, and knowledge, as well as how these are thought to come about through some process of development. Our approach will be dialectic, with concepts first presented straightforwardly from the literature, and then analyzed and questioned. At the close of the chapter we will return once more to these definitions distilled from this literature, and conclude with concrete definitions along with a graphical representation of these concepts to illustrate how we view their transformation within collaborative problem solving, as well as explore their possibilities for measurement within the context of macrocognition in teams.

Data, as characterized in much of the literature, can be construed as raw, unprocessed bits of facts and other phenomenon. Hey (2004) defines data as, 'discrete, atomistic, tiny packets that have no inherent structure or necessary relationship between them' (p. 6). Others have noted that data does not have utility per se in that it is presented without any specific context or interpretation (Gowler, 1953). Data can also be described as symbols that represent properties of objects, events, and their environment, but they are not necessarily of use until applied in some practical context. In short, data can be described as discrete and objective observations or facts; it is considered unorganized and unprocessed and it does not convey much meaning (for example, Ackoff, 1989; Awad & Ghaziri, 2004; Chaffey & Wood, 2005; Pearlson & Saunders, 2004—see also Bellinger et al., 2004).

However, from the human point of view at least, to speak of entities as having absolutely no intrinsic or extrinsic relations is problematic. In one sense, this point brings us back to our previous examination of context, and the observation that even our most basic immediate identification of objects involves understanding them as particular *kinds of things*, and hence entails a network of categories and concepts, an extended semantic context. In this vein it is worth emphasizing as well that *facts*, in their discreetness and concreteness, are themselves articulated entities, composed of parts in relation. The briefest reflection on the nature of facts reveals their articulated character; to state such and such is a fact is to place some thing, some object or entity, or place or person, within some minimal level of relationality: the fact that 'Paris is the capital of France' relates the city of Paris in a specific way to the country of France; the fact that 'the Earth is the third planet out from the Sun' puts the planet Earth in a specific numeric relation to the Sun and the other planets; and so on. This articulation is perhaps what Wittgenstein had in mind when he declared in his *Tractatus Logico-Philosophicus*, 'The world is the totality of facts, not of things.' That is, the world is composed of elements in contextual relation, not of 'things' in empty isolation. Thus, while it may be appropriate in terms of computational implementation to speak of data as 'tiny packets that have no inherent structure or necessary relationship between them,' in the human realm such talk can only serve, at best, as a purely theoretical *asymptote* of sorts, as an extreme idealized point of reference that is never actually reached. However, this is not to say that the basic notion of *objectivity*, in the sense of something being true independently of our thinking or conceptualizing about it, ought to be ignored—quite on the contrary. In the next section, following our descriptions of information and knowledge, we present a 'justified true belief' analysis of knowledge in the hopes that we can clarify this question of *objectivity* in relation to one's attempts to instantiate knowledge.

Information, as the next level in the DIK-T hierarchy, may be said to involve an increase in the level of meaning ascribed to data, based upon two overarching concepts—organization and specified context. Information, then, can be understood as data processed and made meaningful by the specific context in which it is interpreted along with some form of structuring and organizing (Rowley, 2007—see also Ackoff, 1989; Bellinger et al., 2004; Hey, 2004). For example, some argue that information is 'formatted data' (Jessup & Valacich, 2003, p. 7) whereas others state that it is data organized to provide value and meaning (for example, Boddy et al., 2005; Turban et al., 2005). Again, however, insofar as *information* involves the additional formulation of context over and above the level of *data*, this context should properly be understood as a particular *applied* or *practical* context, as opposed to the larger sense of a *semantic* or *conceptual* context or matrix, as previously detailed. So when Rowley (2007) points out that, 'to be relevant and have a purpose, information must be considered within the context where it is received and used' (p. 171), it is the specific context of *use*, of *practice*, that is emphasized. Hence information may be seen as the result of a process whereby meaning is made out of the provided data by viewing it in the context of other data (for example, 'weather data,' 'resources data') relevant to the problem-solving scenario. Thus, transforming data into information requires some form of organization or categorization such that it becomes meaningful within the problem-solving context.

Definitions of what constitutes *knowledge* are in even less agreement, with little or no clear consensus emerging across disciplines. Fitting with our earlier description of expertise and knowledge, Cleveland (1982) defines knowledge as 'expertness—in a field, a subject, a process, a way of thinking, a science, a "technology," a system of values, a form of social organization and authority' (p. 34). Ackoff (1989) defines knowledge as 'know-how, for example, how a system works' (p. 4), which can be contrasted with *knowing-that*, that is, with simple propositional knowledge of facts. Furthermore, Rowley states that, 'knowledge is data and/or information that have been organized and processed to convey understanding, experience, accumulated learning, and expertise as they apply to a current problem or activity' (p. 172). But most relevant to collaborative problem solving, and the definitions with which we closely align, are those put forth by Zaleny (1987) and by Pearlson & Saunders (2004). First, Zelany defines knowledge as holistic and '. . . related to and expressed through systemic network patterns, integrative by definition' (p. 59). In this view, knowledge is contained in an overall organizational pattern and, as such, refers to an 'observer's distinction of "objects" (wholes, unities) through which he brings forth from the background of experience a coherent and self-consistent set of coordinated actions' (p. 61). Pearlson & Saunders (2004) advocate a notion of knowledge requiring reflection on, and synthesis of, information within a particular context and in service of some need. Specifically, knowledge is said to consist of a 'mix of contextual information, values, experience, and rules. . . Knowledge involves the synthesis of multiple sources of information over time' (pp. 13-14).

So, to distill a central thrust from these definitions: knowledge is characterized in its relation and reference to agents and subjects, to subjectivity, broadly speaking, and all that it entails, the assigning of values, the working toward goals, the prevalence of purpose, need, desire, and so forth. As noted by Pearlson and Saunders (2004), the 'amount of human contribution increases along the continuum from data to information to knowledge' (p. 14). This character of knowledge as held by the human, necessarily in reference to the human, is implied as well in the emphasis on pattern and organization in the above definitions; for, as Tsoukas et al. (2001) argue, such organizational and patterned complexity is inherently observer dependent, a second-order function brought to the content in question by the human observer and thinker.

Still, on the whole, questions arise from these characterizations. For instance, to what degree is information merely an intermediary, between the poles of unadorned data and fully elaborated knowledge? And, as noted, to what extent are these poles themselves idealized, functioning as theoretical asymptotes? In the next section we elaborate upon the 'Justified True Belief' analysis of knowledge so as to clarify, if not answer, these questions. As we will see, depending on the stringency of the analysis, the conditions for knowledge may never be entirely satisfactorily met. Nonetheless, consideration of the justified true belief analysis provides the necessary steps in working toward a robust means of measuring the creation of knowledge in the context of macrocognition in teams.

The Philosophical Analysis of Knowledge as Justified True Belief

The above account of the movement from data to information to knowledge may prove valuably illuminated by a philosophical perspective. Here we will outline a potentially fruitful philosophical approach, for questions concerning knowledge can hardly pass by without at least a few words from philosophy. The well-known analysis of knowledge as justified true belief (for example, Chisholm, 1982; Lehrer & Paxson, 1969) provides useful parallels and points of comparison with the discussion of knowledge-building thus far. Its logical notation goes like this, where *S* stands for subject, and *p* for proposition:

S knows that *p* if and only if:
- *p* is true;
- *S* believes that *p*;
- *S* is justified in believing that *p*.

So beginning with the relatively straightforward *truth* condition: in order for someone to know something, that something must be *true*. If someone claims to know something that turns out to be false, it turns out that that person only *thought* he knew that thing, when, in fact, he did not truly know it at all. This condition may be seen as loosely mapping onto the category of data in the DIK-T hierarchy, insofar as data is in some sense understood as pure unadorned objectivity, untouched by human processing. But what at first seemed comparatively uncontroversial now appears problematic, for how can we make sense of something being true without ourselves believing it to be true? That is, how can truth be sensibly separated from belief? Here we encounter a similar problem as that with *data* earlier, in that it seems nonsensical to try to think about something that exists independently of our thinking about it. The philosopher Donald Davidson, though, offers a way out of this apparent impasse, stating, simply, our 'knowledge [is] objective in the sense that [its] truth is independent of [its] being believed to be true' (Davidson, 2001). This amounts to the recognition that believing a thing to be true does not make it true, however difficult this recognition may sometimes be. And so our access to objectivity lies precisely in our fallibility, in the basic fact that we may be wrong, that our beliefs may be true *or* false. This distinction of truth from belief by means of the possibility of being mistaken carries a very clear implication for programs and procedures of knowledge construction: space must be made for error, in the working toward truth, and hence genuine knowledge. For again, whatever sense of objectivity we have lies in the acknowledgement, tacit or otherwise, of our possibly being wrong, of our beliefs possibly being false, as judged against some standard of truth understood as independent of those beliefs. Therefore any process of knowledge creation, particularly in the context of problem solving, should allow for, and even encourage, the making of mistakes, in and through some checking mechanism or feedback loop, whether of trial and error, or discussion and debate.

The *belief* condition holds: for something to be known, it must be known by *someone*, by some knowing subject. That is, it must be *believed*, held in mind. While it might be true that some state of affairs is the case, if no one believes this state of affairs to be true, then, clearly, this state of affairs cannot be known. Again, this point can potentially be complicated; though it does seem safe to say that, in order for there to be knowledge, there must be true belief. The work of Dretske et al. (1994) provides some assistance in this matter. He employs the term *information* by reminding us, in effect, of its verb form: to *inform*. According to him, connections are set up between discreet sites as an event at one *informs* an event at another; hence the *relational* character of information. Dretske goes on to apply the term to the mind, describing it, in effect, as one such site, related to events at other sites precisely by the process of *information*, of the mind and brain being *informed* by whatever is going on in the world. And this relation of information is represented, in this analysis, by the term *belief*. That is, *belief* is a state of the mind as *informed* by something; believing is the state of being informed. Broadly, then, it is the relational character of belief, of belief being a relation between the mind and some state of affairs, that aligns it with the category of *information* in the DIK-T account. And relatedly, the additional implication of the subject, *S*, in the belief condition over that of the truth condition, parallels the aforementioned progressive involvement of subjectivity in the advancement from data to information to knowledge; that is, of the increasing involvement of the human contribution as we move from data to information to knowledge.

But we need to take this analysis further than that of just true belief. The need for the justification condition arises in order to exclude cases of what is called *epistemic luck* (for a discussion see Pritchard, 2004). Imagine a person, for example, who suffers from chronic paranoia concerning her health; indeed, somatic delusions sometimes arise as symptoms of major depression. So this particular person believes she has, among many other ailments, cancer. Yet it turns out, in this case, that she actually does have cancer. Therefore she holds the true belief that she has cancer. Does this qualify as knowledge however? It seems wrong to assert that, in the throes of her delusions, she *knows* she has cancer, despite her believing something that happens to be true. Another example is the classic case of the stopped clock. Say you want to find out what time it is, so you look up at the clock on the wall, which reads 2:53. But, unbeknownst to you, the clock stopped at just that time the day before. And say it just so happens that that is, in fact, the correct time, at the very moment you look at the clock. Do you, then, really know the right time, even though you believe what the stopped clocked reads, which happens to read the true time, just at that moment? In the first case, the response would seem to be, no, she doesn't really know she has cancer, because she is not truly justified in believing it. Her paranoia, in this instance, happens to be true; hence epistemic *luck*. In the second case, since the reading on the clock is not the result of a correctly working mechanism of time-keeping, the tendency, again, would be to say, no, you don't really know the time: you just happened to look at exactly the right time. You would be justified in believing the time only if the clock mechanism were actually working properly.

There is controversy, however, concerning what precisely constitutes *justification*. What is it to be justified in believing something to be true? Some go so far as to allege that, lacking any strict and clear criteria for justification, we can never be sure when this condition is met; thus knowledge, strictly speaking, is never fully securable. Though this is not the place to fully explore this issue, we do suggest that justification is a discursive matter of answering objections to the claims we make, within some collaborative setting in which participants are free to ask questions and challenge the statements of others. Again, we can see here the increasingly inextricable involvement of the subjective component, in the form of multiple subjects participating in a social process of question and answer, call and response, the back and forth of collaboration. We can also note the possible parallel between justification and the centrality of *process* up to this point. So we might say that true belief qualifies as knowledge if it arises from some valid, working, reliable *process* of knowledge creation.

Early theorizing in naturalistic decision making aligns well with this form of critical thinking. In particular, researchers proposed the design of methods and technologies that continually questioned assumptions of the decision maker; forcing individuals or teams to justify their assertions in the hopes of avoiding errors such as false positives (for example, Cohen et al, 1997). But perhaps the purest example of such a process is the paradigmatic scientific experiment. Scientists follow a general procedure that has proven reliable and repeatable in the past, and arrive, by means of this procedure, at some set of findings. If asked to justify their belief in these findings, they would simply describe the process, the method, by which they arrived at those findings; thus the process of justification is the precisely the process of inquiry itself. And, likewise, if the true belief that it is 2:53 arises from an actually working clock that reliably tells the time, then that true belief would be justified by virtue of its connection with a viably regular process, in this case, the mechanism of a correctly working clock. Interestingly, the concept of *information* comes up as a means of explaining and supporting such knowledge-building processes, in terms of the progressive accumulation and formulation of regular and reliable relations between various variables, event sites, or data points (Drestke 1994). Nevertheless, for our purposes, it is enough to say that the process of knowledge building, beginning with basic data, through information, toward knowledge, finds its analogues in philosophical analyses of knowledge.

This distinction has analogs in more practical contexts as well. And, from these, we can illuminate the connections between our opening discussion of the data-information-knowledge transformation to the philosophical treatment of justified true belief. As an example, computer-supported collaborative learning researchers expand on the notion of question generation and argue for the development of prompts or scaffolds for this process in service of understanding. Importantly, and aligning with our earlier distinctions, this research differentiates between fact-oriented and explanation-oriented questions (Lai, 2008). Furthermore, researchers such as Hakkarainen (1998) have created categorization schemes that may contribute to classifying interactions in such a way that knowledge building can be measured. He posited levels of explanation that increased in complexity, starting first with 'isolated facts,' moving next to facts which were 'partially organized,' then those which were 'well-organized' and concluding at higher levels with 'partial explanations' and finally 'explanations'. Fitting, then, with our approach outlined earlier, we could posit that these first three are in alignment with what we described as the transformation from data to information. The latter two, though, are more in line with the information to knowledge transformation. In particular, Hakkarainen's (1998) approach puts pre-eminence on the act of explaining, which, by its very nature, requires creating something over and above that which is given (see also Biggs & Collis, 1982).

Related notions for how to measure knowledge building comes from research on the scientific inquiry process. Here the idea of '*epistemic practices*' provides a conceptual analog to understanding higher levels of knowledge building. In particular, epistemic practices involve activities where one works to make sense of patterns of data, coordinate theory with evidence, and develop representations of this knowledge (see Sandoval et al., 2000). Across these, we see processes in alignment with transitions from data to information to knowledge. Thus, they not only offer more than organization (what we had associated with information), these classification schemes suggest measuring the *interpretive level* where, through the process of explanation and justification, something new arises from that which has been provided.

Integrating the DIK-T with Macrocognition in Teams

Through these views, we can return to our intermediary term of *information*, making up the middle realm of the knowledge-building process. This stage is critical in terms of justification and explaining in that it consists of shifting relations, forming and informing one another, of sifting through and selecting what elements to take up and put together and which to leave out. More importantly, with this emphasis on process, we can return now to the main issue at hand, that of macrocognition in teams. Specifically, we can see its importance in this context in that value-judgments become an inextricable part of the formation and creation of knowledge. Thus, fitting with ideas of information interrogation as put forth in other discussions of macrocognition in teams (for example, Fiore et al., 2008; Letsky et al., 2007), where a critical portion of the problem-solving process involves evaluation and negation of solution alternatives, process, and not outcome, is privileged. In this problem-solving context, facts as discreet outputs, as isolated elements, give way to an appreciation of how they play into the construction and use of elaborated knowledge. Indeed, the admixture of fact and value may be seen as the very signature of knowledge, and reveal its proximity to the hallowed notion of *wisdom*, which is often placed as the culminating term at the top of the hierarchy (see Ackoff, 1989). According to this view, then, collaborative technology should enable the use of information rather than merely support learning information. Thus, 'obtaining, recording, and storing information would become subsidiary functions, designed to serve purposes of knowledge creation' (Scardamalia & Bereiter, 2006, p. 105). In short, information is by no means an end in itself; rather, it is itself a means. Information, in its resonant relationality, in contrast to the discreet isolated quality of data, enables the creation of what becomes a web of knowledge. *Use*, then, acts as the force motivating the process.

In sum, from this review we can begin to see some of the connections between knowledge building as it occurs in collaborating teams. If information, that intermediate entity consisting of data in relation and context, is to be dynamically implemented, it must not flow solely through a single executive faculty, but must be distributed throughout all the individual elements involved. Participants in the knowledge-building process ought all have common access to information in the creation of a space of emergent discourse, rather than having it all pass through the single conduit such as a team leader. Rather, a community's total 'State of Knowledge' is an emergent property of that community, not belonging to any one individual member; moreover, this state of knowledge is not the simple sum or additive aggregate of all the individual members' knowledge. This implies that in the aggregating process, an emergent phenomena takes place, manifesting something that did not exist in the component parts. This quality of emergence is clearly applicable to the building of knowledge in that 'significant advances in knowledge by a research laboratory are obviously emergents; the knowledge didn't pre-exist in anyone's mind' (Scardamalia & Bereiter, 2006, p. 105). We turn now to a more fine-grained treatment of these ideas; that is, a description of how we can more specifically describe the data-information-knowledge transformation as it may occur in macrocognition in teams.

Defining and Measuring the Data, Information, Knowledge Transformation Process in Teams

We started this chapter by providing a definition of macrocognition in teams and we now revisit that in light of the previous discussion of how to conceptualize the knowledge-building process. As stated, the emphasis is on knowledge-based performance where teams are faced with unfamiliar situations which lack pre-existing rules. Here collaborative problem solving requires the generation or adaptation of rules to novel situations and involves individuals and teams generating new knowledge, moving from data to information to knowledge, when faced with unusual problems. In this final section we move our more conceptual discussion of knowledge building to a more practical one. Our goal is to provide candidate definitions of the data-information-knowledge transformation so as to suggest targets for measurement. Toward this end, we focus on both the content and the process of knowledge building. As we have argued, knowledge building is dynamic and is intrinsically wedded to context. As such, to adequately measure the data-information-knowledge transformation, any methodology must take into account, not only what is said (that is, the content), but also how it is said (that is, the process). We first describe our definitions of data, information, and knowledge as we see them in our conceptualization of macrocognition in teams. This represents the type of targets necessary for measuring the progression of knowledge building, that is, for tracking how teams move from data to information to knowledge. We then describe how it may be possible to use measures of communication to assess 'how' things get said during collaboration so as to assess the evaluative or justification component of this process; that is, the factor we suggested drives the transition from information to knowledge.

We turn now to a description of how we have distilled a set of definitions for data, information, and knowledge (refer to Table 12.1) so as to provide a richer conceptualization of our theorizing (cf. Fiore et al., 2008). For illustrative purposes, we use examples from recent research in macrocognition in teams, specifically, the Non-combatant Evacuation Operation (NEO) scenario (Letsky et al., 2008). The

Table 12.1   Definitions and examples of data, information, and knowledge

Concept		Explanation
Data	Definition	Data are disparate statements or facts presented or represented separately and without context.
	Example	'The CH-53 Marine Corps helicopter can hoist 250 feet with a 600 pound lift capacity.'
	Explanation	Here the content is devoid of context and not organized in any way; as such, it is considered only data.
Information	Definition	Information is organized or structured data (i.e., organized or structured statements or facts) that have been related to the problem-solving context.
	Example	'The CH-53 Marine Corps helicopter can carry supplies to the Red Cross workers who have been taken hostage,' or, 'Our three air vehicles are the CH-53, the F-16, and the E2-C.'
	Explanation	The first example represents a transformation in that it involved connecting the piece of data to the problem. The second example represents a transformation in that it involved organizing the data via categorization such that it can serve the problem solving; resources were organized into categories of resources that serve the problem solving.
Knowledge	Definition	Knowledge is the integration of content from two or more categories of information into something which did not explicitly exist before and which has been made actionable by being related to the problem-solving context.
	Example	'The CH-53 Marine Corps helicopter cannot be used to carry supplies because it is foggy over the southwest corridor of Nandor.'
	Explanation	This represents a transformation because vehicle information (a category), was integrated with weather information (another category) in such a way that it serves the problem solving; that is, it was made actionable by explaining and justifying when it could get (or not get) supplies to the hostages.

scenario requires a team to develop plans for evacuating humanitarian workers stranded on an island nation overrun by rebel insurgents. The task requires problem solving amongst a team of specialists with diverse organizational and agency backgrounds along with varied levels of expertise. In this task, an ad hoc team must gather and synthesize information to develop their solution, an evacuation plan for the hostages (Rosen et al., 2008); that is, they must build knowledge from an integration of the task content, the problem context, and their own experience. Our goal here is to illustrate what are the core elements of the DIK-T in such situations. These, then, may be a means through which we could develop measures of knowledge building as it occurs in macrocognition in teams.

First, within the context of a NEO scenario, we can state that data are what is provided by the task materials (for example, 'the CH-53 helicopter has a range of 500 miles'). Data are transformed to information when structured and contextually grounded—that is, it becomes information when it is organized to provide meaning within a particular context (for example, 'we have two air and three land vehicles as resources for use in Nandor'). Information becomes knowledge when it is integrated with other task-related information and/or experience, justified to the team, and made useful for action via synthesization with the problem-solving context. For example, participants may recognize the value of something by noting a piece of information's relevance to another piece of information, or a particular facet of the problem (for example, 'the timing of high tide is important because of we can transfer supplies using only rafts'). For the purposes of supporting measurement of knowledge building, the specific definitions for each stage of the DIK-T process are as follows: (1) *data* are disparate and devoid of context; (2) *information* is organized or structured data within the problem-solving context, but it does not involve integration; and, (3) knowledge involves not only context and organization, but integration—the piecing together of information to produce something actionable and justifiable, that did not explicitly exist before the collaboration began.

Further, given the criticality of 'knowledge' to macrocognition in teams, we more specifically state that, in terms of necessary and sufficient features, to be considered 'knowledge,' it must meet the following criteria: (a) be an integration of pieces of information; (b) have an explication of context for the constructed knowledge that is justified to the team; and, (c) consist of something that did not explicitly exist before in the problem-solving situation. Thus, knowledge has to be *created* by the integration of information generated from the materials and/or from the participants' own experiences, evaluated by and justified to, the team, and actionable for the context of the problem-solving task.

Measuring Knowledge Building in Teams

With this as our foundation, we state that knowledge building is generally defined as the transformation of data, to information, to knowledge. More formally, as illustrated in Figure 12.1 (adapted from Clark, 2004), context represents a critical

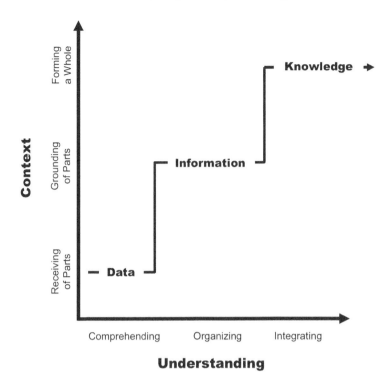

Figure 12.1 Knowledge-building process in macrocognition

component of this process and knowledge building requires an increasingly tight connection to the problem-solving context. Furthermore, as noted earlier, process is an intrinsic part of our conceptualization. As such, we specifically define process, in our conceptualization of knowledge building, as a series of related actions engaged by individuals or the team, and measured over time, that bring about some outcome, result, or product. In this sense, then, knowledge building is a *process* that leads to the *product* of *knowledge*; it is a process where isolated bits of data are organized and then integrated into a tightly coupled network of actionable knowledge for use within a particular problem-solving context. More formally, we define knowledge building as the synthesizing (for example, visualizing numbers via graphs, aggregating, or organizing data using tables, integrating information to create a broader understanding), at the individual and/ or team level, of the relationships among problem relevant content in such a way that it becomes actionable within the problem-solving context. Thus, for data to become information, it must be organized and related to the problem solving. For information to become knowledge, it must be integrated in such a way that it is actionable. It is the expanding volume of problem-relevant knowledge, when taken as a whole, which enables action toward solving a problem.

But, as noted, in addition to an emphasis on the nature of the content to measure, one must also consider the process factors that may be measurable; that is the process of evaluating or justifying the knowledge. So we next include a description of how measurement methods from group communication theory can be considered in light of our definitions and distinctions of knowledge building. In particular, because a core component of what we have put forth is 'process,' and, more specifically, the critical evaluation and justification of information, we suggest that techniques for analyzing group process can help us to assess knowledge building in teams.

First, consider the *Functional Approach to Groups* (Hirokawa, 1980), which holds four core assumptions about groups: 1) groups are goal oriented; 2) performance and behavior within a group varies and can be evaluated; 3) the interaction processes of the group vary and can be evaluated; and 4) various internal and external factors influence the groups performance outcomes through interaction processes (Hollingshead et al., 2005). From this theory comes the notion that decision-making groups must accomplish some set of subtasks, or general functions, in order to successfully arrive at an acceptable decision. But it is the amount and quality of group interactions around these functions which determines the quality of decision-making outcomes. To illustrate the relevance of this approach to the way in which we have conceptualized knowledge building, consider the function where groups work to establish the criteria used for evaluation. Here the group works to define criteria for acceptable solutions so as to set the standards by which decision alternatives will be judged and an acceptable solution selected. Given the primacy of evaluation and explanation processes in our characterization of knowledge building, particularly in the context of justification, we can see how this function is critical. As a target for measurement, then, the amount of communications centered on the generation of such criteria become targets for measurement. For example, to the degree to which teams do not first identify evaluation criteria might be indicative of teams which fail to effectively move from information to knowledge. Alternatively, it might illuminate why it is that teams construct faulty knowledge. Related to this are functions where groups evaluate the hypothetical positive or negative outcomes of their generated knowledge (for example, particular decisions or solutions to a problem). Here groups would explore the relative merits (for example, likely positive outcomes), or the potential problems, associated with decision alternatives. From a measurement standpoint, communications focusing on questioning assumptions behind a particular element of knowledge, or justifying the merit of some knowledge, become targets (Orlitzky & Hirokawa, 2001).

Additional methods for measurement can be gleaned from *Decision Emergence Theory* (Ellis & Fisher, 1994). As with the functional perspective, decision emergence focuses on the quality and nature of the interactions. But the decision emergence perspective tracks the progression of interaction from the introduction of preliminary ideas to the solidification of a group consensus around a solution. Important to our discussion of knowledge-building processes, the decision proposal is the focus of analysis. But, in line with our notions of evaluation and justification as being core processes in moving toward the building of knowledge, each group member's communication is analyzed along the following dimensions: expressing an opinion about the decision proposal (favorable, unfavorable, ambiguous), modifying or clarifying the decision proposal, providing evidence to support an opinion, or agreeing or disagreeing with another group member's stated opinion. Importantly, the measurement focus here is on the substantive content of interaction, that is, the decision proposal. Therefore, it may be informative to take this analytical lens to illuminate how built knowledge may be measured via consideration of members' knowledge proposals.

Finally, it is also possible to consider more complex notions of group interaction such as the multiple sequence models of group decision making proposed by Poole (1981). This theoretical approach argues that groups interweave multiple threads of interaction patterns over time. They differentiate between threads focused on task processes, those on relational processes, and finally topical threads which emphasize substantive issues. These patterns of interaction define a group's *trajectory*, and apropos the current discussion, trajectories bring with them *breakpoints*. Trajectories can be followed, then, to consider a team's attempt to justify some generated knowledge. Similarly, breakpoints may be indicative in failures in justification or when evaluation of generated knowledge leads to dismissal on the part of the team. For example, breakpoints can be interjected intentionally, referred to as called routing statements (Fisher & Stuttman, 1987), and these can be analyzed to determine their relation to evaluation or justification processes (see Poole & Roth, 1989). Importantly, for the purposes of measuring the complexity inherent in macrocognitive contexts, the multiple paths approach allows for nonsequential and recursive processes to be tracked; precisely the form of give and take that would be expected as teams work to interrogate their generated knowledge.

Conclusions

In this chapter we have described knowledge building as a core component of the larger process of complex one-of-a-kind problem solving. We have argued that macrocognition in teams is the process of transforming internalized team knowledge into externalized team knowledge through individual and team knowledge-building processes (see Fiore et al., 2008). Internalized knowledge refers to the collective knowledge held in the individual minds of team members. Individual knowledge building includes actions taken by individuals in order to process and organize data in the context of the problem, and integrate information to create their own knowledge. These processes can take place inside the head (for example, reading, mentally rotating objects) or may involve overt actions (for example, accessing a screenshot, creating graphs, making tables). Externalized team knowledge refers to the integrated information that has been made actionable *and* explicitly agreed upon and justified, or not openly challenged or disagreed upon, by a single or multiple team members. It is this latter process on which we have primarily focused, arguing that knowledge is an emergent property resulting from an active process of evaluation and justification on the part of team members.

In sum, we set out to outline how teams engage in processes that help them construct knowledge so as to begin the process of measuring the nature of the communications driving complex collaborative activity. Although research in team cognition has discussed elements of the theoretical approach we have just articulated (see Salas & Fiore, 2004 for a discussion of these theories), these prior approaches have not focused on knowledge building while investigating macrocognition as an emergent cognitive property of the team's interaction. By presenting, first, conceptual definitions, then more specific definitions, we are reaching for greater descriptive power in service of eventually diagnosing the causal factors associated with collaborative problem solving. For this, we have taken an interdisciplinary approach and have drawn from a variety of literatures, ranging from philosophy to information systems management as well as from group communication theory. Because of the complexity inherent in collaborative environments, it is warranted to draw from these literatures that have set out to understand *interaction* in descriptive and prescriptive ways. We hope that this chapter has taken us at least one step in that direction.

Acknowledgements

The views, opinions, and findings contained in this article are the authors and should not be construed as official or as reflecting the views of the University of Central Florida or the Department of Defense. We thank Anthony Crisafi for his assistance on portions of the literature review for this chapter. Writing this chapter was supported by a Multidisciplinary University Research Initiative Grant from the Office of Naval Research to the University of Central Florida.

References

Ackoff, R. L. (1989). From data to wisdom. *Journal of Applied Systems Analysis, 16*(1), 3-9.

Awad, E. M., & Ghaziri, H. M. (2004). *Knowledge management.* Upper Saddle River, NJ: Pearson Education International.

Bellinger, G., Castro, D., & Mills, A. (2004). Data, information, knowledge, and wisdom. http://www.systems-thinking.org/dikw/dikw.htm, Accessed July 1, 2008.

Biggs, J. B., & Collis, K. F. (1982). *Evaluating the quality of learning—the SOLO taxonomy.* New York, NY: Academic Press.

Boddy, D., Boonstra, A. & Kennedy, G. (2005). *Managing information systems: An organizational perspective.* Upper Saddle River, NJ: Prentice Hall.

Cacciabue, P. C. & Hollnagel, E. (1995). Simulation of cognition: Applications. In J. M. Hoc, P. C., Cacciabue, & E. Hollnagel (Eds.), *Expertise and technology: Issues in cognition and human-computer cooperation* (pp. 55-74). Hillsdale, NJ: LEA.

Cannon-Bowers, J. A., Salas, E., & Converse, S. A. (1993). Shared mental models in expert team decision-making. In N. J. Castellan, Jr. (Ed.), *Individual and group decision making* (pp. 221-246). Hillsdale, NJ: Lawrence Erlbaum.

Chaffey, D. & Wood, S. (2005). *Business information management: Improving performance using information systems.* Upper Saddle River, NJ: Prentice Hall.

Chi, M. T. H., Feltovich, P. J., & Glaser, R. (1981). Categorization and representation of physics problems by experts and novices. *Cognitive science, 5,* 121-125.

Chisholm, R. (1982). *Knowledge as Justified True Belief. The foundations of knowing.* Minneapolis, MN: University of Minnesota Press.

Clark, D. (2004). Understanding and performance. http://www.nwlink.com/~donclark/performance/understanding.html, Retrieved April 10, 2008.

Cleveland, H. (1982). Information as a Resource. *The Futurist,* Dec, 34-39.

Cohen M. S., Freeman, J. T., & Thompson, B. T. (1997). Training the naturalistic decision maker. In C. Zsambok & G. A. Klein (Eds.), *Naturalistic decision making* (pp. 306-326). Mahwah, NJ: Erlbaum.

Davidson, D. (2001). *Subjective, intersubjective, objective.* Oxford, UK: Oxford University Press.

Dretske, F., Clark, A., Wilks, Y., Dennett, D., Chrisley, R. & Cohen, L. J. (1994). The explanatory role of information [and discussion]. *Philosophical Transactions of the Royal Society, 349*(1), 59-70.

Ellis, D. G., & Fisher, B. A. (1994). *Small group decision making: Communication and the group process* (4th edn). New York, NY: McGraw Hill.

Fiore, S. M., Rosen, M., Salas, E., Burke, S., & Jentsch, F. (2008). Processes in complex team problem solving: Parsing and defining the theoretical problem space. In M. Letsky, N. Warner, S. M. Fiore, & C. Smith (Eds.). *Macrocognition in teams: Theories and methodologies.* London, UK: Ashgate Publishers.

Fisher, B. A., & Stuttman, R. K. (1987). An assessment of group trajectories: Analyzing developmental breakpoints. *Communication Quarterly, 35*(2), 105-124.

Gibson, J. J. (1977). The theory of affordances. In R. Shaw & J. Bransford (Eds.), *Perceiving, acting, and knowing: Toward an ecological psychology* (pp. 67-82). Hillsdale, NJ: Lawrence Erlbaum.

Glaser, R. (1989). Expertise and learning: How do we think about instructional processes now that we have discovered knowledge structures? In D. Klahr & K. Kotovsky (Eds.), *Complex information processing: The impact of Herbert A. Simon* (pp. 269-282). Hillsdale, NJ: LEA.

Gowler, H. N. (1953). Plato, Phaedo. In H.N. Gowler (Ed. & Trans.), *Plato 1.* Boston, MA: Harvard University Press.

Hakkarainen, K. (1998) *Epistemology of inquiry and computer-supported collaborative learning.* Unpublished Ph.D. thesis. University of Toronto.

Hey, J. (2004). *The data, information, knowledge, wisdom chain: The metaphorical link.* http://best.me.berkeley.edu/~jhey03/files/reports/IS290_Finalpaper_HEY. pdf, Accessed April 23, 2008.

Hinsz, V. B., Tindale, R. S., & Vollrath, D. A. (1997). The emerging conceptualization of groups as information processors. *Psychological Bulletin, 121*(1), 43-64.

Hirokawa, R. Y. (1980). A comparative analysis of communication patterns within effective and ineffective decision making groups. *Communication Monographs, 47*(4), 312-321.

Hollingshead, A. B., McGrath, J. E., & O'Connor, K. M. (1993). Group task performance and communication technology: a longitudinal study of computer-mediated versus fact-to-face work groups. *Small Group Research, 24*(3), 307-33.

Hutchins, E. (1991). The social organization of distributed cognition. In L. B. Resnick, & J. M. Levine, (Eds.), *Perspectives on socially shared cognition* (pp. 283-307). Washington, DC: American Psychological Association.

Hutton, R. J. E., Miller, T. B. & Thorsden, M. L. (2003). In E. Hollnagel (Ed.), *Handbook of cognitive task design* (pp. 383-416). Mahwah, NJ: LEA.

Jessup, L. M., & Valacich, J. S. (2003). *Information systems today.* Upper Saddle River, NJ: Prentice Hall.

Klein, G., Ross, K. G., Moon, B. M., Klein, D. E., Hoffman, R. R., & Hollnagel, E. (2003). Macrocognition. *IEEE Intelligent Systems, 3*(May/June), 81-85.

Lai, M. (2008). The Role of Argumentation in Online Knowledge Building Activities. In *Proceedings of International Conference on Computers in Education* (ICCE 2008), 317-324.

Larson, J. R. & Christensen, C. (1993). Groups as problem-solving units: Toward a new meaning of social cognition. *British Journal of Social Psychology, 32,* 5-30.

Lehrer, K. & Paxson, T. (1969). Knowledge: Undefeated Justified True Belief. *The Journal of Philosophy, 66*(8), 225-237.

Letsky, M., Warner, N., Fiore, S. M., Rosen, M. A., & Salas, E. (2007). Macrocognition in Complex Team Problem Solving. In *Proceedings of the 12th International Command and Control Research and Technology Symposium* (12th ICCRTS), Washington, DC: United States Department of Defense Command and Control Research Program.

Letsky, M. Warner, N., Fiore, S.M., & Smith, C. (Eds.). (2008). *Macrocognition in teams: Theories and methodologies.* London: Ashgate Publishers.

Levine, J. L., Resnick, L. B., & Higgins, E. T. (1993). Social foundations of cognition. *Annual Review of Psychology, 44*(1), 585-612.

Orlitzky, M., & Hirokawa, R. Y. (2001). To err is human, to correct for it divine: A meta-analysis of research testing the functional theory of group decision-making effectiveness. *Small Group Research, 32*(3), 313-341.

Pearlson, K. E., & Saunders, C. S. (2004). *Managing and using information systems: A strategic approach*. New York, NY: Wiley.

Poole, M. S. (1981). Decision development in small groups I: A comparison of two models. *Communication Monographs, 48*(1), 1-24.

Poole, M. S., & Roth, J. (1989). Decision development in small groups IV: A typology of group decision paths. *Human Communication Research, 15*(3), 323-356.

Pritchard, D. (2004). Epistemic Luck. *Journal of Philosophical Research, 29*(1), 193-222.

Rasmussen, J. (1983). Skills, rules, and knowledge; Signals, signs, and symbols, and other distinctions in human performance models. *IEEE Transactions of Systems, Man, Cybernetics, 13*(3), 257-266.

Rosen, M. A., Fiore, S. M., Salas, E., Letsky, M., & Warner, N. (2008). Tightly coupling cognition: Understanding how communication and awareness drive coordination in teams. *International Journal of Command and Control, 2*(1), 1-30.

Rowley, J. (2007). The wisdom hierarchy: Representations of the DIKW hierarchy. *Journal of Information Science, 33*(2), 163-180.

Salas, E., & Fiore, S. M. (Eds). (2004). *Team Cognition: Understanding the factors that drive process and performance*. Washington, DC: American Psychological Association.

Sandoval, W. A., Bell, P., Coleman, E., Enyedy, N., & Suthers, D. (2000). Designing knowledge representations for learning epistemic practices of science. *Position paper for an interactive symposium of the same name*, presented at the annual meeting of the American Educational Research Association, New Orleans, April 25, 2000.

Scardamalia, M. and Bereiter, C. (2006). Knowledge building: Theory, pedagogy, and technology. In K. Sawyer (Ed.), *Cambridge handbook of the learning sciences* (pp. 97-118), New York, NY: Cambridge University Press.

Schraagen, J. M., Klein, G., & Hoffman, R. R. (2008a). The macrocognition framework of naturalistic decision making. In J. M. Schraagen, L. G. Militello, T. Ormerod, T., & R. Lipshitz (Eds.). *Naturalistic decision making and macrocognition* (pp. 3-25). Aldershot, UK: Ashgate Publishing.

Schraagen, J. M., Militello, L. G., Ormerod, T., & Lipshitz, R. (Eds.). (2008b). *Naturalistic decision making and macrocognition*. Aldershot, UK: Ashgate Publishing.

Tsoukas, H., & M. J. Hatch (2001). Complex thinking, complex practice: The case for a narrative approach to organizational complexity. *Human Relations, 54*(8), 979-1013.

Turban, E., Rainer, R. K., & Potter, R. E. (2005). *Introduction to information technoloy*. New York, NY: Wiley.

Warner, N., Letsky, M., & Cowen, M. (2005). Cognitive model of team collaboration: Macro-cognitive focus. In *Proceedings of the 49th Annual*

   Meeting of the Human Factors and Ergonomic Society. Santa Monica, CA:
   Human Factors and Ergonomics Society.
Zelany, M. (1987). Management support systems: Towards integrated knowledge
   management. *Human Systems Management, 7*(1), 59-70.

PART III
Scenario-based Evaluation Approaches

Chapter 13

Forging New Evaluation Paradigms: Beyond Statistical Generalization

Emilie M. Roth and Robert G. Eggleston

Introduction

A primary motivation for the study of macrocognition is to understand and aid performance in complex naturalistic settings that may involve multiple agents (human and machine), long time scales, uncertainty, and complex dynamics that can pose challenges to macrocognitive functions such as assessing situations, (re)planning, committing to decisions, and executing plans (see Preface). The natural unit of analysis, from this perspective, is the joint cognitive system that encompasses the multiple human and automation/technologies across which macrocognitive functions are distributed to achieve work domain goals (Hollnagel & Woods, 2005; Potter & Rousseau, Chapter 15).

One of the primary challenges is to be able to evaluate whether a new innovation (for example, a software system or new training technology) results in substantial improvement in the joint cognitive system. Effective evaluation is important to justify the financial investment required to implement a new technology. It is also important in regulatory environments where there is a need to establish that a new technology is at least as good (as reliable, as safe) as the legacy technology it is intended to replace (O'Hara, 1999).

Too often emerging technologies are implemented based on optimistic assumptions that they will have well-specified, positive, impacts on overall performance that do not materialize in practice because of a failure to appreciate the full complexities in the work domain and anticipate side effects (Sarter & Woods, 1995; Roth et al., 1997; Vicente et al., 2001; Patterson et al., 2002; Woods & Dekker, 2002; Roth & Patterson, 2005). New support technologies should be regarded as *hypotheses* about what constitutes effective support, and how technological change is expected to shape cognition and collaboration (Woods, 1998; Dekker & Woods, 1999; Potter et al., 2000). Evaluation methods are needed to determine whether the hypothesized benefits of introducing the new technology are realized, as well as to uncover unsupported aspects of performance or unanticipated side-effects of introducing the new technology that need to be addressed, so as to propel further innovation.

Over the last several years we have been engaged in a long-term program to develop work-centered support systems for a military airlift organization. As part of this effort we have had the opportunity to conduct five evaluations of prototype work-centered support systems (Eggleston et al., 2003; Scott et al., 2005, 2009; Roth et al., 2006, 2007, 2009). Our experiences with these evaluations as well as other evaluation efforts (for example, Woods & Sarter, 1993; Lang et al., 2002; Pfautz & Roth, 2006) have convinced us that effective evaluations require a revised theoretical framework that go beyond the standard approaches to evaluation (for example, Charlton & O'Brien 2002; Dumas, 2003). We have coined the term 'work-centered evaluation' for this proposed theoretical framework (Eggleston et al., 2003).

In this chapter we describe limitations of the two most prominent current paradigms for system evaluation, usability testing, and controlled experiments, that make them poorly suited for informing and propelling the design of joint cognitive systems intended to enhance macrocognitive performance. We then describe the elements and theoretical underpinnings for an alternative, work-centered evaluation paradigm, that more closely meets the goals and pragmatic constraints of conducting evaluations of complex joint cognitive systems in naturalistic environments.

Limitations of Formative vs. Summative Evaluation Dichotomy

Traditionally a distinction has been made between two classes of evaluation: *formative evaluation* and *summative evaluation* (Nielsen, 1993). *Formative evaluations* are intended to collect data for the purpose of improving a system design. The main goal of a formative evaluation is to provide feedback to developers with respect to what aspects of the system design work well and which can be improved as part of an iterative design process. There are a variety of approaches to formative evaluation including discount methods such as heuristic evaluations conducted by usability experts and more formal usability tests that collect performance and feedback data from representative users. In contrast, *summative evaluations* are intended to provide an overall assessment of a system design. Summative evaluations are typically conducted after the completion of a system to evaluate the efficacy of the final system design or to compare competing design alternatives. Charlton & O'Brien (2002) refer to these as 'operational test and evaluation' that are conducted to determine if the new system can meet the requirements and goals identified during the system definition phase. Summative evaluations often involve comparing the new system to some comparison system (for example, the current system that serves as a 'baseline') using an experiment design methodology (O'Hara, 1999; Snow et al., 1999).

While the dichotomy between formative and summative evaluation is conceptually clear, in practice it is more productive to develop evaluations that combine formative and summative objectives. This is particularly so for evaluations of complex joint cognitive systems in real-world settings, where evaluation opportunities are typically scarce and resource intensive. Each evaluation 'look,' whether early or late in the design, should be capitalized on as an opportunity to assess whether the aid is meeting important design objectives (summative evaluation) as well as to identify problems to be corrected and opportunities for improvement (formative evaluation).

Limitations of the Usability Test Paradigm

One of the most prominent formative evaluation approaches is the usability test paradigm (Nielsen, 1993; Dumas, 2003). Representative users are asked to use the aiding technology to perform specified tasks, typically in a controlled usability laboratory environment. A variety of measures are collected and used to identify problems and opportunities for improvement. Nielsen (1993) defines elements of usability to include:

- Learnability: Is the system easy to learn?
- Efficiency: Once the user has learned the system can they perform assigned tasks efficiently?
- Memorability: Is it easy for the user to remember how to perform system functions?
- Errors: Do users make errors in trying to use system functions?
- Satisfaction: Do users like the system?

While the usability test paradigm is a powerful tool for conducting formative evaluations, it has a number of limitations that make it a poor match for evaluating complex joint cognitive systems. First, as the list of usability test objectives provided above suggests, it is vulnerable to focusing too narrowly on evaluating the *usability* of a new technological aid—how easy is it to learn and use the features of the aid. In evaluating a joint cognitive system, our concern is not only how easy a new aid is to learn and use but also how *effective* the joint system is in achieving work-domain goals. As a consequence a 'work-centered' evaluation needs to broaden the scope of the evaluation beyond what is traditionally addressed by usability studies.

A complete evaluation also needs to consider the *usefulness* of the new technology. To assess usefulness it is necessary to ground the evaluation in an understanding of the range of cognitive activities, work contexts, and sources of complexities that are likely to arise in the domain and to insure that these aspects of the work context are adequately represented in the evaluation. Traditionally this has not been a point of emphasis in the usability paradigm.

In addition a complete evaluation needs to also consider *impact*. Impact addresses the question of whether the new technology substantially contributes to achieving the mission goals of the larger organization within which this work is embedded. For example, a new aid may enable a worker to complete a task faster, but this might not translate into substantive improvements on dimensions that the organization values (for example, improved quality of product, improved customer satisfaction, reduced cost, improved safety). A question often asked at the organizational level is whether the benefits to be accrued from introducing this technology are sufficient to warrant the required investment.

Limitations of the Experimental Paradigm

Traditional approaches to conducting summative evaluations often adopt a classic controlled experiment paradigm where two or more conditions are compared (Charlton & O'Brien, 2002; O'Hara, 1999; Snow et al., 1999). The test conditions might be a system that is currently in place versus the new system being developed or two or more competing systems. Performance is compared and standard inferential statistics are employed to assess whether there are statistically significant differences in performance between the test conditions.

While we believe that it is important to incorporate summative evaluation elements into work-centered evaluations, there are a number of limitations of the classic experimental paradigm that make it an inappropriate model for work-centered evaluations that are intended to inform design.

Limited diagnostic power   The logic of the experimental paradigm is best suited for addressing narrow questions where there is only a single, or at least a small number, of (independent variable) differences across conditions (Cook & Campbell, 1979; Monk, 1998). If a statistically significant difference is found between the test conditions, the difference can be causally attributed to the independent variable(s) that differed across conditions. However, when comparing two or more systems of interest (for example, a new system and an existing system) it is hardly ever the case that they differ on only a single variable. In almost all cases systems will differ on a large number of dimensions ranging from differences in the interface 'look and feel,' to differences in information content, to differences in aspects of work supported. If the new system is shown to result in statistically significantly improved performance when compared to an existing baseline system, the specific aspects of the new system that led to the improvement in performance cannot reliably be identified.

Thus, when comparing multiple complex systems the experimental paradigm is not diagnostic in pointing to the causes of differences in performance observed. As a consequence, it is not possible to easily draw generalizations with applicability beyond the specific systems being examined (for example, to the next systems development build cycle or to similar systems operating in other contexts). This limits the ability of the study to contribute to a broader scientific base. Similarly, if performance with the new system is found to be poorer than the existing system, the reason for the poorer performance cannot be identified. The results of the study will not be sufficiently diagnostic to drive future design improvements.

An important criterion for a work-centered evaluation is that it not only answers the question, 'Does the new design lead to better performance than the old design?' but also '*Why* does it lead to better performance (or not)?' This can be achieved by incorporating additional measures that are specifically intended to be diagnostic with respect to why the observed differences in performance were obtained. Diagnostic measures can include process trace measures (for example, eye movement measures, analysis of team communication, situation awareness measures), expert observer assessments, or participant feedback via post-test verbal debriefs and questionnaires (Woods & Sarter, 1993; Endsley, 1995; Lang et al., 2002).

Emphasis on internal validity over external validity Another limitation of the experimental paradigm relates to the inherent tradeoff between factors that contribute to *internal validity* and *external validity*. Campbell & Stanley (1963) were the first to define these two elements of validity. *Internal validity* refers to 'the approximate validity with which we infer that a relationship between two variables is causal or that the absence of a relationship implies the absence of cause' and *external validity* refers to 'the approximate validity with which we can infer that the presumed causal relationship can be generalized to and across alternate measures of the cause and effect and across different types of persons, settings, and times' (Cook & Campbell, 1979, p. 37).

Internal validity depends on minimizing extraneous variability within and between conditions. This is needed to support drawing 'cause and effect' conclusions, as well as to reduce variability in performance that can make it more difficult to draw statistical conclusions from the results. The practical consequence of this concern is to limit the range of different types of individuals (for example, gender, age, experience levels) and the range of different types of test situations (for example, routine cases, non-routine cases) included in a study.

In contrast, external validity depends on expanding the range of different types of cases included in the study. The more different types of test participants and test situations included in the study, the higher the confidence can be for the generality of the obtained results. If the results hold across the different types of individuals or situations then you have more confidence that you can generalize across them. If the results appear to hold for some cases but not others, that is valuable information about the boundary conditions to which the results apply.

The perspective on tradeoff between internal and external validity is very different when conducting a joint cognitive system evaluation than when conducting an experiment in support of basic research. From the point of view of evaluating a joint cognitive system for potential implementation in the field, the importance of being able to assess the generality of results across a representative range of individuals and situations to which the system will be applied outweighs concerns to minimize variability for the purposes of statistical conclusion validity. We compensate for this loss of statistical conclusion validity through the use of triangulation which is discussed later.

When designing a work-centered evaluation we deliberately expand heterogeneity with respect to test participants and test situations so as to increase *external/ecological* validity. The objective is to increase confidence in the generality of the findings across different types of individuals and situations as well as to identify edge cases and boundary conditions beyond which effectiveness of the joint cognitive system is likely to degrade (c.f., Potter & Rousseau, Chapter 15). Expanding the range of test participants and test cases is done explicitly, recognizing that by doing so we are increasing within condition variability, thus inflating the error terms in statistical analyses and reducing statistical power.

Statistical power requires large sample sizes Another limitation of the experimental paradigm as a model for joint cognitive system evaluation is that it is often not possible to meet the statistical power requirements. In a typical psychology laboratory experiment it is relatively easy to run enough test participants and trials to achieve sufficient statistical power to obtain statistical significance, given a reasonable size treatment effect. The pool of candidate test participants is generally large, making it relatively easy to recruit a sufficient number of test participants (typically ten or more participants per test condition) for the study. Trial lengths are typically short, in the order of seconds to minutes, making it relatively efficient to run a large number of test trials in a reasonable period of time (for example, a one to two hour test session), that is needed to achieve reasonable statistical power in the assessment. Further, the responses obtained from test participants as the dependent variable tend to be highly constrained (for example, force choice response times), limiting the inherent variability in performance across test participants and test materials.

The situation is often very different in the case of evaluating a new joint cognitive system. In many cases the total population of potential users is relatively small, limiting the number of participants that can be included in the study. For example, a power plant may only employ eight to ten crews of control room operators. This puts a severe limit on the total number of crews that can be run in a study to evaluate a new control room design. Similarly, realistic test scenarios can take a substantial amount of time to run. For example, realistic scenarios for normal and emergency power plant operations can take between two and six hours each to run. This imposes a pragmatic limit on the total number of scenarios that can be run in a study. As an example, in a recent evaluation study of a power plant control room, five crews were tested on each of ten test scenarios (Lang et al., 2002). Testing required a full week of testing per crew. Similar examples can be cited for the development of military weapon systems and other areas of large scale systems development (for example, Jauer et al., 1987). This is an expensive proposition that is warranted for final summative evaluation of a new system for a high-risk domain, but is not a practical model for system evaluation as part of an iterative design process.

Furthermore, the performance of interest in evaluating a new system tends to be relatively complex because the focus of interest is the ability of the system to support realistically complex units of work in response to representatively complex situations. For example, in the power plant case, test participants might be presented with an abnormal plant condition and the measure of interest is their ability to recognize the plant condition and take appropriate action. As a consequence, the degrees of freedom available in responding is large, leading to wide variability in response—both with respect to the elements of the response (what information is looked at, what hypotheses are considered, what action is taken and in what order) and the time to complete the response. All these factors, the limited number of test participants, the complexity of the test 'trials' and the complexity of the 'target responses' that are the dependent measures of interest, place severe constraints on the ability to achieve sufficient statistical power to detect a treatment effect statistically, placing in question the appropriateness of adopting a experimental paradigm that relies on statistical generalization.

Beyond Statistical Generalization

One way to overcome the limitations associated with the experimental paradigm is to look for alternative strategies for gaining confidence in the evaluation results that do not depend as heavily on the logic of experimentation and the requirements for statistical power to draw generalizations.

Importance of Analytic Generalization

Important generalizations are often not dependent on the logic of *statistical generalization* but rather the logic *of analytic generalization* that relies on previously developed theory (Yin, 2003). This is true for all types of studies including controlled experiments. A controlled experiment inevitably includes a limited number of test conditions, test situations, and test participants. For example, a psychology study may use undergraduate students who are enrolled in an introductory psychology class as test participants. To the extent that the students in the study can be characterized as a random sample of undergraduate students taking an introductory psychology class, then the logic of *statistical generalization* allows the researchers to draw conclusions about the population of undergraduates in introductory psychology classes from the performance of the test participants in the study. In practice, however, the researchers are interested in drawing conclusions about larger populations—all undergraduates, all adult English speakers, all human beings. The logic of statistical generalization does not provide the basis for such generalizations. To generalize beyond the population from which one samples requires reasoned arguments that are theoretically grounded (Yin, 2003). For example, if the study involved perceptual processes, one can argue, on theoretical grounds, that perceptual processes should not vary across different subpopulations of adults and that thus the results of the study should apply to all adults. Meaningful generalization invariably requires a leap beyond what is justified by the logic of statistical generalization alone.

Given the inherent tradeoff between the requirements of statistical generalization (*statistical conclusion validity*) and the requirements of external (or ecological) validity), in developing the work-centered evaluation paradigm, we have adapted the theoretical perspective to generalization that has been pioneered by Yin (2003) for generalizing from case study research. The emphasis is on building a rich web of interconnected, converging findings that support *analytic generalization* by connecting to theory.

Hollnagel et al. (1981) were among the first in the cognitive engineering literature to argue for generalization based on mapping to theory. Their ideas have been extended by Woods & Hollnagel (2006) and most recently by Roth & Patterson (2005) in relation to drawing generalizations from observational data obtained in cognitive field studies. Roth & Patterson (2005) have argued that *conceptual frames* drawn from prior empirical studies or theory (for example, the theory of distributed cognition, lessons learned across domains with respect to supervisory control of automation) serve to guide the interpretation and aggregation of findings to draw broader generalizations from observational studies. Yin (2003) makes the same point with respect to analytic generalization from case studies.

In explaining the basis for analytic generalization, Yin (2003) writes '. . . a previously developed theory is used as a template with which to compare the empirical results of the case study. If two or more cases are shown to support the same theory, replication may be claimed. The empirical results may be considered yet more potent if two or more cases support the same theory but do not support an equally plausible, *rival* theory.' (pp. 32-33).

Several points can be drawn from this line of argument. First, it argues for the importance of *theory* as a basis for guiding and generalizing from observation. This point also holds for generalization from evaluation studies. Data from evaluation studies can be generalized beyond the specific cases tested to the extent that it makes contact with well-established theory. Second, it highlights the importance of explicitly searching for, addressing, and eliminating *plausible rival theory*.

Donald Campbell, in his foreward to Yin's book, reinforces this point. He points out that the inference logic underlying the case study approach, and quasi-experimentation more generally, relies on the ability to specify and eliminate plausible rival hypotheses. The same point holds for design of and generalization from evaluation studies. In drawing generalizations from evaluation studies, one needs to take into account possible confounds that could account for the observed findings. For example, a steep learning curve might cause test participants to perform less well with a new system than they would perform with more experience. The possible impact of a learning curve on performance would need to be explicitly considered and addressed in drawing conclusions from an evaluation study (for example, by explicitly looking at changes in performance over time, or utilizing test participants who have equivalent level of experience with various systems being compared).

Importance of Ecological Validity

Another important criterion for enabling generalization is establishing the ecological validity of the test. Ecological validity stresses the importance of creating test conditions that are representative of the work content and context to which the joint cognitive system will be applied. This includes the types of test cases presented and performance measures collected.

Newman (1998) has argued eloquently for the importance of understanding and capturing real-world task goals, methods, and context in designing representative tasks for use in an evaluation study. He argued that it is not enough to simply look for frequently performed tasks; the overall *structure* of the work needs to be understood and reflected in the selection of representative tasks to include in the study.

Newman (1998) provides a poignant example illustrating the need for such an ecological approach to design of test cases. He describes the case of a family practitioner interacting with patients and writes, 'For example, if the system is to support a family practitioner, studies should focus at the very least on gathering data on a sequence of consultations and on recurring structures within consultations, and individual tasks should then be selected only if their contribution to the whole consultation is understood. Then we might see fewer situations of the kind that pertains in the United Kingdom, where family doctors make notes on paper during the consultation only to type them into their computers between each consultation and the next' (p. 320).

Work-centered evaluations emphasize the use of test cases that are representative of the range of tasks, cognitive activities, work contexts, and sources of complexities that are likely to arise in the domain and ecologically valid performance measures. There are a variety of field observation and cognitive analysis methods that can be used to uncover the ecology of work and provide the basis for defining test cases that are broadly representative of the demands of work and ecologically valid performance measures (c.f., Bisantz & Roth, 2008; Evenson et al., 2008; Patterson, et al., Chapter 14).

Importance of an Explicit Model of Support

The design of an evaluation also needs to be strongly informed by the hypothesized *model of support* that underlies the new aid to be evaluated. System developers have expectations about how the aid is expected to influence performance. This includes what aspects of macrocognition are expected to be facilitated and what range of situations the aid is intended to apply to (and which are beyond its' capabilities). These implicit assumptions about how the aid is expected to influence macrocognition and work constitute a hypothesized *model of support.*

The *model of support* needs to be made explicit so that it can be effectively tested. This includes a characterization of the activities/decisions/cognitive functions/teamwork elements that the aid is intended to support as well as the range (and bounds) of situational contexts and complicating factors for which it is expected to provide effective support (Roth et al., 2002). The model of support is analyzed along with the work context to motivate specific hypotheses regarding particular aspects of cognitive or collaborative work that are likely to be positively affected as well as the types of situations where performance improvements are likely to be realized. These specific hypotheses can then be used to define test cases and diagnostic measures that need to be included in the evaluation for a meaningful test.

Test situations should include a range of conditions for which the aid is explicitly designed to address so as to test the model of support. It may also include situations explicitly intended to identify the boundary conditions for support—conditions beyond which performance is expected to breakdown (c.f., Patterson et al., Chapter 14).

Importance of Triangulation and Converging Evidence

Another approach to strengthen the accuracy and persuasiveness of generalizations is to use multiple sources of evidence that converge on the same conclusion. In the qualitative research literature this is referred to as the *principle of triangulation* (Seale, 1999; Yin, 2003). Any finding or conclusion of an evaluation study is likely to be much more convincing and accurate if it is based on several different sources of information. The objective of triangulation is to develop converging lines of inquiry, so that the same fact or phenomenon is corroborated by multiple means.

The term triangulation is designed to evoke an analogy with surveying or navigation in which people discover their position on a map by taking bearings from multiple landmarks, the lines from which will intersect at the observer's position.

There are four types of triangulation discussed in the qualitative research literature:

- *Data source triangulation*, which involves using diverse sources of data so that one seeks out instances of a phenomenon in several different settings, at different points in time or space.
- *Investigator triangulation,* which involves leveraging the perspectives of multiple researchers. The premise is that by engaging in continuing discussion of their points of differences and similarity, personal biases can be reduced.
- *Theory triangulation*, which involves approaching data with several hypotheses in mind, to see how each fares in relation to the data. Theory triangulation reinforces the point raised earlier with respect to the importance of actively searching for and addressing how well rival hypotheses account for observations.
- *Methodological triangulation*, which involves using multiple methods, whose conclusions converge.

While all four elements of triangulation are important for bolstering the validity of conclusions from evaluation studies, methodological triangulation is particularly central to generalization validity. Methodological triangulation is explicitly called out by Cook & Campbell (1979) as an important tool for addressing construct validity. The idea is that abstract concepts such as intelligence, aggression, and situation awareness may be difficult to capture with any single measure. Confidence that the construct of interest is being tapped is increased if multiple measures are used and they produce converging results. This approach was recently used in a study by Roth et al. (2002) to examine the impact of control room technology on situation awareness. The study employed three different operational measures of situation awareness: self report ratings; a variant of the Situation Awareness Global Assessment Technique (Endsley, 1995); and an open-ended question technique. The pattern of results across the three measures increased confidence that meaningful aspects of situation awareness were being tapped.

Work-centered evaluations draw upon the logic of evidence triangulation in deriving generalizations. Multiple, partially overlapping measures are used to assess the usability, usefulness, and impact of a new technological aid for supporting work. These include both objective measures of performance under a range of representatively complex task conditions as well as elicitation of subjective opinions and suggestions from the test participants informed by their experiences in using the aid under a range of conditions.

Across studies, a wide variety of methods can be used that include analysis of objective measures of user interaction with a prototype, field observations of how a new system is adopted (and adapted) in an operational setting, as well as interviews with and surveys of 'users' at multiple levels within the organizational structure.

As Oviatt (1998) points out, individual methods seldom stand alone when answering real-world questions. Methodological diversity is among our most powerful design techniques for bolstering confidence in the validity of conclusions from evaluation studies.

Importance of Replication

Perhaps the most important tool available to a research community for increasing the confidence in a generalization (external validity) is *replication*—demonstrating the same result across cases. The replication can occur within a given study (for example, same result obtained over multiple participants or multiple test cases), or through multiple studies across space (for example, in different labs, in different domains) or across time (for example, at different stages in the system development process). If an important finding is replicated, then the likelihood that the results in question could have co-occurred by chance alone is reduced dramatically. Thus, concerns regarding statistical conclusion validity that cannot be addressed by demonstrating statistically significant results in any single study can largely be addressed through replication across studies.

In the final analysis, no single study by itself can ever address all possible threats to validity or potential biases of researchers. In practice it takes multiple studies, run by different people in different settings, at different points in time, to address concerns with threats to validity, eliminate rival explanations, and increase the confidence in the robustness of conclusions.

In the work-centered evaluation paradigm, replication is achieved both within and across studies. Within a study it is achieved by including a broadly representative range of test participants and test cases. The work-centered approach also relies on across-study replication by performing multiple evaluations or work-centered 'looks' at different points throughout the iterative design process.

Toward a Work-Centered Evaluation Framework

As we have tried to show above, the goals and pragmatic constraints of conducting evaluations of joint cognitive systems often mismatch the assumptions underlying traditional usability and experimental paradigms. We have tried to point to alternative theoretical frames that can provide more appropriate foundation for 'work-centered' evaluations. Table 13.1 summarizes the main mismatches between traditional evaluation paradigms and the goals and pragmatic constraints of joint cognitive system evaluations and how they are addressed in the work-centered evaluation approach.

One of the key elements of a work-centered evaluation is that it simultaneously addressed both *formative* and *summative* questions. From a summative perspective the aim is to assess whether the proposed design concepts, as embodied in the prototype, have the positive effects predicted by the system developers. The aim is to evaluate the model of support embodied in the prototype. The model of support is framed in terms of the macrocognitive elements of work that the system is intended to support as well as the *context of work* where the support is likely to be most needed and thus most effective. The objective is to assess how well the aid provides the hypothesized cognitive and collaborative support under *a realistic range of complex situations* that can arise in the domain.

The objective of a summative evaluation is not to establish that the aid being evaluated is 'better' than some competing design (for example, the current system) but rather to establish that the aid impacts performance in the manner predicted by the 'model of support' that underlies the design. As a consequence, an experimental paradigm that is aimed at comparing overall performance across conditions (for example, performance with the aid under evaluation versus performance with the current system or a competing system) is neither necessary nor sufficient. The aid under evaluation can turn out to result in better (or even worse) performance for a variety of reasons that have little to do with the underlying model of support that underlies the design. The focus of a work-centered evaluation is to be *diagnostic* in assessing whether the aid is supporting the specific macrocognitive aspects of work predicted by the model of support. This emphasis on diagnosticity contributes significantly to the ability of the work-centered evaluation paradigm to support design decisions throughout the design and development process.

Table 13.1 Mismatches between traditional evaluation paradigms and the goals and pragmatic constraints of joint cognitive system evaluations and how they are addressed in the work-centered evaluation approach

Characteristics of Traditional Evaluation Paradigms	Mismatches with Joint Cognitive System Evaluation Goals and Pragmatic Constraints	Work-Centered Evaluation Approach for Addressing Joint Cognitive System Evaluation Goals and Pragmatic Constraints
Clear Dichotomy between Formative and Summative Evaluation.	To capitalize on scarce evaluation opportunities, each evaluation cycle should combine elements of formative and summative evaluation.	Address formative and summative questions in each evaluation cycle. • Evaluate the model of support underlying the aid to determine whether the hypothesized benefits are realized. • Collect information on performance obstacles and boundary conditions to propel further innovation.
Usability Paradigm: Prone to narrowly focus on usability issues.	Desire to also assess usefulness and impact across the range of cognitive activities, work contexts, and sources of complexities that are likely to arise in the domain.	• Explicitly include test cases that are representative of the range of cognitive activities, naturalistic work contexts, and sources of complexities in the domain. • Collect usability, usefulness and impact measures.
Experimental Paradigm: Limited Diagnostic Power. Focus is on establishing reliable differences in performance across conditions—but not why those differences arose.	Desire for diagnostic tests that can be used to evaluate understand why results were obtained so as to advance theory and propel further design innovation.	Include diagnostic measures such as: • Process trace measures. • Expert Observer assessments. • Participant feedback via post test debriefs and questionnaires.
Experimental Paradigm: Emphasis on internal validity over external validity. Experimental logic requires restricting variability through tight control.	Desire to generalize as broadly as possible across individuals and situations in natural contexts. Desire to explore the bounds of generalization by searching for edge cases and boundary conditions.	• Emphasize ecological validity and breadth of test cases. • Include a broadly representative range of users as test participants. • Probe for boundary conditions and edge cases.
Experimental Paradigm: Statistical power requiring large sample sizes.	The sample size is usually severely limited by participant availability, time and cost factors.	Utilize alternative strategies to draw conclusions and build confidence in results: • Logic of triangulation and converging measures. • Replication across multiple studies over time and space. • Theoretically grounded analytic generalization.

At the same time a work-centered evaluation needs to serve a *formative* function. As in the classic usability paradigm, part of the goal of a work-centered evaluation is to uncover usability problems that need to be addressed prior to final implementation. In addition, the work-centered evaluation aims to uncover additional demands and unanticipated requirements at the level of *work support* so as to propel further work-centered design innovation. An important aim is to probe for the boundaries of effectiveness of the aid and the breakdown conditions where the aid no longer provides effective support. To achieve this formative objective, it is important to broadly sample the range of work contexts that reflect different kinds of complexities that can arise in the field of practice that complicate the cognitive and collaborative aspects of work, requiring flexibility and adaptation in response. It is also important to include test participants that reflect the range of variability of the target user population (for example, different experience levels; different positions). It should be noted that the objective of broad sampling is not to support *statistical generalization* but rather to support *analytic generalization*. The objective of sampling is neither to provide a random sample nor to provide a sample that minimizes uncontrolled variability. The intent is to include a broadly representative range of participants and test conditions that are based on the results of the work ecology model as well as theoretical grounds so as to enable discovery of the boundary conditions of support. The results can then be used to drive further design efforts and expand the theoretical base for macrocognition.

References

Bisantz, A. & Roth, E. M. (2008). Analysis of cognitive work. In D. A. Boehm-Davis (Ed.), *Reviews of human factors and ergonomics volume 3* (pp. 1-43). Santa Monica, CA: Human Factors and Ergonomics Society.

Campbell, D. T., & Stanley, J. C. (1963). *Experimental and quasi-experimental designs for research*. Chicago: Rand McNally.

Charlton, S. G. & O'Brien, T. G. (2002). The role of human factors testing and evaluation in systems development. In S. G. Charlton & T. G. O'Brien (Eds.), *Handbook of human factors testing and evaluation.* (2nd edn). Mahwah, NJ: Lawrence Erlbaum Associates, Inc.

Cook, T. D. & Campbell, D. T. (1979) *Quasi-experimentation: Design and analysis issues for field Settings*. Chicago, IL: Rand McNally College Publishing Company.

Dekker S. & Woods D. D. (1999). Extracting data from the future: Assessment and certification of envisioned systems. In S. Dekker & E. Hollnagel (Eds.*), Coping with computers in the cockpit* (pp. 2-27). Aldershot, UK: Ashgate Publishing.

Dumas, J. S. (2003). User-based evaluation. In J. Jacko & A. Sears (Eds.), *The human-computer interaction handbook: Fundamentals, evolving technologies, and emerging applications* (pp. 1094-1117). Mahwah, NJ: Lawrence Erlbaum Assoc.

Eggleston, R. G., Roth, E. M. & Scott, R. (2003). A framework for work-centered product evaluation. In *Proceedings of the Human Factors and Ergonomics Society 47th Annual Meeting*. Santa Monica, CA: Human Factors and Ergonomics Society.

Endsley, M. R. (1995). Measurement of situation awareness in dynamic systems. *Human Factors, 37*(1), 6-84.

Evenson, S., Muller, M. & Roth, E. M. (2008). Capturing the context of use to inform system design. *Journal of Cognitive Engineering and Decision Making, 2*(3), 181-203.

Hollnagel, E., Pederson, O., & Rasmussen, J. (1981). Notes on human performance analysis. *Technical Report Riso-M-2285*. Roskilde, Denmark: Riso National Laboratory.

Hollnagel, E. & Woods, D. D. (2005). *Joint cognitive systems: Foundations of cognitive systems engineering*. Boca Raton, FL: Taylor and Francis.

Jauer, R. A., Quinn, T. J., Hockenberger, R. L., & Eggleston, R.G. (1987). Radar aided mission/aircrew capability exploration: Full task simulation study, *Report No. AAMRL-TR-86-015*. Wright-Patterson Air Force Base, OH: Armstrong Aerospace Medical Research Laboratory.

Lang, A. W., Roth, E. M., Bladh, K. & Hine, R. (2002). Using a benchmark-referenced approach for validating a power plant control room: Results of the baseline study. In *Proceedings of the 46th Annual Meeting of the Human Factors and Ergonomics Society* (pp. 1878-1882). Santa Monica, CA: Human Factors and Ergonomics Society.

Monk, A. F. (1998). Experiments are for small questions, not large ones like 'What Usability Evaluation Method Should I Use?' *Human Computer Interaction: Special Issue: Experimental Comparisons of Usability Evaluation Methods, 13*(3), 296-303.

Newman, W. M. (1998). On simulation, measurement, and piecewise usability evaluation. *Human-Computer Interaction: Special Issue: Experimental Comparisons of Usability Evaluation Methods, 13*(3), 317-323.

Nielsen, J. (1993). *Usability engineering*. San Diego, CA: Morgan Kaufman.

O'Hara, J. (1999). A quasi-experimental model of complex human-machine system validation. *International Journal of Cognition, Technology, and Work, 1*(1), 37-46.

Oviatt, S. L. (1998). What's science got to do with it? Designing HCI studies that ask big questions and get results that matter. *Human-Computer Interaction: Special Issue: Experimental Comparisons of Usability Evaluation Methods, 13*(3), 303- 307.

Patterson, E., Cook, R. & Render, M. (2002). Improving patient safety by identifying side effects from introducing bar coding in medication administration. *Journal of the American Medical Informatics Association, 9*(5), 540-553.

Patterson, E. S., Roth, E. M., &Woods, D. D. (this volume). Facets of complexity in situated work.

Pfautz, J. & Roth, E. M. (2006). Using cognitive engineering for system design and evaluation: A visualization aid for stability and support operations. *International Journal of Industrial Ergonomics, 36*(5), 389-407.

Potter, S., & Rousseau, R. (this volume). Evaluating the resilience of a human-computer decision-making team: A methodology for decision-centered testing.

Potter, S. S., Roth, E. M., Woods, D. D. & Elm, W. (2000). Bootstrapping multiple converging cognitive task analysis techniques for system design. In J. M. Schraagen, S. F. Chipman & V. L. Shalin (Eds.), *Cognitive Task Analysis* (pp. 317-340). Mahwah, NJ: Lawrence Erlbaum Associates, Inc.

Roth, E. M., Gualtieri, J. W., Elm, W. C., & Potter, S. S. (2002). Scenario development for decision support system evaluation. In *Proceedings of the Human Factors and Ergonomics Society 46th Annual Meeting.* (pp. 357-361). Santa Monica, CA: Human Factors and Ergonomics Society.

Roth, E. M., Malin, J. T., & Schreckenghost, D. L. (1997). Paradigms for intelligent interface fesign. In M. Helander, T. Landauer & P. Prabhu (Eds.), *Handbook of human-computer interaction* (2nd edn) (pp. 1177-1201). Amsterdam: North-Holland.

Roth, E. M. & Patterson, E. S. (2005). Using observational study as a tool for discovery: Uncovering cognitive and collaborative demands and adaptive strategies. In H. Montgomery, R. Lipshitz, & B. Brehmer (Eds.), *How professionals make decisions* (pp. 379-393). Mahwah, NJ: Lawrence Erlbaum Associates.

Roth, E. M., Scott, R., Whitaker, R., Kazmierczak, T., Forsythe, M., Thomas, G., Stilson, M. & Wampler, J. (2007). Designing decision support for mission resource retasking. In *Proceedings of the 2007 International Symposium on Aviation Psychology*, April 23-26, 2007, Dayton, OH.

Roth, E. M., Scott, R., Whitaker, R., Kazmierczak, T., Truxler, R., Ostwald, J., & Wampler, J. (2009). Designing work-centered support for dynamic multi-mission synchronization. In *Proceedings of the 2009 International Symposium on Aviation Psychology*, April 27-30, 2009, Dayton, OH.

Roth, E. M., Stilson, M., Scott, R., Whitaker, R., Kazmierczak, T., Thomas-Meyers, G. & Wampler, J. (2006). Work-centered design and evaluation of a C2 visualization aid. In *Proceedings of the Human Factors and Ergonomics Society 50th Annual Meeting* (pp. 255-259). Santa Monica, CA: Human Factors and Ergonomics Society.

Sarter, N. B. & Woods, D. D. (1995). How in the world did we ever get into that mode? Mode errors and awareness in supervisory control. *Human Factors, 37*(1), 5-19.

Seale, C. (1999). *The quality of qualitative research.* Thousand Oaks, CA: Sage Publications.

Snow, M., P., Reising, J. M., Barry, T. P. & Hartsock, D. C. (1999). Comparing New Designs with Baselines. *Ergonomics in Design, 7*(1), 28-33.

Scott, R., Roth, E. M., Deutsch, S. E., Malchiodi, E., Kazmierczak, T., Eggleston, R. G., Kuper, S. M., &Whitaker, R. (2002). Using software agents in a work centered support system for weather forecasting and monitoring. In *Proceedings of the Human Factors and Ergonomics Society 46th Annual Meeting* (pp. 433-437). Santa Monica, CA: Human Factors and Ergonomics Society.

Scott, R., Roth, E. M., Deutsch, S. E., Malchiodi, E., Kazmierczak, T., Eggleston, R., Kuper, S. M. & Whitaker, R. (2005). Work-centered support systems: A human-centered approach to intelligent system design. *IEEE Intelligent Systems, 20*(2), pp. 73-81.

Scott, R., Roth, E. M., Truxler, R., Ostwald, J. & Wampler, J. (2009). Techniques for effective collaborative automation for air mission replanning. In *Proceedings of the Human Factors and Ergonomics Society 53rd Annual Meeting.* Santa Monica, CA: Human Factors and Ergonomics Society.

Vicente, K. J., Roth, E. M., & Mumaw, R. J. (2001). How do operators monitor a complex, dynamic work domain? The impact of control room technology. *International Journal of Human Computer Studies, 54*(6), 831-856. Available online at: http://www.idealibrary.com

Woods, D. D. (1998). Designs are hypotheses about how artifacts shape cognition and collaboration. *Ergonomics, 41*(2), 168-173.

Woods, D. & Dekker, S. (2002). Anticipating the effects of technological change: a new era of dynamics for human factors. *Theoretical Issues in Ergonomics Science, 1*(3), 1-11.

Woods, D. D. & Hollnagel, E. (2006). *Joint cognitive systems: Patterns in cognitive systems engineering.* Boca Raton, FL: Taylor and Francis.

Woods, D. D. & Sarter, N. B. (1993). Evaluating the impact of new technology on human-machine cooperation. In J. Wise, V. D. Hopkin, & P. Stager (Eds), *Verification and validation of complex systems: Human factors issues* (pp. 133-158). Berlin: Springer-Verlag.

Yin, R. K. (2003). *Case study research: Design and methods.* (3rd edn). Thousand Oaks, CA: Sage Publications.

Chapter 14

Facets of Complexity in Situated Work

Emily S. Patterson, Emilie M. Roth, and David D. Woods

Introduction

One of the major challenges in designing effective 'person-in-the-loop' software evaluations is to sufficiently capture the complexity inherent in real-world operations. If the test scenarios are not sufficiently complex during an evaluation, fielded software systems may fail to achieve their desired benefits, miss opportunities to support critical functions, and/or introduce negative unintended consequences ('side effects'). This chapter focuses on one of the most challenging aspects of construction of meaningful evaluations—how to *design* evaluation scenarios that adequately probe the boundaries (cf., testing the 'edges' of system capability in Chapter 15) of effectiveness of the support system. In this chapter, we describe an approach to scaling up the complexity of existing scenarios based upon a working model of what facets make up complexity in supervisory control domains. We discuss the multiple levels of description that need to be considered when designing scenarios, and present a domain-independent set of cognitive and collaborative 'complicating factors' that can be used to guide the design of evaluation scenarios. These complicating factors can be embedded individually or in clusters into existing scenarios so as to exercise the software under conditions that accurately reflect the types of complexities that may arise in the field. We view these factors as comprising a working definition of the facets that contribute to complexity in dynamic, high-risk, real-world settings where cognition is distributed across human agents supported by sophisticated artifacts.

Software systems intended to support individual and team macrocognitive processes are rapidly emerging across a wide variety of domains from healthcare, to process control to military command and control. These support systems range from new visualizations and multimedia communication devices intended to foster common ground and facilitate collaboration, to automated decision-support tools intended to support situation assessment and planning, to entire computerized command and control centers intended to facilitate the work of multiple individuals engaged in a broad range of activities.

One of the major challenges for the macrocognition research community is to develop guidance for how to design effective 'person-in-the-loop' evaluations to ensure that the resulting software systems achieve their desired benefits without introducing unintended negative side effects. Typically representative users from the target domain are asked to exercise the software under a range of plausible scenarios in order to assess whether the software meets its intended objectives (for example, as measured by improved performance or positive user assessment). Roth (this volume) and Potter & Rousseau (this volume) describe some of the challenges to design of effective evaluations and introduce promising evaluation paradigms.

In this chapter, we discuss how to design evaluation scenarios that adequately probe whether the software systems are providing the benefits envisioned by the system designers (and are not introducing new sources of difficulty). Too often evaluation scenarios are selected based on convenience (for example, scenarios that were previously developed for training purposes; or a 'data-snapshot' that was previously captured to support design activities) with the result that they do not necessarily constitute an effective evaluation. In this chapter we argue for the importance of explicitly designing the evaluation scenarios to insure that they reflect the range of task complexity that users in the domain are likely to confront. We describe an approach to scaling up the complexity of existing scenarios. Our approach distinguishes levels at which scenarios can be described, including complexity, and provides a detailed set of 'complicating factors' which can be embedded individually or in clusters into existing scenarios so as to better reflect the range of complicating factors that arise in the domain.

Starting Points for Evaluation Scenario Design

Scenarios are a specification of a sequence of circumstances and events that depict a realistic situation that individuals and teams may confront in a target domain. Scenarios may take multiple forms ranging from a simple verbal or text description of the circumstances and events being postulated to an elaborate simulation of the events using a high fidelity dynamic simulator.

Scenarios are commonly used for multiple purposes including analysis, design, training, and evaluation. For example, scenarios simulating emergency situations (such as a chemical spill in a heavily populated urban environment) can be used to examine the readiness level of an emergency response organization to respond to unanticipated events. They can also be used to train members of the organization to work more effectively as a team or to evaluate how the introduction of a new communication technology or protocol alters how the organization functions when responding to an emergency.

Scenarios are routinely used to motivate and guide design (Carroll & Rosson, 1990; Rosson & Carroll, 2002). In scenario-based design, scenarios are used to capture and communicate important characteristics of the users, the typical and critical tasks they engage in, the tools they use, and their organizational context. They are also used to communicate the vision of how the new support system is likely to change the user experience and task performance. Software demonstrations often feature scenarios embodied in storyboards as a way to communicate to the potential client the intent of the underlying design concept, how that concept has been implemented in the system, and how users are expected to interact with the system to accomplish their goals. Some have applied variations of this 'envisioned world' scenario technique in order to elicit alternative stories from users about how software might be useful in an envisioned setting (Dekker & Woods, 1999; Evenson et al., 2008). In addition, for some design teams, 'use cases' (Cockburn, 2001) serve as a communication vehicle for how a software program is intended to meet design requirements in order to foster a shared understanding among disparate stakeholders and facilitate negotiations in cases of conflicting goals.

Scenarios that have been developed for other purposes can often be leveraged in the design of evaluation scenarios. One clear advantage of starting with already existing scenarios is that they tend to have high 'face validity' in the sense that the surface details of the scenario (for example, the names, locations, and relationships) readily resonate as realistic with domain practitioners. However, because the scenarios have been developed for other purposes, they typically do not include the range of domain complexities that are required to serve as an effective evaluation scenario set 'as is.'

Scenarios that have been developed to obtain stakeholder buy-in or to motivate design will tend to provide a biased sample from the perspective of software evaluation because they generally focus on situations where the benefits of the software tool are most likely to be realized. To be effective, evaluation scenarios need to cover a broader range of situations and complicating factors that are likely to confront users in actual operation. For this reason, the software evaluation team will need to expand the range of scenarios included in the set and/or increase the representativeness of the scenarios by introducing additional complicating factors.

Another common source for scenarios is scenarios that have previously been developed for training purposes. A large corpus of face valid training scenarios are often available in military, aviation, and industrial domains where simulation-based training is routinely conducted. The military in particular often develops rich, complex 'overarching' scenarios, typically involving 'fictitious' countries, that include elaborate descriptions of the historical, political, military, social, and economic situation, complete with maps and timelines. These overarching scenarios can provide an invaluable resource to leverage for creating more specific targeted vignettes within the broader scenario context. While training scenarios can provide a rich starting point for evaluation scenario design, they often tend to represent straightforward, well-practiced situations that do not tap the range of cognitive and collaborative demands that can arise in the domain (Mumaw & Roth, 1992).

In general, existing scenario sets usually cover relatively simple, routine tasks using the designed system as intended because these best meet the prior needs for scenarios. During software evaluations, it is important to include not only routine or textbook cases, but also cases that challenge both cognition and collaborative processes and reflect the real complexities that can arise in the domain (Mumaw & Roth, 1992; Roth et al., 2002). This is particularly important in settings with high consequences of failure, where significant loss of life, property, and economic assets could result from implementing a system that does not adequately handle the range of complexity inherent in the real-world setting.

Levels of Scenario Design

Designing evaluation scenarios is one of the most challenging and important aspects of software evaluation development. In many ways, constructing a set of scenarios is like crafting a puzzle that must simultaneously satisfy constraints at multiple levels, and changes at one level interact with the design at other levels. When designing a crossword puzzle, one must consider the impact of word selections on both the rows and columns. Similarly, in designing scenarios one must consider the scenario at multiple levels of description and craft a set of scenarios that collectively satisfy the required constraints at each level.

Table 14.1 presents the various levels of scenario description that need to be taken into consideration when crafting a set of scenarios. These levels of description represent different perspectives that need to be considered in design of scenarios. The table provides a definition of each level of scenario description, an *integrity* target that specifies the target characteristics that a set of scenarios should meet (individually or as a collective set), and a description of objective indicators that can be used to assess whether sufficient integrity across evaluation scenarios has been achieved. While the focus of this chapter is on the level of scenario *representative complexity*, it is important to consider all of the different levels of description when creating new scenarios or augmenting existing scenarios. Since changes at one level can impact other levels, the scenario designer constantly needs to be cognizant of their inter-relationship so as to insure that scenario constraints at all levels are simultaneously met.

At the most basic level, the scenarios need to display *surface validity*. The scenarios need to display verisimilitude with respect to the domain details. The terminology used, and rules of operation and relationships specified, need to 'ring true' to domain practitioners or the evaluation will lack credibility in eyes of the evaluation participants. At the same time, the scenarios need to reflect important underlying domain complexities—*representative complexity*. This requires mapping the scenario characteristics to a more abstract-level domain-independent characterization of the underlying domain complexities and macrocognitive challenges. This is a fundamental step required to insure that the set of scenarios covers the range of anticipated complexity in the field setting.

The relevant sources of complexity in a domain are typically uncovered by a cognitive task or cognitive work analysis (Evenson et al., 2008; Bisantz & Roth, 2008; Roth, 2008). The uncovered complexities can then be weaved into scenarios. Typically this is accomplished by working closely with domain experts and involves translation back and forth between the language of the domain and more abstract, domain-independent descriptions of cognitive demands. A domain-

Table 14.1   Levels of scenario descriptions

Level of Description	Definition	Integrity Target	Indicators of Sufficient Integrity
Surface validity	Detailed scenario elements are face valid to real-world practitioners, including tasks, events, software support, communication support, names, locations, relationships, rules of operation.	The scenario details follow domain conventions and are readily accepted as realistic by domain practitioners.	• Professionals representing the target user population are engaged in the scenario activities and consider them face valid. • Participants are willing to perform tasked activities during the software evaluation. • Participants can easily explain why scenario elements are appropriate for their work setting. • Participants recommend participation in the evaluation to senior, respected colleagues.
Representative complexity	A mapping of scenario characteristics to underlying domain complexities and macrocognitive challenges.	The scenario set covers the range of anticipated complexity in the field setting, from nominal cases to challenging cases.	• The scenarios reflect the range of complicating situations that have been identified by prior analyses (e.g., ethnographic field studies; cognitive task analyses).
Model of support	A mapping of scenario characteristics to the activities/decisions/ cognitive functions/ teamwork elements that the software aid is intended to support.	The primary justifications for the design of the software are explored with the scenario.	• The scenario reflects complex situations where the software is expected to improve performance. • The scenario reflects a range of complexities to explore the boundaries of support—where the support system may break down. • The scenario supports a qualitative, holistic assessment by participants and evaluation team members about whether the software would be useful in their real-world setting.

Table 14.1 *Concluded*

Level of Description	Definition	Integrity Target	Indicators of Sufficient Integrity
Scenario difficulty	A characterization of scenario in terms of level of task difficulty it will pose to study participants (and justification).	The scenarios are at a level of difficulty that creates the potential for level of performance to vary across study conditions so as to insure that the evaluation will be diagnostic.	• There is a range of performance across study conditions, making it possible to evaluate the contribution of the software tool to performance.
Performance observability	A mapping of scenario characteristics to observable individual or team response that is anticipated and the conditions under which these observable behaviors will occur (e.g., observable behavior anticipated if study participants make correct situation assessment).	The scenario includes situations/ probes/target events that create the opportunity for cognitive and collaborative activities of interest to be exercised in an observable manner.	• The scenarios enable objective observers to reliably determine the quality of cognitive and collaborative performance via externally observable indicators (e.g., participant utterances or observable actions).
Value/ impact potential	A mapping of scenarios to value metrics of importance to the customer organization (e.g., operational effectiveness improvement, risk reduction, cost reduction).	The scenarios enable collection of performance metrics along value dimensions of importance to the organization so that the impact of implementing the software can be assessed for cost/benefit analyses purposes.	• The scenarios reflect complex situations where improvements in performance would be anticipated to reap significant benefits to the organization (e.g., measurable economic or safety benefits) as well as to the end-user (e.g., system is easier and faster to use).

independent set of complicating factors that can be used in establishing this mapping is provided in the next section.

The focus on sampling a range of domain-complicating factors that can stress cognitive and collaborative performance contrasts with traditional approaches to scenario design that emphasize high 'surface validity' from the point of view of containing rich operational details. A set of situations can be realistic from the point of view of capturing operational details (the surface features of the situation), but not reflect the range of cognitive, collaborative, and decision-making demands that arise in the domain (the deep structure of the situations). For example, the situations can reflect a wide range of mission areas but be all routine 'textbook' cases that are straightforward to handle. In our approach to scenario design, while an attempt is made to create situations that encompass a range of realistic operational settings (that is, situations that have face validity at the surface level), as much or more effort is spent ensuring that the situations adequately sample the range of complicating factors that arise in the domain.

In crafting a set of scenarios it is also important to remain cognizant of how the scenarios map to the *model of support* underlying the design of the software system. The *model of support* provides a specification of the activities/decisions/ cognitive functions/teamwork elements that the visualization or decision aid is intended to support (Roth et al., 2002; Roth & Eggleston, Chapter 13). Evaluation scenarios need to exercise these cognitive/collaborative elements of performance in order to determine whether the decision aid affects performance as intended. Scenarios should include a range of conditions for which the software system was explicitly designed to address so as to test the model of support. It may also include situations explicitly intended to identify the boundaries of support—conditions under which performance is likely to breakdown.

The scenario designers must also consider the level of difficulty of the scenarios. In order for the evaluation to be diagnostic, there needs to be a potential for level of performance on the scenarios to vary (for example, the level of performance between participants who use the new software tool and participants who use an existing legacy tool). It is important to avoid 'floor' effects where the scenario is so challenging that performance is poor irrespective of support system (that is, whether the user is using the new software aid or existing legacy tool) and 'ceiling effects' where the scenario is so straightforward that performance is high irrespective of support system. Having the potential for spread in performance is critical for being able to distinguish the usefulness of an innovation.

Another level of description that needs to be considered in designing scenarios relates to *performance observability*. The scenarios need to include situations/ probes/target events that create the opportunity for cognitive and collaborative activities of interest to be overtly demonstrated (Woods & Sarter, 1993; Johnston et al., 1995; Smith-Jentsch et al., 1998) so that impact of the software aid on performance can be objectively assessed. For example, to assess the impact of a new software system on situation awareness, one could include explicit target events in the scenarios and record whether study participants detected them.

A final level of description that needs to be considered in designing and characterizing scenarios relates to characterizing the *value 'impact' potential* to the customer organization if performance on this scenario were to improve. One of the primary considerations in evaluating a new software tool from the customer decision-maker perspective relates to whether the new software tool provides enough benefit on value dimensions of importance to the organization to justify the cost expenditure to implement the software. Relevant value dimensions include improved operational effectiveness, improved safety, and reduced operating cost. In developing evaluations, it is important to create scenarios where different levels of user performance can be straightforwardly translated into quantitative 'savings' on the value dimensions of interest to the organization (Eggleston et al., 2003). For example, in designing an evaluation of a new software aid for route planning for an air transportation organization, one might include evaluation scenarios where differences in study participant performance could be translated into fuel usage savings (and therefore cost savings).

Complicating Factors

In this section, we provide a domain-independent characterization of complicating factors that often arise in dynamic high-risk worlds that pose challenges to individual and team cognitive and collaborative processes. These domain-independent characterizations can be used as starting probes to facilitate dialogue between cognitive analysts and domain experts so as to streamline the domain discovery and scenario design processes. We believe that it is more efficient, reduces the reliance on the ability of domain participants to be reflective practitioners (Schon, 1983), and can reduce the scenario design 'trial and error' process when domain-independent descriptions of what makes cognition challenging in complex, sociotechnical settings are used as the basis for dialogue.

The literature on complexity in general spans nearly every discipline (see Campbell, 1988, for a literature review of task complexity). For the purposes of this chapter, we limit our references to the pioneering work of Jens Rasmussen, Eric Hollnagel, David Woods and colleagues, who initiated work domain analysis methods to uncover domain characteristics that contribute to cognitive and collaborative challenges in dynamic, high-risk worlds (Rasmussen & Lind, 1981; Rasmussen, 1986; Woods, 1988; Woods et al., 1988; Woods & Hollnagel, 2006). This work was done primarily under the label of Cognitive Systems Engineering (CSE), which was a new approach to 'coping with complexity,' primarily in response to an increased use of automated controllers, computerization, and digital data storage and display formats in supervisory control domains (Hollnagel & Woods, 1983).

Efforts to capture characteristics of dynamic, high-risk, sociotechnical systems that introduce complexities and challenges in domain-independent terms continue. Most recently Nancy Leveson (manuscript in preparation) has developed an extensive framework for capturing domain complexities that can contribute to cognitive and collaborative challenges in complex sociotechnical systems and increase risk. Some of the dimensions of sociotechnical systems that she identifies include:

- Interactive complexity: interaction among system components.
- Dynamic complexity: changes over time.
- Decompositional complexity: inconsistency between the structural decomposition and the functional decomposition.

• Non-linear complexity: non-intuitive relationships between cause and effect.

These dimensions of complexity, while framed in terms of characteristics of domains and technologies, can result in challenges to cognitive and collaborative functions. For example, in domains where there are interactions among system components, such as nuclear power plants where changes in the primary system can impact parameters in the secondary system and vice versa, the complex dynamics that result can complicate a range of macrocognitive functions.

Interacting systems can challenge monitoring and situation awareness because when operators make a change in one system, they need to continue to monitor and maintain awareness of impacts on the other system. It can challenge situation assessment and fault diagnosis because disturbances in one system can produce symptoms in another (effects at a distance). It can challenge action planning because operators need to anticipate the impacts of actions across multiple systems to avoid unanticipated side effects of actions. It can also complicate team processes because operators that are each responsible for a different system that interact with one another need to maintain awareness of the goals and actions of the other team members that operate interacting systems so as to avoid unanticipated interactions (Roth & Woods, 1988).

Descriptions of domain complexities that increase cognitive difficulty have also come out of the cognitive science literature (for example, Burleson & Caplan, 1998; Smith et. al., 1991; Feltovich et al., 1997, 2004). Of particular note is the work of Chi and his colleagues (1981). They have identified eleven dimensions of domains that impact cognitive difficulty. These include: static vs. dynamic; discrete vs. continuous; separable vs. interactive; sequential vs. simultaneous; homogeneous vs. heterogeneous; single vs. multiple representations; mechanism vs. organicism; linear vs. nonlinear; universal vs. conditional; regular vs. irregular; surface vs. deep.

Interestingly, the characterizations of domain complexity drawn from the cognitive engineering literature and the cognitive science literature identify very similar elements of complexity. Both call out complex interactions among parts of the domain, and dynamics as factors that increase complexity. The commonality across the two lists is particularly noteworthy, given that they were drawn from observations and analyses of very different domains (learning and performance in medical education in the case of Feltovich and his colleagues (1997); and dynamic, high-risk technologies in the case of Hollnagel et al. (2006)).

Abstract-level characterizations of dimensions of domain complexity provide a useful foundation for identifying characteristics of a particular domain that may create cognitive or collaborative challenges. One could start by characterizing a particular domain in terms of the dimensions identified in either one (or both) of the complexity dimension characterizations provided above. Scenarios could then be created that exploit particular elements of complexity exhibited in the domain. For example, if a domain was characterized by multiple interacting systems, then the scenario designer could work with domain experts to create scenarios where problems in one system are manifested in symptoms in another interdependent system to create situation assessment challenges. Similarly one could create scenarios that create planning challenges where the study participants would need to think about potential side effects that actions on one system would have on another system.

While abstract-level characterizations of domain complexity are useful, they require the cognitive analyst and scenario designer to infer how the dimensions of domain complexity will impact particular macrocognitive functions.

There has been a growing body of research that has attempted to identify and characterize more specific examples of complicating factors that naturally arise in complex, sociotechnical settings and that are more clearly linked to macrocognitive functions (Roth et al., 1994; Woods et al., 1994; Woods & Patterson, 2001; Watts-Perotti & Woods 2007). These factors have generally been identified through descriptive ethnomethodologically-informed analyses of real-world cognitive and collaborative processes. Examples include 'effects at a distance' that can influence situation assessment processes (for example, Roth et al., 1994) and goal-conflict situations that can influence planning and decision making (Woods et al., 1994).

A collection of complicating factors is presented in Table 14.2. This list represents the most recent embodiment of a working effort to capture characteristics of domains that pose challenges to macrocognitive functions in a domain-independent fashion. These factors are loosely organized around a set of macrocognitive functions (see Preface for more information), which is not necessarily a fixed or comprehensive list:

- *Detecting:* Noticing that events may be taking an unexpected (positive or negative) direction that require explanation and may signal a need or opportunity to reframe how a situation is conceptualized (sensemaking) and/or revise ongoing plans (planning) in progress (executing). (Related terms: detecting problems, monitoring, observe, anomaly recognition, situation awareness, problem detection, reframing.)
- *Sensemaking:* Collecting, corroborating, and assembling information and assessing how the information maps onto potential explanations; includes generating new potential hypotheses to consider and revisiting previously discarded hypotheses in the face of new evidence. (Related terms: orient, analysis, assessment, situation assessment, situation awareness, explanation assessment, hypothesis exploration, synthesis, conceptualization, reframing.)
- *Planning:* Adaptively responding to changes in objectives from supervisors and peers, obstacles, opportunities, events, or changes in predicted future trajectories; when ready-to-hand default plans are not applicable to the situation, this can include creating a new strategy for achieving one or more goals or desired end states. (Related terms: replanning, flexecution, action formulation, means-ends analysis, problem solving.)

Table 14.2   Factors that increase scenario complexity

Factor	Description	Primary Macrocognitive Functions Impacted	Example
Data overload (buried information)	Problems that need to be detected and addressed are buried in a large amount of potentially relevant information.	Detecting	Critical information about an impending attack by insurgents is buried in a large dataset (e.g., email notification buried in a large 'inbox,' phone call easy to defer, information embedded in continuously streaming media data on a background channel). A large number of alarms come on simultaneously in a control center.
Signal-noise relationship (false alarms)	Detecting a signal from background noise is difficult because the signal is close to the background noise distribution. Examples include environments with high false alarm rates, especially where there are negative consequences for acting on false alarms. Information is prone to be discounted if the indicators are perceived to be unreliable or have a high false alarm rate.	Detecting	It is challenging for sonar operators to distinguish an enemy submarine from schools of fish and 'friendly' watercraft. If insurgents are constantly issuing threats, it may be difficult to distinguish and act upon indicators of an actual threat situation.
Attention demands (attention bottlenecks)	Requirements for rapid attention shift.	Detecting	Situations where information and requests are coming in from multiple sources requiring rapid shift in attention. Examples include monitoring and detecting tasks where there is a need to detect and track targets of interest.

Table 14.2 *Continued*

Factor	Description	Primary Macrocognitive Functions Impacted	Example
Distributed information across individuals/ organizations (scattered puzzle pieces)	Information distributed across participants and/or roles is required for recognition of a coherent pattern. Particularly challenging scenarios include unique access (only one individual has particular information), distributions that do not match the responsibility of the roles participants play, information that requires prior knowledge or specialized expertise in order to appreciate the significance, information of mixed levels of sensitivity (e.g., top secret classification documents mixed with publicly available documents) and when the participants are incentivized not to share information (e.g., sharing information puts them at personal legal risk).	Detecting Sensemaking Coordinating	Multiple participants have different pieces of evidence that when combined would allow definitive identification of the perpetrator of a crime. Multiple army participants have unique knowledge about available resources, capabilities of vehicles, fuel cost, and safety concerns that needs to be shared to devise the safest, least expensive, and fastest route for transporting troops and supplies in preparation for a strategic offensive.

Table 14.2 *Continued*

Factor	Description	Primary Macrocognitive Functions Impacted	Example
False prime explanations (garden path)	Initial evidence is strongly suggestive of the false prime ('garden path') explanation. Over time, evidence is presented in increasingly hard-to-ignore formats that suggest that the false prime explanation is inaccurate. A highly similar approach has a prime explanation that is not provably inaccurate, but is less supported than an alternative ('emerging path') when all of the available evidence is considered.	Sensemaking	False prime explanation: Nuisance sniper attacks on a checkpoint on a major supply route are routinely being conducted by untrained civilians. The civilians are unhappy that their country is being occupied by another country's military. They are using easily accessible weaponry (e.g., sniper rifles that are standard issue for armies in many countries). This leads participants to initially assume that untrained civilians are responsible for current attacks.

Emerging path explanation: Sniper attacks are being conducted by military personnel of a neighboring country. These personnel are hiding among an ethnic sub-population that is known to be sympathetic to the neighboring country. Their activities are intentionally deceptive in that they are intended to mask planning for an imminent large-scale invasion by the neighboring country's military, which will use tanks to cross the border. |
| Context change (not in Kansas anymore) | Requirements, plans, actions, decisions, or social norms that make sense in one context do not translate to another one. | Sensemaking | An intelligence analyst with experience in one region where power is centrally located in the elected president for the nation must re-evaluate default assumptions about command and control structures and procedures when transferred to another region where power is centered in tribal elders for multiple ethnic groups. |
| Missing information (information gap) | Information that is needed for an accurate assessment is missing (e.g., due to lack of sensors or failed sensors, lack of system update, lack of informants on the ground, or poor communication). | Sensemaking | A classic example is the case of a missing leading indicator scenario in a nuclear power plant domain. One of the earliest and clearest indicators of a steam generator tube rupture event is the detection of radiation in the secondary side of the plant (where normally there is no radiation.) If the radiation detection sensors fail, then diagnosing the steam generator becomes more challenging. |

Table 14.2 *Continued*

Factor	Description	Primary Macrocognitive Functions Impacted	Example
Misleading indicators (low predictive value)	Information may be inaccurate (e.g., the information may not have been updated in a timely manner, the source may be inherently unreliable, or inaccurate information may be provided due to degraded sensors, problems in communication or intentional deception).	Sensemaking	Misleading information may be provided due to inherent limitations of reports (e.g., stale information, inherent limitations of predictions, distortions resulting from indirect reports, secondary sources, translations) or explicit intent to deceive through misinformation. It can also result from reliance on indirect indicators that are usually correlated with the information of interest, but not in that situation. A classic example occurred during the Three-Mile Island accident, where level in the pressurizer was assumed to indicate that there was sufficient coolant in the reactor vessel, when in fact this was not the case.
Multiple simultaneous influences (more than one explanation required)	There are multiple independent 'influences' that are simultaneously present and in combination explain the observed evidence.	Sensemaking	A software flaw in the Ariane 5 rocket design accounted for the rocket during the maiden Ariane 501 launch having no guidance data mid-flight, the rocket swiveling abnormally during take-off, and a simultaneous primary and backup inertial reference system shutdown. Although it was initially assumed that an unexpected roll torque was also explained by the software flaw, the identical problem was also observed on the subsequent Ariane 502 launch. The best explanation was that this problem was due to an unrelated mechanical failure of a small part, which was replaced with a stronger design before the next successful launch, Ariane 503. In process control domains there may be multiple thermodynamic influences that combine to produce plant behavior. For example, in a steam generator tube rupture a leak of fluid from the primary side will tend to cause level in the steam generator to go up. However, there may be countervailing forces such as a simultaneous decrease in temperature that causes level in the steam generator to appear to be going down, making it more difficult to detect the steam generator tube rupture.
Ambiguous cues (no obvious answer)	There are multiple, alternative, explanations for the pattern of symptoms observed.	Sensemaking	Classic examples come from medical diagnosis where a set of presented symptoms could be explained by multiple alternative diseases. For example, someone could display gastro-intestinal symptoms because of an allergic reaction to something eaten, food poisoning, or a gastro-intestinal flu.

Table 14.2 *Continued*

Factor	Description	Primary Macrocognitive Functions Impacted	Example
Uncertain information (unreliable data)	The accuracy of the information cannot be definitively ascertained.	Sensemaking	Information can be uncertain because the information is old or because it is from an inherently unreliable or untrustworthy source. For example, an automated system may categorize an entity as 'threat' or 'non-threat' but be unreliable because of a high false alarm rate.
Complex or counterintuitive dynamics (non-intuitive)	A process changes over time in a complex manner making it difficult to develop an appropriate mental model and to anticipate/project the impact of changes over time.	Sensemaking	Counterintuitive dynamics make it difficult for operators to understand and the current state and/or predict future states. Complex thermodynamic processes in process control applications where adding cold water to a system will first cause level to decrease (because of shrink effects) and then to slowly start to increase.
Stereotype violations (not the usual suspect)	In most situations, there is a stereotypical explanation for a set of data. In this scenario, the stereotype ('usual suspect') explanation is not the best explanation.	Sensemaking	Classic examples come out of healthcare. A set of symptoms may be 'prototypical' of a common disease but in this case may be caused by a more obscure disease. For example, a person presenting the classic symptoms of a flu (a fever and cough) during flu season may be diagnosed with the 'flu' when in fact they may have a more serious respiratory ailment.
Implied relationship (red herring)	Findings that appear to be relevant to the situation to be explained, may in fact be unrelated and misleading.	Sensemaking	Multiple data elements are presented that are implied to be related, based on co-occurrence in time, are not related. One of the data elements (red herring) can easily be falsely assumed to be explained by explanations that account for some of the other data elements. A concrete example from an intelligence analysis application is a series of reported events that suggests growing violence by anti-democracy forces (e.g., mob activity at rallies; small explosions causing property damage). Included in the series is a report that there was a house fire that killed a prominent pro-democracy judge that has provoked fear among the populous and pro-democracy lawyers and judges that anti-democracy forces might attack them. Upon further investigation, it is learned that the fire was caused by faulty wiring (and thus the report of the house fire is a red herring).

Table 14.2 *Continued*

Factor	Description	Primary Macrocognitive Functions Impacted	Example
Hidden coupling (effects at a distance; cascading effects)	A source of a problem is difficult to detect because of cascading secondary effects that make it difficult to connect the observed symptoms to the originating source.	Sensemaking	A leak in a tube at a nuclear power generation facility generates numerous alarms about cascading problems in multiple physically dispersed systems farther down the line, making it difficult to trace the interconnections among the symptoms and uncover the underlying 'root cause.'
Distributed information across time (overturning updates)	Situations that require integrating information over time periods.	Sensemaking	Situations can arise where individual pieces of information come in at different points in time and must be integrated across time to form a coherent picture. Individuals may fail to recall prior information, connect the information distributed in time, or revise their assessment based on the new incoming information. For example, an adverse event investigation concluded that the primary cause of death of a nursing home patient was fecal impaction, which could have been avoided with a stronger dose of medication to alleviate constipation. After the investigation was closed, an autopsy report was filed that stated that the primary cause of death was a cardiac event (specifically atheroemboli from shaggy atherosclerotic plaques on the aorta wall).
Inadequate guidance (poor seed)	The participant is provided guidance (e.g. from a peer/colleague) that would predictably result in suboptimal performance if followed. This guidance could include an inaccurate or poorly supported explanation or a plan that is suboptimal or inadequate.	Planning	An anesthesiologist is told by an outgoing anesthesiologist (who is actually a confederate researcher) during a handover update to administer a medication during the upcoming simulated surgery to which the patient has been highly allergic in the past (e.g., a patient had anaphylactic shock during a prior surgery when he received that medication).

Table 14.2 Continued

Factor	Description	Primary Macrocognitive Functions Impacted	Example
No predefined plan or procedure (novel situation)	A participant is presented an unfamiliar situation that has not been trained for, and for which no predefined guidance, plan or procedure is available.	Planning	An aircraft pilot may confront a novel malfunction for which training has not been provided, and no predefined guidance is provided.
Unintended effects (managing side effects)	An action encouraged to be taken by the participant has unintended effects which are challenging to predict in advance (without training or case-specific expertise).	Planning	Participants simulating the response of firefighters to a basement fire are likely to enter through the back door of a house since that is the most direct route to the basement. The next participants on the scene (who are actually 'confederate' researchers) erroneously assume that they are first on the scene because they cannot see that the first responders entered through the backdoor and so enter through the front door, causing a dangerous 'blowback' situation that has to be managed. The unintended effect of using the most direct, but difficult to see for the next on the scene, route is to make direct communication necessary about who has already entered the building.
Mismatch between predefined plans or procedures and the situation confronted (wrong plan)	Individuals are provided predefined plans or procedures to use in responding to a situation. However, the situation deviates from the assumptions underlying the plan/procedure and if followed verbatim will not achieve the desired goals.	Planning	Preplanned response procedures are common in many domains. Situations can arise where the assumptions underlying the predefined plan are not met (e.g., the plan assumed that multiple forces would attack synchronously, but some of the forces have been delayed resulting in less desirable combat power ratio than was assumed in the plan.) The challenge is to recognize the mismatch between the state of the world assumed by the plan and the actual state of the world and to modify the plan so as to achieve the desired objective.

Table 14.2 *Continued*

Factor	Description	Primary Macrocognitive Functions Impacted	Example
Impasses and opportunities (unworkable plan)	A participant must revise a plan when it can no longer be executed (impasse). A similar tactic is to set up a possibility of taking advantage of an unexpected resource or event to greatly improve upon an existing plan (opportunity) at the participant's initiative.	Planning	In conducting the operation against the insurgents, troops and supplies have to be moved safely from one place to another. The planned route becomes unavailable after an attack on a bridge destroys it.
Shifting objectives (moving target)	The tasks originally given to the participants change over time, due to shifts in objectives, detection by others of violated assumptions, or the need to synchronize with other personnel who did not accomplish everything according to the original plan. This requires a revision in plan to meet original goals/intent.	Planning	The fuel that was going to be used to move all of the wounded troops was blown up by the enemy forces. In the meantime, the participants must use the available fuel to transport only the most seriously wounded to a different, closer hospital. Then the participants must transfer those that require more extensive treatment to a larger hospital when the replacement fuel arrives in approximately two hours.
Demands on prospective memory (delayed follow-up activity)	Requirements to perform a disconnected activity in the future for which there is no strong memory cue under high attention and working memory load conditions.	Executing	Classic examples are instances where an individual forgets to take the original sheet of paper from the copier machine after making copies, and an individual forgetting to take their bank card out of an automatic teller machine after taking out their cash.

Table 14.2 *Continued*

Factor	Description	Primary Macrocognitive Functions Impacted	Example
Incomplete advice (leave them hanging)	An automated algorithm provides partial advice to solve a problem.	Executing	A Global Positioning System (GPS) provides navigational guidance until a state boundary is crossed, at which time it displays 'Download data for West Virginia.'
Timing issues in following procedures (out of sequence steps)	The expected order for procedural steps is deviated from due to issues in timing activities.	Executing	A software package requires entry of all demographic patient information when a patient record is opened for the first time, a necessary prerequisite step for ordering medications and procedures, but the primary diagnostic code for a patient used in the billing procedure at a later time is not available until the physician reports it to the participant, and the physician is not currently available.
Interruptions (memory bottleneck)	Interruptions make it easy to forget to do unresolved tasks and prioritize tasks appropriately.	Executing	Frequent requests are made by media personnel (who are confederate researchers) to learn updates about the status of a coordinated response by an Emergency Operations Center director to a hurricane.
Mismatch between stimulus and response (mode confusion)	An intended response to a stimulus probe requires an unnecessary translation to a different side, modality, or navigation to a different physical or conceptual space.	Executing	A flashing alert on the right hand side of a touchscreen display requires pushing a button on the left side of the screen to respond.
Negative transfer between tasks (not used to doing it this way)	Identical or similar tasks done in different settings, modes, or procedural sequences require different approaches.	Executing	A person with years of experience driving in the left front seat of a vehicle is required to drive in the right front seat.

Table 14.2 *Continued*

Factor	Description	Primary Macrocognitive Functions Impacted	Example
Similar to different task (capture error)	A new task begins in a way that is similar to an overlearned task.	Executing	Driving to the grocery store begins with the same route as driving to the workplace.
Overconstrained task (can't do it all)	A participant is required to do tasks that cannot be accomplished without relaxing one or more constraints (and sacrificing the associated goals).	Deciding	A nurse is tasked to administer medications within a 30-minute time window, respond to requests from simulated patients to increase their comfort, respond to phone calls from simulated physicians that were previously initiated about urgent patient events, and document the results of responses to pain medication in less time than would be required by top performers.
Doublebind situations (dilemmas)	Doublebinds where two (or more) choices all have highly undesirable elements. With authority-responsibility doublebinds, the participant has the responsibility but not the authority to take an action that is needed to reduce risk or accomplish a goal. With early-late doublebinds, the participant is required to choose between intervening early to head off potential problems at the cost of junior	Deciding	A participant acting as a co-pilot is provided ambiguous cues that indicate that the pilot (who is actually a confederate researcher) has entered a flight plan that would result in their plane crashing into a mountain.

Table 14.2 *Continued*

Doublebind situations (dilemmas) *(cont.)*	personnel losing training opportunities and reputation or at the cost of senior or more powerful personnel disagreeing with the need to intervene, or intervening later when problems might escalate out of control, at which time they cannot be fully addressed. With specialist-generalist doublebinds, the person with generalist knowledge has a difficult time identifying when the benefits of accessed specialized expertise to detect problems and/or greatly improve ongoing plans surpass the costs of adding an interdependency which requires coordination to manage, thereby delaying action and increasing uncertainty. With differential cost-benefit doublebinds (sometimes referred to as 'Grudin's law' situations, Grudin 1994), the participant must choose between helping another person accomplish a task or accomplishing tasks for which they are directly responsible.		

Table 14.2 *Continued*

Factor	Description	Primary Macrocognitive Functions Impacted	Example
Goal conflicts (impossible task)	A participant is tasked to meet two or more incompatible goals simultaneously. Variations on this theme include learning that apparent goal conflicts can be resolved without sacrificing any goals and having hidden goal conflicts emerge over time.	Deciding	A nurse is tasked to scan the wristband of a series of simulated patients in order to actively identify the patients to reduce wrong patient errors. One simulated patient has been diagnosed with an infectious disease (methicillin-resistant Staphylococcus aureus - MRSA). The hospital requires that handheld devices, including barcode scanners, not enter rooms for patients diagnosed with MSRA ('contact precautions'). Therefore, the participant must choose between violating the 'wrong patient' policy or the infectious disease policy.
Workload (time pressured)	Requirements for multiple mental or physical actions that need to be accomplished within a limited period of time.	Deciding Executing Coordinating	A command and control task requires navigating across multiple displays to aggregate required information and execute required control actions under time-pressured conditions.
Stress, fatigue (tunnel vision)	Emotional incentives are created to respond in a manner that differs from desired practices. Fatigue and fear can be induced, making it difficult to follow slow, methodical practices that are likely to avoid mistakes with potentially high consequences. Fear of personnel in positions in higher authority can encourage undesirable	Deciding	An angry commander of a different unit (who is actually a confederate researcher) demands immediate action on a relatively low priority task, collecting weapons from a known cache site, that he believes will save the lives of his men. If the participants respond immediately without questioning the order, lives are lost from not following procedures to verify that the cache is not booby-trapped.

Table 14.2 *Continued*

Stress, fatigue (tunnel vision) (*cont.*)	priority orderings. Fear for the safety of close peers can make it difficult to meet strategic and safety objectives of larger groups of people. Fatigue, stress, and a fast-paced tempo can induce 'tunnel vision,' where it is more challenging to question the framing of tasked activities; in particular, it is hard to step back and 'see the forest for the trees' to make sure that there is not a need to reformulate the overall problem space.	
Weak leadership (in-fighting)	Competing factions undermine the potential effectiveness of the individual participant as well as the overall team. The leader does not intervene, or even actively encourages competitive behavior, amongst competing factions.	Coordinating	An air force intelligence officer is required to brief his superior on the most recent attacks on the armed forces, but the army intelligence officer refuses to share relevant information prior to the briefing.
Unreliable communication systems (poor communication)	Communication tools are unreliable and there are noisy, ambiguous signals.	Coordinating	The pager for a surgeon participant trying to decide if he needs to leave an operation to deal with another patient is unreliable and the text information is hard to read without contaminating his surgical gloves.

Table 14.2 *Concluded*

Factor	Description	Primary Macrocognitive Functions Impacted	Example
Decreased access to team members (remote teams)	Team members are physically distant, distant in time (e.g., time lags for responses from team members in other time zones), or have reduced richness of the communication medium content as compared to real-time face-to-face interactions (e.g., only text or audio interaction capability).	Coordinating	A participant has audio (e.g., SKYPE) communication capabilities, but not video or face-to-face interactions, with team members to perform tasks.
Interdependencies among roles (coordination bottlenecks)	The costs of coordinating work among multiple participants is increased when there are increased interdependencies, including sequencing constraints (A happens before B), resource sharing (A is needed by both B and C and cannot be used by two people at the same time), and synchronization (A and B have to happen at the same time).	Coordinating	A nurse cannot administer an ordered intravenous (IV) medication until a specialist (who is actually a confederate researcher) obtains IV access to the patient's vein.

- *Executing:* Converting a prespecified plan into actions within a window of opportunity (related terms: adapting, implementation, action, act); this includes adapting procedures based on incomplete guidance to an evolving situation where multiple procedures need to be coordinated, procedures which have been started may not always be completed, or when steps in a procedure may occur out of sequence or interact with other actions.
- *Deciding:* A level of commitment to one or more options that may constrain the ability to reverse courses of action. Decision making is inherently a continuous process conducted under time pressure that involves re-examining embedded default decisions in ongoing plan trajectories for the predicted impact on meeting objectives, including whether to sacrifice decisions to which agents were previously committed based on considering tradeoffs. This function may involve a single 'decision maker' or require consensus across distributed actors with different stances toward decisions. (Related terms: decision making, decide, choice, critical thinking, committing to a decision.)
- *Coordinating:* Managing interdependencies across multiple individuals acting in roles that have common, overlapping, or interacting goals. (Related terms: collaboration, leadership, resource allocation, tracking interdependencies, communication, negotiation, teamwork.)

The factors can be embedded within existing scenarios semi-independently to scale up complexity. The goal is to increase the potential for simulating challenges that will generate erroneous or suboptimal human performance without the aid of the planned software artifact (or training or organizational innovation) so as to be able to diagnostically assess under what conditions the software artifact improves performance and where the boundary conditions are beyond which the effectiveness of the software aid breaks down. We view these factors as comprising a working definition of what constitutes facets of cognitive complexity in real-world settings.

Clusters of Complicating Factors

Although we have collaborated with domain experts to augment the complexity of existing scenarios by sequentially going individually through a list of complicating factors such as is provided in Table 14.2, in some cases we have also taken short-cuts in that we embedded probes that naturally contain clusters of these factors. In Table 14.3, we illustrate this approach with a non-exhaustive set of clusters used in a few studies (Roth et al., 1994; Patterson et al., 2001, 2004).

Table 14.3 Clusters of complicating factors

Cluster Probe	Description	Complicating Factors
Problem that requires diagnosis but which also simultaneously impacts capabilities of supporting systems.	Diagnostic activities need to be conducted in parallel with 'safing' activities to keep the system safely operating (e.g., tube rupture in a nuclear power plant). Action is required before a definitive diagnosis can be made.	Multiple hypotheses, stereotype violations, implied relationship, hidden coupling, data overload, signal-noise relationship, interdependencies among roles, over-constrained task, doublebind situations, goal conflicts, and unintended effects.
Leading indicators that are usually relied upon in diagnosis are unavailable.	Indicators that usually provide relatively definitive diagnoses with little ambiguity are no longer available, and so study participants have to use a combination of indicators with which they have less experience to infer what is happening, some of which require coordinating with others to learn the information.	Multiple hypotheses, stereotype violations, implied relationship, data overload, signal-noise relationship, distributed information, interdependencies among roles, and unreliable communication systems.
Accident.	An event with a highly undesired outcome is embedded in the scenario. A flurry of initial reports contain fragmented, inaccurate, emotional information that is hard to synthesize into a coherent narrative. Later reports contain more coherent narratives that could be highly biased and are missing some of the details available in earlier reporting. Many requests for updates are made of study participants in roles of authority, frequently interrupting their activities.	False prime explanations, implied relationship, hidden coupling, data overload, distributed information, over-constrained task, doublebind situations, shifting objectives, stress, and interruptions.
Late change in a plan.	Planning activities conducted during relatively low workload periods need to be reconsidered at the onset of an escalating period of activities due to a late change of plan.	Interdependencies among roles, over-constrained task, doublebind situations, impasses and opportunities, unintended effects, shifting objectives, unreliable communication systems, and decreased access to team members.
Interruption about a prior task that was believed to have been completed while working on a new primary task.	The synchronization of tasks was intended such that a prior task would be completed prior to the initiation of a new task. The participant is required to unexpectedly reorient to a prior task while also maintaining responsibility for a new ongoing task.	Interdependencies among roles, over-constrained task, doublebind situations, goal conflicts, impasses and opportunities, shifting objectives, and decreased access to team members.
Unexpected change of role.	During the course of the scenario, the participant responsible for a role is unexpectedly replaced with a different participant who was not anticipating assuming the role.	Data overload, distributed information, interdependencies among roles, inadequate guidance, goal conflicts, impasses and opportunities, unintended effects, shifting objectives, and stress.

Conclusion

In this chapter, our primary contribution was a list of domain-independent complicating factors, categorized by macrocognitive function, which can be embedded within scenarios so as to scale up complexity and increase representativeness of evaluation scenarios. If the evaluation scenarios are not sufficiently complex during an evaluation, fielded software systems may fail to achieve their desired benefits. Without examining the usefulness of a proposed system and the boundaries for effectiveness, significant loss of life, property, and economic assets could result. In our experience participating in software evaluations across a variety of complex domains, this approach has proven useful for quickly designing complex scenarios for effective software evaluations that provide a richer scenario toolkit than is typically used.

In addition to reducing the risk of a software system failing to be useful after implementation, this approach reduces the risk of costly evaluations failing to demonstrate revolutionary improvements in decision effectiveness. Only by fully exercising software designs can revolutionary improvements in decision effectiveness be demonstrated. In an evaluation, at least some percentage of the scenarios need to predictably and repeatably generate erroneous or suboptimal human performance without the aid of the planned software aid (or training or organizational innovation), so as to increase the diagnostic power of the evaluation study in establishing the potential impact of the software aid on performance. Without testing the 'edges' of system capability, it is also not possible to defend claims about the boundary of conditions within which the aid provides effective support.

Even if the detailed complicating factors are not incorporated in software evaluation scenarios, our overall perspective on the importance of sampling a representative as well as a challenging 'problem space' and adopting a skeptical stance toward predicted benefits of a proposed system on performance may be of value in itself. Specifically, this chapter may be used to remind software developers and 'educated consumers' to consider these kinds of questions prior to system implementation (c.f., Woods & Sarter, 1993; Smith-Jentsch et al., 1998; Roth et al., 2002; Roth, this volume, Potter & Rousseau, this volume):

- What is the model of support (design hypothesis) embodied in the software being evaluated? Do the scenarios provide opportunities to exercise this model of support?
- Do the evaluation scenarios capture the range of complicating factors that arise in the actual operational environment so as to assess extent and boundaries of effectiveness of the aiding concept?
- Have the performance issues (for example, potential vulnerabilities, biases, errors, and breakdown points) that can impact the decisions and related cognitive and collaborative activities of interest been identified? Do the test scenarios create opportunities to assess the impact of the software on these potential performance deficiencies?
- Have situations/probes/target events been embedded in the scenarios so as to create the opportunity for cognitive and collaborative activities of interest to be exercised in an *observable* manner?

- Does the study serve as a tool for discovery—providing a vehicle to uncover additional domain demands, and unanticipated requirements for support, to propel design innovation?

Perhaps most importantly, this chapter also provides a theoretical contribution, by explicitly and operationally defining the facets that contribute to complexity in dynamic, high-risk, real-world settings where cognition is distributed across human agents supported by sophisticated software systems. We view these factors as comprising a working definition of what constitutes (macro)cognitive complexity in real-world settings, which lays the groundwork for developing complexity metrics that could be useful in a variety of applications.

Acknowledgements

This research was supported by the Air Force Research Laboratory (S110000012) and the Office of Naval Research (GRT00012190). The views expressed in this article are those of the authors and do not necessarily represent those of the Air Force or Navy.

References

Bisantz, A. & Roth, E. M. (2008). Analysis of cognitive work. In D. A. Boehm-Davis (Ed.), *Reviews of human factors and ergonomics volume 3* (pp. 1-43). Santa Monica, CA: Human Factors and Ergonomics Society.

Burleson, B. R., & Caplan, S. E. (1998). Cognitive complexity. In J. C. McCroskey, J. A. Daly, M. M. Martin, & M. J. Beatty (Eds.), *Communication and personality: Trait perspectives* (pp. 233-286). Creskill, NJ: Hampton Press.

Campbell D. J. (1988). Task complexity: A review and analysis. *The Academy of Management Review, 13*(1), pp. 40-52

Carroll, J. M. & Rosson, M. B. (1990). Human-computer interaction scenarios as a design representation. In *Proceedings, Vol II of HICSS-23. 23rd Hawaii International Conference on System Sciences, Software Track*, B. D. Shriver (Ed.), (pp, 555-561). Los Alamitos, CA: IEEE Computer Society Press.

Chi, M. T. H., Feltovich, P. J. & Glaser, R. (1981). Categorization and representation of physics problems by experts and novices. *Cognitive Science 5*(2), 121-152.

Cockburn, A. (2001). *Writing effective use cases.* Boston, MA: Addison-Wesley Longman Publishing.

Dekker, S. & Woods, D. D. (1999). Extracting data from the future: Assessment and certification of envisioned systems. In S. Dekker & E. Hollnagel (Eds.), *Coping with computers in the cockpit* (pp. 7-27). Aldershot, UK: Ashgate.

Eggleston, R. G., Roth, E. M. & Scott, R. A (2003). A framework for work-centered product evaluation. In *Proceedings of the Human Factors and Ergonomics*

Society 47th Annual Meeting. (pp. 503-507). Santa Monica, CA: Human Factors and Ergonomics Society.

Evenson, S., Muller, M. & Roth, E. M. (2008). Capturing the context of use to inform system design. *Journal of Cognitive Engineering and Decision Making, 2*(3), 181-203.

Feltovich, P. J., Hoffman, R. R., Woods, D. D., & Roesler, A. (2004). Keeping it too simple: How the reductive tendency affects cognitive engineering. *IEEE Intelligent Systems, 19*(3), 90-94.

Feltovich, P. J., Spiro, R. J., & Coulson, R. L. (1997). Issues of expert flexibility in contexts characterized by complexity and change. In P. J. Feltovich, K. M. Ford & R. R. Hoffman, *Expertise in context: Human and machine* (pp. 135-146). Menlo Park, CA: AAAI Press.

Grudin, J. (1994). Groupware and social dynamics: Eight challenges for developers. *Communications of the ACM, 37*(1), 92-105.

Hollnagel, E. & Woods, D. D. (1983). Cognitive systems engineering: New wine in new bottles. *International Journal of Man-Machine Studies, 18*(6), 583-600.

Hollnagel, E., Woods, D. D., & Leveson, N. (Eds.) (2006). *Resilience engineering: Concepts and precepts.* Aldershot, UK: Ashgate Publishing.

Johnston, J. H., Cannon-Bowers, J. A., & Smith-Jentsch, K. A. (1995). Event-based performance measurement system. In *Proceedings of the Symposium on Command and Control Research and Technology*, June (pp. 268-276).

Leveson, N. (in preparation). Engineering a Safer World. http://sunnyday.mit.edu/book2.pdf

Mumaw, R. J. & Roth, E. M. (1992). How to be more devious with a training simulator: Redefining scenarios to emphasize cognitively difficult situations. *1992 Simulation Multi-Conference: Nuclear Power Plant Simulation and Simulators*, Orlando, FL, April 6-9.

Patterson, E. S., Rogers, M. L., & Render, M. L. (2004). Simulation-based embedded probe technique for human-computer interaction evaluation. *Cognition, Technology, and Work, 6*(3), 197-205.

Patterson, E. S., Roth, E. M., & Woods, D. D. (2001). Predicting vulnerabilities in computer-supported inferential analysis under data overload. *Cognition, Technology, and Work, 3*(4), 224-237.

Rasmussen, J. (1986). *Information processing and human-machine interaction: An approach to cognitive engineering.* Amsterdam: North-Holland.

Rasmussen, J. & Lind, M. (1981). Coping with complexity. In H. G. Stassen (Ed.), *First Europena annual conference on human decision making and manual control.* New York, NY: Plenium.

Rosson, M. B & Carroll, J. M. (2002). *Usability engineering: Scenario-based development of human-computer interaction.* San Francisco, CA: Morgan Kaufmann Publishers.

Roth, E. M. (2008). Uncovering the requirements of cognitive work. *Human Factors, 50*(3), 475-480. (Golden Anniversary Special Section on Discoveries

and Developments). http://www.ingentaconnect.com/content/hfes/hf/2008/00
000050/00000003/art00022

Roth, E. M., Gualtieri, J. W., Elm, W. C., & Potter, S. S. (2002). Scenario
development for decision support system evaluation. In *Proceedings of the
Human Factors and Ergonomics Society 46th Annual Meeting* (pp. 357-361).
Santa Monica, CA: Human Factors and Ergonomics Society.

Roth, E. M., Mumaw, R. J., & Lewis, P. M. (1994). *An empirical investigation
of operator performance in cognitively demanding simulated emergencies.*
Washington DC: US Nuclear Regulatory Commission (NUREG/CR-6208).

Roth, E. M. & Woods, D. D. (1988). Aiding human performance: I. Cognitive
analysis. *Le Travail Humain, 51*(1), 39-64.

Roth, E. M., Woods, D. D. & Pople, H. E. (1992). Cognitive simulation as a tool for
cognitive tasks analysis. *Ergonomics: special issue on Cognitive Engineering,*
35(10), 1163-1198.

Schon, D. A. (1983). *The reflective practitioner: how professionals think in action.*
London, UK: Temple Smith.

Smith, P. J., Galdes, D., Fraser, J., Miller, T., Smith, J. W., Svirbely, J. R., Blazina,
J., Kennedy, M., Rudmann, S., & Thomas, D. L. (1991). Coping with the
complexities of multiple-solution problems: A case study. *International
Journal of Man-Machine Studies, 35*(3), 429-453.

Smith-Jentsch, K. A., Johnston, J. H. & Payne, S. (1998). Measuring team-related
expertise in complex environments. In J. A. Cannon-Bowers & E. Salas (Eds.),
Making decisions under stress: Implications for individual and team training.
Washington, DC: American Psychological Association.

Watts-Perotti, J., & Woods, D. D. (2007). How anomaly response is distributed
across functionally distinct teams in Space Shuttle Mission Control. *Journal of
Cognitive Engineering and Decision Making, 1*(4), 405-433.

Woods, D. D. (1988). Coping with complexity: The psychology of human behavior
in complex systems. In L. P. Goodstein, H. B. Andersen, & S .E. Olsen (Eds.),
Mental models, tasks and errors (pp. 128-148). London: Taylor & Francis.

Woods, D. D. & Hollnagel, E. (2006). *Joint cognitive systems: Patterns in cognitive
systems engineering.* New York, NY: Taylor & Francis.

Woods, D. D., Johannesen, L., Cook, R. & Sarter, N. (1994). *Behind human error:
Cognitive systems, computers and hindsight.* Dayton, OH: Crew Systems
Ergonomic Information and Analysis Center, WPAFB.

Woods, D. D., Roth, E. M. & Pople, H. Jr. (1988). Modeling human intention
formation for human reliability assessment. *Reliability Engineering & System
Safety, 22*(1-4), 169-200.

Woods, D. D., & Patterson, E. S. (2001). How unexpected events produce an
escalation of cognitive and coordinative demands. In P.A. Hancock & P.A.
Desmond (Eds.), *Stress workload and fatigue.* (pp. 290-304). Hillsdale, NJ:
Lawrence Erlbaum Associates.

Woods, D. D. & Sarter, N. (1993). Evaluating the impact of new technology on human-machine cooperation. In J. Wise, V. D. Hopkin, & P. Stager (Eds.), Verification *and validation of complex systems: Human factors issues*. Berlin: Springer-Verlag.

Chapter 15

Evaluating the Resilience of a Human-Computer Decision-making Team: A Methodology for Decision-Centered Testing

Scott S. Potter and Robert Rousseau

Introduction

In developing new decision support systems, it is essential to evaluate the impact of the new system on the decision making it was designed to support. Embracing the perspective of macrocognition, defined in complementary ways as (1) the cognitive functions that are performed in natural decision-making settings (Cacciabue & Hollnagel, 1995) or (2) how cognition adapts to complexity (Klein, 2007; Rasmussen & Lind, 1981) imposes significant challenges for evaluation of new decision support systems (Roth & Eggleston, Chapter 13). This perspective requires using measures that go beyond typical performance metrics such as the number of subtasks achieved per person per unit of time and the corresponding simple baseline comparisons or workload assessment metrics. For instance, for a new naval command and control system evaluation, typical performance measures might be the percent of correct identifications and proximity of engagement to optimal probability of kill within the period of watch. This procedure is likely to involve standard scenarios to facilitate comparisons between systems, but does not address the resilience of the Joint Cognitive System (JCS) in the face of novel situations outside the boundaries. However, in order to impact these types of evaluations, macrocognition-grounded evaluation frameworks must provide guidance with respect to defining the characteristics of the evaluation and corresponding metrics.

In response to the need to evaluate the decision-making effectiveness of a practitioner teamed with a new decision support system, we have developed an evaluation methodology based on principles from Cognitive Systems Engineering (CSE). Decision Centered Testing (DCT; Rousseau et al., 2005) aims at testing the effectiveness of operators teamed with Decision Support Systems (DSS) in any challenging work domain. DCT is grounded in a CSE framework, where the concept of a Joint Cognitive System (JCS) is central. We define a JCS as the combination of human problem solver and automation/technologies which must act as co-agents to achieve goals and objectives in a complex work domain (cf. Hollnagel & Woods, 2005 for their cyclic model of the function of JCS).

An implication of our CSE foundation is that a JCS needs to be evaluated for its effectiveness in performing the complex cognitive work requirements. This JCS perspective implies that the system must be designed and evaluated from the perspective of the teaming of the practitioner and the DSS. Metrics will have to reflect such phenomena as resilience of a JCS. This places new burdens on evaluation techniques and frameworks, since metrics should be generated from a principled approach and based on fundamental principles of interest to the designers of the JCS.

DCT involves explicit design and analysis of tests based on the key decision-making demands within the naturalistic work domain. The result is an explicit test design describing the cognitive problem under test, the hypothesized edge or latent potential weaknesses in the JCS, as well as the events that need to be included in the scenario. In DCT, test scenarios are developed to specify a progressive evolution of events that would be expected to stress the defined edge within the JCS. This decision-centered approach to testing has proven effective in discovering fundamentally new ways for evaluating the net decision-making effectiveness of the joint human-technology decision-making team. Ultimately, DCT is about evaluating the resilience of the JCS.

This chapter describes the DCT methodology and the initial application of the methodology in a competitive game micro-world domain. This application enabled the definition of appropriate test metrics and the construction of unique test scenarios to exercise the decision-making effectiveness. From this application, insights were gained regarding the construction of an evaluation framework for assessing the net decision-making effectiveness of a JCS.

The Decision Centered Testing (DCT) Methodology

Overview

The critical aspects of this technique are:

- The evaluation needs to **address the attributes/characteristics that a joint cognitive system must have** in order to effectively conduct cognitive work in any goal-directed task (Woods & Hollnagel, 2006) rather than more pure decision-making abilities of the human agent. These include requirements such as observability, directability, teamwork with agents, and directed attention. For example, testing for observability might entail assessing the insights into a process that the user was able to gain through the representation provided by the DSS.
- The evaluation needs to be **focused on the fundamental cognitive demands of the natural decision-making setting**. To meet this need, we have found significant leverage from an analytical model of cognitive demands of the work domain derived from a cognitive work analysis. This has provided a

pragmatic approach to focus the evaluation on the most profitable functions (since an exhaustive evaluation is infeasible).

- In order to evaluate the resilience of the JCS, it is important to **define potential weaknesses, or fissures in the JCS** where decision-making effectiveness may break down under pressure. Within the DCT methodology, these weaknesses are referred to as 'edges.' An edge is any discontinuity within any of the relationships within a JCS. This is in contrast to approaches which simply manipulate task difficulty until natural limits are reached.
- Once the target of the DCT has been identified, **scenarios need to be defined explicitly on the analytical basis in order to exercise the desired cognitive demands**. Scenarios can be derived from the specific edge under investigation. For example, a DCT focused on selection between competing goals would include scenarios specifically designed to emphasize the different goals.
- **Cognitive pressure needs to be defined to explicitly stress the edges** and therefore assess the resilience of the JCS. Cognitive pressure is not simply swamping or overloading the JCS by the number of stimuli or amount of cognitive work to be performed. Rather, it is a carefully crafted manipulation of controllable aspects of the work environment related to JCS support requirements, cognitive demands of the work domain, and/or edges in the JCS.

Phase—Describing the Work Domain (or World)

As the starting point to DCT, it is essential to focus the evaluation on a particular target within the JCS. To start this process, the work domain must be understood. The work domain is the set of constraints, boundary conditions, forcing functions, and so on, and resulting events that occur. If humans desire to influence the work domain so as to achieve one or more desired goals, then there is usually a set of decisions and actions that must take place, often in a timely manner and sequence, in order to make that influence happen. These decisions and actions are demands that are required by the work domain without respect for the capabilities, or lack thereof, of the humans that wish to influence the events in the work domain.

For the purposes of DCT, the demands and constraints of the work domain must be understood in order to appropriately specify the context for the evaluation. This can be accomplished very effectively by a work domain analysis. While typically this analysis serves as the context for a decision support system to be designed, in DCT it is also used to identify potential target areas for the evaluation. For example, particularly complex functional nodes or relationships between functional nodes indicate exceptional demands on the decision making of the human plus decision support system.

Phase—Defining the Test Objective/JCS Support Requirement

Once the particular target area within the work domain has been identified, it is possible to define the particular test objective. Because the CSE JCS Support Requirements identify the attributes or characteristics of an effective JCS, they define the objectives of DCT and provide focus for the detailed test design. This CSE framework makes it possible to identify JCS support requirements (Christoffersen & Woods, 2002; Elm et al., 2005). These requirements describe the attributes/characteristics that a joint cognitive system must have in order to effectively conduct cognitive work in any goal-directed task.

Previous research in CSE and our own experience led us to identify a set of generic support requirements that apply to cognitive work by any cognitive agent or any set of cognitive agents, including teams of people and machine agents (Billings & Woods, 1994; Dekker & Woods, 1999: Christoffersen & Woods, 2002; Elm et al., 2005). These include:

- *Observability*—the ability to form insights into a process (either a process in the work domain or in the automation), based on feedback received. Observability overcomes the 'keyhole' effect and allows the practitioner to see sequences and evolution over time, future activities and contingencies, and the patterns and relationships in a process.
- *Directability*—the ability to direct/redirect resources, activities, and priorities as situations change and escalate. Directability allows the practitioner to effectively control the processes in response to (or in anticipation of) changes in the environment.
- *Teamwork with agents*—the ability to coordinate and synchronize activity across agents. This defines the type of coordination (for example, seeding, reminding, critiquing) between agents. Teamwork with agents allows the practitioner to effectively redirect agent resources as situations change.
- *Directed attention*—the ability to re-orient focus in a changing world. This includes issues like tracking others' focus of attention and their interruptability. Directed attention allows the human-system team to work in a coordinated manner, resulting in increased effectiveness.
- *Resilience*—the ability to anticipate and adapt to surprise and error. This includes issues such as failure-sensitive strategies, exploring outside the current boundaries or priorities, overcoming the brittleness of automation, and maintaining peripheral awareness to maintain flexibility.

The selection of a decision-making focus needs to include the selection of the JCS support requirement as the test objective. That said, the task of selecting which support requirement to choose as the test objective for a particular test is not straightforward. This effort has found that there is an appropriateness factor involved in the selection, that is, not all of the support requirements are relevant or appropriate to a particular decision-making requirement. As a result, the selection of a decision-making focus needs to include the selection of the JCS support requirement as the test objective. For example, a decision in the work domain that requires that aspects of the work domain be made observable in order to assure an effective JCS may not mean that the same decision is at all dependent on work domain entities being directable or that there is any relationship between team work with other agents and effective cognitive coupling for that decision, and so on. As a result, effective test design is something of an iteration process to ensure that the decision-making focus is appropriately difficult and that the test objective is applicable in order to establish effective cognitive coupling for that decision.

Phase 3—Defining the Decision making under Investigation

For the DCT Methodology to be applied, it is essential that the specific decision making to be assessed be clearly identified and defined. Obviously, the decision making under investigation must be within the scope of support of the DSS. In the context of the DCT methodology, the term 'cognitive work' or 'decision making' refers to the set of activities that include (from Rasmussen, 1986):

- data collection due to alerts related to problems, or as a result of casual monitoring;
- state identification;
- comparison against desired states or goals;
- planning for remedial action if the current state and the desired state do not match satisfactorily;
- remedial action execution; and
- evaluating the feedback resulting from the action against goal achievement.

This decision-making definition can be achieved by the application of a variety of cognitive analysis approaches, but the important issue is that the demands and constraints that make the identified cognitive work difficult also be identified as well. In this way, insights into potential manipulations in the test conditions can be derived. For example, if the decision making under investigation in a military anti-missile defense domain is 'determine the engagement sequence against a set of incoming missiles,' the demands on the decision making include temporal (that is, time to decide) and volume (that is, number of incoming missiles) aspects. However, these demands by themselves are insufficient to truly exercise the decision-making challenges of the JCS.

Phase 4—Specifying the Decision Support System (DSS) to be Evaluated (and the Joint Cognitive System (JCS) that is Created)

In order to fully specify the JCS (and thus conduct a DCT), the support tools to be used by the decision maker must be specified. Though not necessarily limited to computer-based implementations, the decision-making aids that are considered in this effort include a wide range of multi-functional computer-based DSSs. Therefore, it is important to define the specific characteristics of the DSS to be evaluated. This included the types of representations provided, the degree of automation, the nature of the user interaction requirements, and the scope of responsibility (that is, the interactions that are supported and those that are not supported).

This description of the DSS is performed in order to provide the basis for the identification of potential edges in the JCS (to be described in the next section). The critical issue is that there is a wide spectrum of JCSs that are created based on the characteristics of the work domain and the DSS. These characteristics must be described in order to define the focus for the DCT.

Phase 5—Identifying Edges in the Joint Cognitive System (JCS)

Given that the focus of the evaluation is on the decision making of the JCS as a single unit, it is important to define potential weaknesses, or fissures, in the JCS where decision-making effectiveness may break down under pressure. Within the DCT methodology, these weaknesses are referred to as 'edges.' An edge is any discontinuity within any of the three relationships in the cognitive triad. This could include:

- events or changes in the work domain that are not represented well by the DSS;
- automatic control actions taken by the DSS but not communicated to the user;
- transition points within an algorithm that are not represented to the user.

DCT attempts to exercise these edges by assessing the decision-making effectiveness of the JCS as it works across an edge. A successful transition across an edge is a characterization of a resilient JCS, while decision-making breakdown at an edge is the mark of an ineffective JCS.

It is important to note (as indicated in Figure 15.1) that an edge is not equivalent to simply overloading the JCS by domain events until a decision-making breakdown occurs (for example, the number of simultaneous incoming missiles). While the focus of an evaluation is often to identify any improvement in the maximum overall capacity of the JCS, DCT is focused on the latent, brittle edges in the JCS design. Thus, an edge in an anti-missile defense domain might be the combination of heterogeneous bearing/distance/speed of incoming missiles to which the operator is unable to solve (in real time) a solvable multi-threat situation. Note that this is not simply increasing the number of incoming missiles until the defensive position is hit, as a large number of missiles on the same bearing at the same speed and distance could be effectively engaged without difficulty.

Phase 6—Determining Cognitive Pressure to Apply to the Edge(s)

Once the particular edge has been identified, it is important to determine the most appropriate pressure to apply against the edge. In the DCT methodology, this is referred to as cognitive pressure. DCT stresses the JCS through the application of cognitive pressure on the three basic relationships in the cognitive triad (humans, technology, work domain) in an attempt to cause differences in decision-making effectiveness. Within a particular experiment, cognitive pressure is the independent variable that is manipulated. Cognitive pressure is not simply swamping or overloading the JCS by the number of stimuli or amount of cognitive work to be performed. Rather, it is a carefully crafted manipulation of controllable aspects of the work environment related to JCS support requirements, cognitive demands of the work domain, and/or edges in the JCS. In the anti-missile defense domain previously mentioned, the cognitive pressure would be variations in bearing, distance, speed, and number of missiles centered around an equivalent time-to-impact point. One of the critical aspects of cognitive pressure in DCT is that the manipulation is based on the hypothesized edge and therefore is more focused than traditional evaluation approaches. As indicated in Figure 15.1, applying pressure around a particular edge can result in performance decrements in a significantly different point in the spectrum than simply overloading the JCS. In the figure, moving from right to left on the 'Cognitive Pressure' dimension indicates an increase in pressure. The

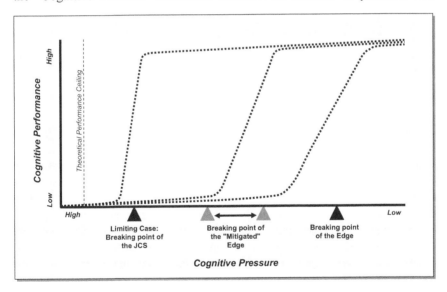

Figure 15.1   **Applying cognitive pressure against a hypothesized edge vs. generic cognitive pressure. The breaking point at the edge will be significantly different from the limiting case (not at an edge)**

hypothesis of DCT is that testing a particular 'edge' will result in performance decrements far sooner than the limiting case. The limiting case is indicated by the left performance curve, the breaking point of the edge is indicated by the right curve, and the potential performance improvements achieved by mitigating the edge is indicated by the middle curve. Note how the theoretical performance ceiling is represented by the greatest amount of cognitive pressure that can be applied before performance drops.

Phase 7—Defining Scenarios

The most difficult and also most important contribution of DCT is in defining scenarios to instantiate the desired cognitive pressure to stress the edge under test. Scenarios can include events and behaviors within the work domain, DSS, or operator. Scenarios can be derived from the specific JCS support requirement being tested. For example, in a DCT focused on directed attention, there will need to be event(s) in the scenario to establish attention away from the focus area in order to then assess the effectiveness of the JCS in re-orienting attention to the proper focus area. In addition, scenarios can be derived from the specific decision-making demands of the work domain being tested (Roth et al., 2002; see Patterson et al. (Chapter 14) for a discussion of 'complicating factors' that can serve as a key basis for scenario design).

As another example, a DCT focused on the cognitive work of inferring degree of threat of various contacts in a military command and control (C2) domain will need to include characteristics of what makes this decision difficult (for example, missing or inconsistent data) in the scenarios. Also, scenarios can be derived from the specific edge under investigation. For example, a DCT focused on coordination across automatic mode shifts by the DSS will need to include events that trigger mode shifts in the scenarios.

Phase 8—Defining Cognitive Performance

Defining cognitive performance within the DCT methodology focuses on measuring the cognitive work under investigation. These metrics must be indicative of JCS decision-making effectiveness. As noted earlier, the focus of DCT is not on assessing the overall outcome of the decision making, but rather the effectiveness of the decision making in achieving this outcome. Thus, for a given scenario, two different JCSs could result in the same outcome but with very different measures of decision-making effectiveness. For example, in the anti-missile defense domain, the performance metrics would address decisions about selecting the most optimal engagement sequence for the multiple incoming missiles (that is, did the subject engage with the most tactically efficient sequence?) rather than assessing the overall outcome (that is, was the subject able to engage all of the missiles?).

An Application of the Methodology

Test Summary

The DCT methodology was applied to a competitive game micro-world work domain where two opponents are competing to be the first to sum scores to 15 points, requiring both offensive and defensive maneuvers to be considered during each turn (see Elm et al., 2003 for a complete description and analysis of the domain). The JCS support requirement under test was observability. Specifically, it is observability provided by the DSS of multiple goals (offensive and defensive) supported by the decision making. That is, any move impacts both offensive and defensive goals. If a player has better observability into these supported goals, it is more likely that a better decision can be made more quickly.

Figure 15.2 illustrates the relationship between goals for this micro-world work domain. The offensive and defensive competition for the use of 'select your next piece' is explicitly evident in the two high-level goals (the goal on the left is 'maximize own win potential' and the goal on the right is 'minimize adversary's win potential') defining competing goals for the selection of any game piece. In playing the game it is often observed that players without effective decision support will fixate on one goal or the other, sometimes missing the opportunity to pick the winning piece, or failing to pick the piece that their opponent uses to win on their next turn. Players have even failed to recognize that they have a win available among their collection of pieces.

Three different DSSs were tested. The first was essentially an unaided JCS—the DSS did not provide any real observability. The second was a partial representation design, in which some of the fundamental concepts of the work domain were represented in the DSS. The third was a full representational design, with an elegant mapping of the fundamentals of the work domain captured in an intuitive DSS.

The edge under investigation for this DCT was the lack of observability of the multiple goals (offensive and defensive) provided by the representation of the work domain. That is, selecting any game piece impacts both offensive and defensive win potential. If a player has better observability into these supported goals, it is more likely that a better decision can be made more quickly. This edge is truly a characteristic of the JCS. First, it is an intrinsic aspect of the work domain (as reflected in the Functional Abstraction Network from Figure 15.2), as indicated by one function supporting multiple higher-level goals. This is a reflection of the fact that each piece carries, in addition to its numerical value, a contribution to both offensive and defensive objectives; a concept referred to as 'win potential.' Second, it is a potential characteristic of the artifact (decision-support tool), as the representation may or may not encode information about offensive and defensive win potential for the practitioner.

In this experiment, cognitive pressure was manipulated in two ways. First, scenarios were designed to explicitly represent the two supported goals within

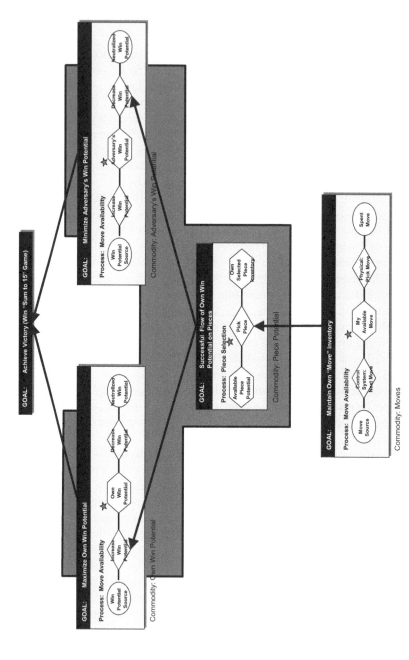

Figure 15.2 Specific region of the Functional Abstraction Network for the micro-world work domain highlighting the multiple supported goals in need of observability

the work domain. Second, the amount of time available for the participants to make their selection was varied. These times were chosen based on pilot studies to encompass the entire range from 'not enough time to make the selection' to 'plenty of time to make the selection.' Cognitive performance was determined to be the ability to make the best selection for the given scenario. This was a result of the scenarios providing a best selection in each case and the remaining selection options all being suboptimal.

Three functionally distinct types of scenarios were designed. For each, the test was set up so that the participant would encounter the possibility to satisfy at least one supported goal. That is, the game board was set up so that the participant had the chance to select a game piece that would either win the game (scenario type #1), prevent the opponent from winning the game (scenario type #2), or contain an opportunity to win and prevent the opponent from winning (scenario type #3). In each scenario, the game was a mid-game situation, with only half of the game pieces already selected. However, each offensive scenario was also an end-game situation, with the potential to win the game on the current turn.

Forty-nine subjects served as participants in this study. Sixteen of these served as calibration and pilot subjects to determine the appropriate response durations for each of the DSSs. The 33 subjects who served in the actual evaluation were randomly assigned to one of the three DSSs for the experiment. The subjects were college students who volunteered for this experiment. Each subject received a standardized set of instructions prior to beginning the experiment. As part of this training, they each played an offline version of the game with the experimenter to assess that they understood the rules. The first eight trials of the experiment were considered practice to acclimate the participants to the variable response durations they were to experience.

After training and practice, each subject was shown a series of game conditions (test stimuli) with six different response durations and eight instances of the three scenarios. The eight instances were designed so that each of the game pieces was involved in the instances. This experimental design resulted in $6 \times 8 \times 3 = 144$ trials. These 144 trials were presented in a completely randomized manner.

Results

The DCT test results were plotted using the two Cognitive Pressure manipulations as the independent variables and the Cognitive Performance metric as the dependent variable. For each of the subjects, the percent correct for each of the response durations and scenarios were computed and plotted. Then, a regression polynomial was fit for the entire set of data for each of the DSSs (as shown in Figure 15.3). This was used to estimate the cognitive performance threshold. The response threshold used for this experiment was the midpoint between chance performance (20 percent in this case since there were five possible selections for each trial) and perfect performance (100 percent); therefore the threshold was 60 percent correct responses.

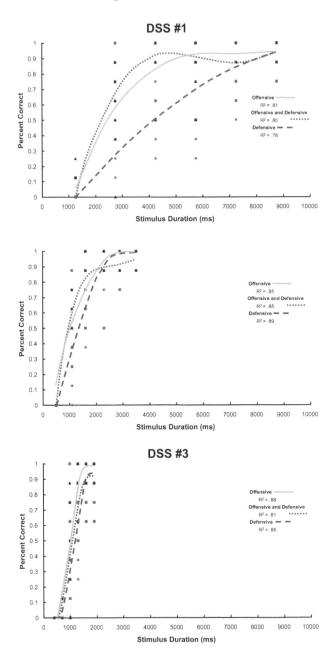

Figure 15.3 **Decision-making results for the three DSSs. Percent correct for each of the scenarios is plotted for each of the durations used in the experiment. The three graphs are vertically aligned to permit visual comparisons across the DSSs**

For each of the DSSs, a third order polynomial was used as the best fit for the data. These polynomials resulted in R^2 values of 0.859, 0.863, and 0.906 for the three DSSs, respectively. All of these regression equations are excellent fits for the data. These correlation coefficients indicate that the amount of variability decreased as the representational quality of the DSS increased.

The performance thresholds were used as the primary metric for comparing cognitive performance between the two DSSs (Gesheider, 1997). As indicated in Table 15.1, the overall thresholds for the three DSSs were 3175 ms, 1200 ms, and 1175 ms for DSSs 1, 2, and 3, respectively. This results in a difference in thresholds between DSSs #1 and #2 of 1975 ms, between DSSs #2 and #3 of 25 ms, and between DSSs #1 and #3 of 2000 ms. These results indicated that, compared to DSS #1, subjects were able to achieve equivalent cognitive performance in approximately 37 percent of the time with either DSS #2 or DSS #3. In addition, 95 percent confidence intervals at these response thresholds were calculated as a means of comparing the differences between the response thresholds for the three DSSs.

Table 15.1 **Response thresholds for the three DSSs and the three scenarios. These values are all based on the polynomial models presented in Figure 15.3**

	DSS		
	1	2	3
Offensive	3125	1050	1100
Defensive	4950	1550	1250
Offensive + Defensive	2950	1250	1175
Overall	3175	1200	1175

Discussion—Decision Support Insights

1. Differences in observability of supported goals between the three DSSs This evaluation provides support (as hypothesized) for the strong degree of observability of the offensive and defensive goals provided by the representations of both DSSs #2 and #3. This was evidenced by the differences in threshold values between the DSS. With this increased observability (achieved by explicit representation of the higher-order goals), the decision-making threshold was dramatically reduced. This dramatic increase in cognitive performance in this DCT provides strong support for success in identifying a key fundamental property of the work domain as the essential component in the design of the two DSSs.

These differences in observability were assessed by calculating 95 percent confidence intervals for each of the response threshold estimates. These results are indicated in Figure 15.4. This analysis indicates that DSSs #2 and #3 were significantly different from DSS #1 but not different from each other (based on the midpoints of the confidence intervals for both DSSs #2 and #3 being outside of the confidence interval of DSS #1). Because of the dependence on the representational properties of the DSS, effective performance required an effective representative design concept that capitalizes on human perceptual capabilities. It is important to note that the decision-making benefit of DSSs #2 and #3 representations was demonstrated without any explicit instructions or descriptions of the potential shift from a cognitive to perceptual task. Participants were only told of the layout of the game pieces and how the possible winning combinations are arranged in the DSSs.

2. Differences in decision-making heuristics   As indicated in Figure 15.3, the results provided some interesting insights on the decision-making heuristics that appear to be used with the different DSSs. Cognitive performance thresholds are consistently lower for the two offensive scenarios with DSS #1. This difference was evidenced to a much lesser degree with DSS #2 and virtually nonexistent in DSS #3 representation. This pattern of results was observed in all subjects.

This suggests a coping strategy with DSS #1 in which an offensive move was investigated first. Then, if none existed, a defensive move was pursued. At the

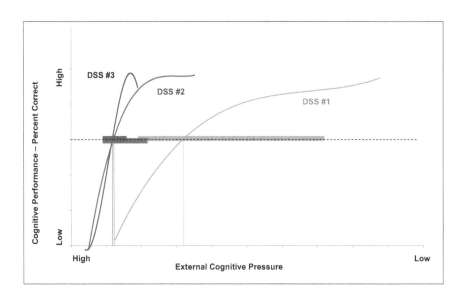

Figure 15.4   Confidence intervals for the three DSSs. These were computed at the performance threshold (60% correct). Note that the confidence bars are staggered in the graph to aid in discrimination

moderately short response durations, subjects were able to respond correctly most of the time in offensive scenarios but only minimally in defensive scenarios.

This sequential strategy would appear to be a result of the lack of decision support—the adaptive and flexible human problem solver developed a workaround based on the characteristics of the work domain. However, with proper decision support, the human problem solver was able to respond to either offensive or defensive with equal efficiency.

This is an interesting example of a superficial decision-making strategy that develops as a result of ineffective decision support. The essential issue from a DSS design perspective is that the heuristic can and should not be the focus. Designing to support this coping strategy would only lead to marginal improvements in decision-making effectiveness.

3. Reduced variability in cognitive performance   Also evident in the figures is a reduction in the variability of the responses from DSS #1 to both DSS #2 and #3. This was evidenced in three ways. First, the R^2 values for the regression estimate for the cognitive performance with DSSs #2 and #3 were higher than for DSS #1. Second, the confidence intervals at performance threshold for both DSSs #2 and #3 were substantially smaller than for DSS #1. Third, the slope of the cognitive performance versus. pressure curve for DSSs #2 and #3 is substantially steeper than that of DSS #1. While these three findings were not subjected to statistical tests, they were all evident based on visual inspection of the data.

These results indicate that not only did the increased Observability of the supported goals dramatically improve cognitive performance, but it also made performance more consistent across subjects. This is most likely explained by individual differences across subjects—some people were quicker at performing the computations required to determine if a win or block was feasible. The slight difference between DSSs #2 and #3 is explained by the reduced visual scanning requirements of DSS #3, as DSS #2 contained redundancy in its representation game state.

This is an interesting example of DCT-based evidence of an effective DSS. As effective support is provided for the fundamental decision-making requirements, there should be less variability in the decision-making effectiveness from individual differences. In addition, as indicated in Figure 15.1, the breaking point of the JCS (that is, the amount of cognitive pressure required to result in performance degradation) is hypothesized to be a much more significant drop-off in cognitive performance (as the limits of human abilities are reached) than the breaking point resulting from an edge (where performance is much more variable). This was certainly evidenced in the current DCT—small changes in response duration (cognitive pressure) made a significant impact on percent correct (cognitive performance) for DSS #3 (elegant representation design) as the upper limits of human abilities were reached.

Discussion—Evaluation Methodology Insights

1. Edge definition   For this experiment, identifying the edge to be investigated provided significant insights for constructing the DCT. Focusing the experiment on the one-to-many mapping between decision making and goals supported grounded the entire experiment in the fundamental decision-making difficulty in the micro-world work domain. This was the critical issue in defining the different scenarios to be used, the cognitive pressure, as well as the cognitive performance metrics.

This insight provides a useful analytical pattern that can guide future DCTs as well. This type of structure in previous Cognitive Work Analyses has been an indicator of constraints on cognitive work that have proven to be difficult. Therefore, exploiting this pattern should prove to be valuable for targeting evaluations around the essential cognitive demands.

2. Scenario generation   Defining scenarios based on edges within the JCS, defined by characteristics of the work domain's functional structure, is a tremendous insight provided by DCT. In this experiment, the scenarios were very different than typical evaluations in this type of decision-making domain. Rather than conducting complete games and deriving some decision-making score for the selections made by the participants, our three scenarios defined a best selection for each instance, making for remarkably straightforward analysis of the response data.

In this DCT, the different scenarios were explicitly based on the edge definition. This provides an evaluation methodology insight that should be an essential aspect of all DCTs. With this explicit mapping, the DCT was able to provide explicit insights into the observability of the multiple goals to be supported.

3. Cognitive pressure   In this experiment, cognitive pressure proved to be extremely effective at highlighting the identified edge within the work domain. This evaluation required quite extensive pilot studies to identify the breakpoint around the edge under investigation. By then using that as the base pressure and then manipulating the available response duration, cognitive performance was able to be varied from 0 percent to 100 percent for almost all of the subjects. This permitted a clear identification of the performance breaking point due to the edge and a clear comparison between the DSSs.

As discussed previously, applying cognitive pressure against the hypothesized edge is an essential component of any DCT. By having the hypothesized edge defined, it provides a prescriptive methodology for defining the types of manipulations to determine the effectiveness of the JCS at dealing with the edge.

4. Cognitive performance   The critical issue in defining our cognitive performance metric for this evaluation was the creation of scenarios around the edge to be investigated. With these three scenarios, we were able to define a best response for each instance of the scenarios. This avoided any ambiguity in the interpretation of, or scoring of, responses. This allowed cognitive performance to be able to be assessed with a simple transformation of the individual responses into a percentage correct. This permitted data analysis based on psychophysics experiments to be applied to the results, which resulted in a clear metric for comparing the three JCSs.

Conclusions

This DCT methodology application provides strong support for the framework for defining DCT evaluations based on characteristics of the JCS as identified by a cognitive analysis of the work domain. The insights gained by using this methodology provided unique assessment of the decision-making differences between the three JCSs. However, future DCTs need to evaluate the decision-making effectiveness in a wide variety of situations in order to evaluate the robustness of the methodology to different DSSs. For example, evaluations need to be conducted of the different JCS support requirements, different types of edges in the JCS, and scenarios to instantiate the cognitive pressure. By further exploring the evaluation space of DCT, a more robust methodology can be developed to evaluate the decision-making effectiveness of the JCS.

References

Billings, C. E. & Woods, D. D. (1994). Concerns about adaptive automation in aviation systems. In M. Mouloua & R. Parasuraman (Eds.), *Automation and human performance: Recent research and trends* (pp. 264-269). Hillsdale, NJ: Erlbaum.

Cacciabue, P. C. & Hollnagel, E. (1995). Simulation of cognition: Applications. In J. M. Hoc, P. C. Cacciabue, & E. Hollnagel (Eds.), *Expertise and technology: Cognition and human-computer cognition* (pp. 55-73). Mahwah, NJ: J Lawrence Erlbaum Associates.

Christoffersen, K. & Woods, D. D. (2002). How to make automated systems team players. In *Advances in human performance and cognitive engineering research, Vol. 2,* (pp. 1-12). Bridgewater, NJ: Elsevier Sciences Ltd.

Dekker, S. W. & Woods, D. D. (1999). To intervene or not to intervene: The dilemma of management by exception. *Cognition, Technology and Work, 16*-96.

Elm, W. C., Potter, S. S., Gualtieri, J. W., Roth, E. M., & Easter, J.R. (2003). Applied cognitive work analysis: A pragmatic methodology for designing revolutionary cognitive affordances. In E. Hollnagel (Ed.), *Handbook of cognitive task design* (pp. 357-382). Mahwah, NJ: Lawrence Erlbaum Associates.

Elm, W., Potter, S., Tittle, J., Woods, D., Grossman, J., & Patterson, E. (2005). Finding decision support requirements for effective intelligence analysis tools.

In *Proceedings of the Human Factors and Ergonomics Society 49th Annual Meeting*. Santa Monica, CA: HFES.

Gesheider, G. A. (1997). *Psychophysics: The fundamentals*. Mahwah, NJ: LEA.

Klein, G. (2007). Personal communication.

Hollnagel, E. & Woods, D. D. (2005). *Joint cognitive systems: Foundations of cognitive systems engineering*. Boca Raton, FL: Taylor and Francis.

Rasmussen, J. (1986). *Information processing and human-machine interaction: An approach to cognitive engineering*. Amsterdam: North-Holland.

Rasmussen, J. & Lind, M. (1981). Coping with complexity. In H. G. Stassen (Ed.), *First European annual conference on human decision making and manual control*. New York, NY: Plenum.

Roth, E. M. & Eggleston, R. G. (Chapter 13). Forging new evaluation paradigms: Beyond statistical generalization. In E. S. Patterson & J. Miller (Eds.), *Macrocognition metrics and scenarios: Design and evaluation for real-world teams*. Farnham, UK: Ashgate Publishing.

Roth, E. M., Gualtieri, J. W., Elm, W. C., & Potter, S. S. (2002). Scenario development for decision support system evaluation. In *Proceedings of the Human Factors and Ergonomics Society 46th Annual Meeting* (pp. 357-361). Santa Monica, CA: Human Factors and Ergonomics Society.

Rousseau, R., Easter, J., Elm, W., & Potter, S. (2005). Decision-Centered Testing (DCT): Evaluating joint computer cognitive work. In *Proceedings of the Human Factors and Ergonomics Society 49th Annual Meeting*. Santa Monica, CA: HFES.

Woods, D. D. & Hollnagel, E. (2006). *Joint cognitive systems: Patterns in cognitive systems engineering*. Boca Raton, FL: Taylor and Francis.

Chapter 16

Synthetic Task Environments: Measuring Macrocognition

John M. Flach, Daniel Schwartz, April M. Courtice, Kyle Behymer, and Wayne Shebilske

Introduction

> I claim that many patterns of Nature are so irregular and fragmented, that, compared with Euclid—a term used in this work to denote all of standard geometry—Nature exhibits not simply a higher degree but an altogether different level of complexity. . . .
>
> The existence of these patterns challenges us to study those forms that Euclid leaves aside as being 'formless,' to investigate the morphology of the 'amorphous.' Mathematicians have disdained this challenge, however, and have increasingly chosen to flee from nature by devising theories unrelated to anything we can see or feel. (Mandelbrot, 1983, p. 1)

Consistent with Mandelbrot's comments comparing the 'cold' geometry of Euclid with the patterns of nature, there seems to be a growing dissatisfaction with the ability of classical experimental approaches to human information processing to capture the complexity of cognition 'in the wild' (for example, Hutchins, 1995). While many cognitive scientists have 'disdained this challenge' and have fled from the apparently amorphous patterns of everyday work to study sterile, logical puzzles, a few have been plagued by a nagging fear that this research may not be representative of how people experience life. In some sense, the construct of macrocognition reflects the challenge to address the patterns of cognition as they appear in natural work contexts. In addressing this challenge, questions are raised about the very nature of cognition and thus about the appropriate ways to measure it. There is a growing consensus that the cold geometry of micro-laboratory tasks (for example, reaction time, tracking, logical puzzles) is not capturing the complexities of human performance in natural settings.

This chapter first considers how assumptions about cognition from a 'macro' perspective may differ from assumptions that have been made from 'micro' perspectives. In doing this, we raise important questions about measurement. The chapter concludes by considering the use of synthetic task environments as one means for wrestling with both the theoretical assumptions and the measurement challenges of a 'macro' approach to cognition.

What's the System?

One of the first decisions researchers must make is to identify the phenomenon or system of interest. That is, they must identify what dimensions are endogenous to the phenomena—essentially the 'state dimensions.'

As the following quote from Marr (1982) suggests, from the start researchers interested in humans as information processors, recognized that the system must include both the agent and the task demands:

> . . . the critical point is that understanding computers is different from understanding computations. To understand a computer, one has to study the computer. To understand an information-processing task, one has to study that information-processing task. To understand fully a particular machine carrying out a particular information-processing task, one has to do both things. Neither alone will suffice. (Marr, 1982, p. 5)

This suggests that the system must be defined in a way that includes both properties of the cognitive agents (that is, constraints on awareness) and properties of the task or problem space (that is, constraints on situations). Although the pioneers in the field (for example, Marr, 1982; Newell & Simon, 1972) understood this, one gets a distinct impression from the research literature on human cognition and its application in terms of human factors that the task or problem space is arbitrary. This literature creates a distinct impression that the system of interest is 'in the human's head.' And that the research goal is to characterize the internal 'limitations' within isolated stages of information processing so that these limitations can be accommodated in the design of complex systems. The primary motivation for choosing one laboratory task or another is the ability to isolate specific stages of processing (for example, encoding or decision making) or specific internal constraints (for example, a bottleneck or resource limit). There is little discussion of how well such tasks represent the demands of natural situations.

Thus, an important difference between micro- and macro-approaches to cognition is that microcognition defines the system as a process internal to an agent, whereas macrocognition defines the system to include both the agent and the problem demands. In other words, macro approaches consider cognition to be a process that is 'situated' or 'distributed.' It also assumes that understanding the 'situation' is a critical component to a computational theory of cognition. In fact, to a large extent the 'computation' is an adaptation to the problem or situation constraints. The implication for measurement is that we need to consider how to measure both situations and awareness and we must measure them in a way that we can index the fitness of one relative to the other. They are not two separate systems, but two facets of a single system. The computational system (that is, the cognitive phenomenon) depends on the fit between awareness and situations.

Measuring Situations

When considering measuring situations, it is important to begin with fundamental lessons about the nature of information. The information value of an event (for example, selecting the number 42 from a jar) cannot be determined unless the possibilities (for example, the other numbers in the jar) are specified. In simple choice decision paradigms (for example, Hick (1952) and Hyman(1953)), the other possibilities are well-defined in terms of the number and probabilities of alternatives. However, how do you specify the possibilities when the task is controlling a nuclear power plant, flying a modern aircraft, or directing air operations during battle; much less when the question has to do with the life of meaning or the meaning of life (Adams, 1979)?

From the micro-perspective on cognition, there is a natural tendency to extrapolate from the measures that provided well-defined performance functions in the laboratory. This has stimulated initiatives to reduce events in a nuclear power plant and other complex situations to probabilities and durations that can be integrated using linear techniques like THERP or Monte Carlo style simulations such as MicroSAINT.

From the macro perspective, there is great skepticism about whether the dynamics (or possibility space) of many natural situations can be captured using event probabilities or time-based measures alone. The alternative is to describe the constraints that shape the space of possibilities in terms of goals/values, general physical laws, organizational constraints, and specific physical properties. This perspective is well illustrated by Rasmussen's (1986) Abstraction Hierarchy. The Abstraction Hierarchy is significant as one of the first clear specifications of the different classes of constraints that limit possibilities in natural work environments. In relation to the focus of this chapter, it is important to recognize that measurement is a form of abstraction. Thus, the Abstraction Hierarchy is a statement about measurement. In fact, we suggest that it could easily have been termed a Measurement Hierarchy, where each level suggests different ways to index constraints within the work domain.

In the context of specifying or measuring situations, the Abstraction Hierarchy provides a useful guide for thinking about the various levels or types of constraints that shape the field of possibilities within a work domain. It stimulates discussion about relations within and across these levels. Thus, when considering how to measure situations, one must consider both the need to describe the constraints in ways that reflect significant relations both within and across the various levels. Vicente (1999) illustrates this very clearly with the DUal REServoir System (Duress) example of a process control micro-world simulation. One caution is that while Duress is a great illustration—it represents a task with fairly well-defined goals and constraints—in many domains the constraints will be far more amorphous and discovering the right metrics to characterize the significant relations among the constraints is a significant challenge. But it is a challenge that must be engaged if there is to be any hope of understanding the computations involved in cognitive work.

Let's consider some of the levels associated with situation constraints and the issues associated with measuring them. First, consider goals and values. Even in micro-laboratory tasks (for example, reaction time or signal detection) it is evident that tradeoffs between goals (for example, speed versus accuracy or hits versus false alarms) have a significant role in shaping performance. Natural work domains are typically characterized by multiple goals and success often depends on balancing the demands associated with these goals (for example, setting priorities or precedence). An important question for measurement is how to index performance with respect to multiple goals so that the data can be integrated across the goal dimensions in a way that will reflect whether performance is satisfactory with regard to the aspirations for the system. Measures should allow some classification (satisfactory versus unsatisfactory) or ordering (better or worse) of performance with respect to these aspirations.

Brungess' (1994) analysis of the Suppression of Enemy Air Defenses (SEAD) is an important example of someone who is explicitly wrestling with the problem of how to measure performance relative to goal constraints. For example, he writes:

> SEAD effectiveness in Vietnam was measured by counting destroyed SAM sites and radars. Applying that same criteria to SEAD technologies as used in Desert Storm yields a confused, possibly irrelevant picture. SEAD weapons and tactics evolution has outpaced the development of criteria to measure SEAD's total contribution to combat (Brungess, 1994, pp. 51-52).

At another level, it should be quite obvious how general physical laws (for example, thermodynamics or laws of motion) provide valuable insight into the possibilities of natural processes (for example, feedwater regulation or vehicle control). These laws suggest what variables are important for specifying the state of the system (for example, mass, energy, position, velocity). These variables are critical both to the cognitive scientist interested in describing the computational processes and to the active control agents (whether human or automated) in terms of feedback (that is, observability and controllability). Note that for the variables to be useful in terms of feedback, they must be indexed in relation to both goals and control actions. That is, it must be possible to compare the information fed back about the current (and possibly future) states with information about the goals, in a way that specifies the appropriate actions. Thus, questions about controllability and observability require indexes that relate goals, process states, and controls (for example, Flach et al. 2004b). This is a clear indication of the need to choose measures that reflect relations within and across levels of the Measurement Hierarchy.

In addition to considering relations across levels in the Situation Measurement Hierarchy, it is important not to lose sight of the fact that the measures should also help to reveal important relations to constraints on awareness. Remember the system of interest includes both the problem and the problem solver. Understanding general physical laws can be very important in this respect, because these laws suggest ways to organize information to allow humans with limited working memory to 'chunk' multiple measures into a meaningful unit. This is a recurrent theme in the design of 'ecological' displays—to use geometric relations to specify constraints (for example, physical laws) that govern relations among state variables (for example, Vicente, 1999; Amelink et al., 2005).

For the purposes of illustrating the general theme of measuring the situation using indexes that reveal relations both across levels in the hierarchy and in relation to constraints on awareness, we only describe this one level in the Situation Measurement Hierarchy. For more discussion of other levels in the abstraction/measurement hierarchy see Flach et al. (2004a).

Measuring Awareness

Micro-approaches to cognition tend to focus on identifying awareness dimensions that are independent of (or invariant) across situations. Thus, they tend to address issues such as perceptual thresholds, memory capacities, bandwidths or bottlenecks, and resource limits—as attributes of an internal information processing mechanism. However, within this research literature, it is not difficult to find research that attests to the adaptive capacity of humans. For example, basic work on signal detection suggests that performance is relatively malleable as a function of the larger task context (for example, expectancies and values). Even the sensitivity parameter (d) is defined relative to signal and noise distributions. So, there is ample reason for skepticism about whether any attribute of human performance can be specified independently from the larger task context.

Whether or not it is possible to characterize constraints on human information processing that are independent of the task context, few can argue that humans are incredibly adaptive in their ability to meet the demands of natural situations. Macro-approaches tend to focus on this adaptive capacity and this raises the question of measuring awareness relative to a domain (for example, skill or expertise). There is little evidence to support the common belief that expert performance reflects innate talent. Rather the evidence suggests that expert performance reflects skills acquired through extended, deliberate practice in a specific domain (Erikson & Charness, 1994). In fact, Erikson & Charness (1994) conclude that 'acquired skill can allow experts to circumvent basic capacity limits of short-term memory and of the speed of basic reactions, making potential limits irrelevant' (p. 731).

Again, we feel that credit goes to Rasmussen (1986) as one of the first to explicitly recognize the flexibility of human information processing and to introduce a conceptual framework specifying important distinctions that must be addressed by any program to quantify human performance in natural contexts (that is, the decision ladder and the SRK distinction between Skill-, Rule-, and Knowledge-based processing). The decision ladder explicitly represents the shortcuts that might allow experts to 'circumvent' basic capacity limitations. The SRK distinction provides a semiotic basis for relating the properties of the situation (for example, consistent mapping) to the potential for utilizing the various shortcuts (Flach & Rasmussen, 2000).

Rasmussen (1986) illustrates how the decision ladder can be utilized to visualize qualitatively different strategies for fault diagnoses. This is clearly an important form of measurement that helps to index performance in relation to potential internal constraints on awareness and in relation to the demands of situations. Furthermore, it allows these strategies to be compared to normative models of diagnoses.

Consistent with the basic theory of information, it is important not to simply ask what experts 'know' and what strategy experts typically use; we must explore the possibilities about what experts 'could know' and about what strategies might be effective in a given situation. Note that the awareness of experts will be constrained by the types of representations that they have been exposed to. For example, pilots and aeronautical engineers utilize very different forms of representations for thinking about flying. Thus, there can be striking contrasts for how these different experts think about flight performance. Exploring the differing forms of awareness can be important for differentiating more and less productive ways for thinking about a problem such as landing safely (Flach et al., 2003). It is important to keep in mind that the best operators (for example, pilots or athletes) often do not have the best explanations for how and why they do what they do.

Considering alternative representations across different experts can suggest possibilities for shaping 'awareness' through the design of interfaces. As Hutchins' (1995) work clearly illustrates, the choice of a specific technique for projecting the world onto the surface of a map has important implications for the cognitive processes involved in navigation. Again, this is a key theme behind the construct of ecological interface design—to shape the nature of awareness to facilitate information processing (in terms of allowing shortcuts and supporting multiple pathways to satisfactory outcomes) (Rasmussen & Vicente, 1989; Vicente & Rasmussen, 1990). The point is not to simply match existing mental models, but to design representations that help shape the mental models to enhance awareness and resilience.

Measuring Performance

At the end of the day, one of the most important measurement challenges is to be able to index the quality of performance. For example, to be able to index whether one interface, training protocol, incentive system, leadership style, organization plan, and so on leads to 'better' performance than another. Micro-approaches to cognition prefer to focus on one dominant measure (for example, time to completion or percent correct) to index quality. Even when there is clear evidence of the potential for tradeoffs (for example, speed versus accuracy), micro-style research tends to frame the task to clearly emphasize one dimension (for example, 'go as fast as possible with zero errors'). These approaches typically assume that performance functions are monotonic (for example, 'faster is better'). This is typically generalized to research in human factors, where two designs might be evaluated in terms of which design produces a statistically significant advantage in response time. But whether a statistically significant difference in response time leads to a practical gain in work performance is difficult to address with the micro-approach. The General or Chief Executive Officer (CEO) who asks whether the 'improved' system will be worth the cost in terms of achieving the objectives that are important to him (for example, greater safety or a competitive advantage) rarely gets a satisfactory answer.

In the everyday world, there is rarely a single index of satisfaction. As discussed in the section on measuring situations, there are typically multiple goals that must be balanced—a good system should be fast, accurate, safe, and not too expensive. This requires either multiple performance measures or at least an explicit integration of indexes associated with the various goals to yield a single 'score' for ranking goodness or at least for distinguishing between satisfactory and unsatisfactory performance. Rarely are the quality indices monotonic. That is, success typically depends on responding at the right time (not too early or too late). At least for closed-loop systems there will always be a stability boundary that limits the speed (that is, gain) of response to stimuli. Thus, response speed is rarely monotonic—a system that is too fast can become unstable (for example, pilot-induced oscillations).

It is impossible to address questions about the right information, the right place, the right person, or the right time without considering the specific problem that is being solved (that is, the work domain or task). 'Right' is highly context dependent. It cannot be addressed by a micro-research program that is designed to be context independent. This is an important motivation for a macro-approach to cognition and work—to specify the specific criteria for satisfying the demands of specific work domains.

In order to know whether a difference in response time is practically significant, it can be useful to compare this against landmarks that reflect the optimal or best case situation. Here is where analytic control models (for example, the optimal control model) or Monte Carlo simulations (for example, Microsaint) can be very useful. Not as 'models' of human information processes, but as ways to explore the boundary conditions of performance. What is the best possible performance assuming certain types of processes? How do changes at one step in a process (for example, to speed or accuracy) or in properties of a sensor (for example, signal-to-noise ratio) impact system performance? Where are the stability limits? In this sense, the models are being used to explore boundaries (or limits) in the workspace. These boundaries may provide important insights into what are realistic targets for improvement and into the practical value of specific improvements. In essence, these models can suggest 'normative' landmarks against which to assess actual performance.

An example of a case where this type of insight might be useful is a recent initiative on the part of the US Air Force to reduce the response time for executing dynamic targets to single digit minutes. This is motivated by the threat of mobile missile systems (for example, Scuds) that can fire a missile and then move to cover within about ten minutes. Few have raised the question about whether this response time is realistic given the unavoidable lags associated with acquiring the necessary information and communicating with the weapons systems. We fear that the blind pursuit of this single digit minute goal may lead to instabilities and unsatisfactory solutions to the overall goals of the Air Force. Rather than 'reacting' faster, the solution to the Scud missile problem may depend on improving the ability to predict or anticipate launches (for example, Marzolf, 2004). Thus, the solution to Scud missiles may rest with the design of the air battle plan, to include dedicated aircraft to patrol areas where launchers are likely to be hidden. This solution does not require speeding the dynamic targeting process for dealing with events not anticipated in the air battle plan.

Another important consideration for measuring performance is the distinction between process and outcome. In complex environments, an optimal process can still result in a negative outcome due to chance factors that may be completely beyond control. For example, a coach can call the perfect play that results in a touchdown and have it nullified by a penalty flag, incorrectly thrown by a poor referee. Thus, it is important to include measures of process as well as measures of outcome. It is important to have standards for measuring process as well as outcome. For example, most military organizations have doctrine that provides important standards for how processes should be conducted (see Chapter 8 for an example on communicating commander's intent to subordinates).

The astute reader should realize that as we talk about performance measurement, we are covering some of the same ground that was discussed in terms of measuring situations. We are talking about ends (goals and values) and means (processes). In classical micro-approaches that define the system of interest as inside the cognitive agent, the situation is typically treated as an independent variable and performance measures are treated as dependent variables. This creates the impression that these are different kinds of things. However, in natural work ecologies, understanding the situation requires consideration of both means and ends. So, using the Abstraction Hierarchy to think about situations will go a long way toward addressing questions about performance measures. It should also help to frame these questions in terms that are meaningful to the problem owners (those who have a stake in success).

Thus, consistent with our discussion about situations, we believe that it is useful to think about a nested hierarchy of performance measures, where higher levels in the hierarchy reflect global criteria for success (for example, how do you know whether you are winning or losing); and where lower levels address subgoals and process measures that reflect the means to higher-level goals (for example, showing patterns of communication or organization). It is important to keep in mind that the primary goal of measurement is to reveal the patterns of association between process and outcome. In other words, a key objective is to connect the micro-structure associated with the design and organization of work activities to qualitative changes associated with global indexes of quality!

Synthetic Task Environments

In contrasting micro- and macro-approaches to cognition, our intent is not to eliminate micro-level research, but rather to make the case that this is only one element of a comprehensive research program. We fear that a research program that is exclusively framed in terms of low-dimensional tasks will not satisfy our goals to understand cognition in natural contexts or to inform the design of tools to support cognitive work. Thus, the goal of a macro-approach is to enrich the coupling between the laboratory and the natural world. It is in this context that we would like to suggest that research employing synthetic task environments can be an important means for bridging the gap between basic experimental research and natural cognition.

We use the term synthetic task environment to represent experimental situations where there is an explicit effort to represent the constraints of a natural work domain. This is in contrast to experimental paradigms designed around the parameters of a particular analytic model (for example, choice reaction time or compensatory tracking) or designed to isolate a specific stage of an internal process (for example, visual and memory search tasks). It is also in contrast to micro-world research that attempts to represent the complexity of natural domains, without representing the constraints of specific actual domains (for example, space fortress or other research using computer games). In a synthetic task, the work domain has to be more than a cover story. The 'task' must be representative of some natural work—even though the implementation is synthetic, typically utilizing a simulation.

For example, research using a flight simulator may or may not satisfy our definition for synthetic task research. If the focus is on flight performance, perhaps in relation to a specific training protocol, to compare alternative interfaces, or to evaluate different procedures, then this is consistent with our definition of synthetic task research. However, if the focus is on cognitive workload and the flight task is just one aspect of a multiple task battery, then we would not consider this to be synthetic task research. Again, this does not mean that such research is not valuable, but we simply want to emphasize that for synthetic task research, the focus should be on the impact of specific natural constraints of the work on cognitive processes. The key to synthetic task research is NOT the use of a simulator, but the framing of research questions with respect to properties of the natural task or the natural work domain!

A second important facet of synthetic task environments is the ability to measure performance at multiple levels, as discussed in previous sections of this chapter. The synthetic task environment should allow performance to be scored relative to global objectives (for example, was a landing successful; was the mission completed successfully?). It should also allow direct measures of the work processes (for example, time history of control and communication activities and of system state, such as the actual flight path).

One of the important goals for synthetic task research is to provide empirical data with respect to the coupling of global metrics (goals and values) and micro-metrics (work activities and movement through the state space). The goal is to empirically relate variations at one level of the measurement hierarchy to variations at the other levels. Thus, for example, this may allow the question about whether a significant difference in reaction time is practically significant with respect to the global intentions for the system to be addressed empirically. Do quantitative differences in response time to a particular class of events lead to increased probability of successfully completing the mission?

Another consideration with respect to synthetic task research is the question of fidelity. How much is enough? This is a bit tricky, because this is one of the questions that we are asking when we frame questions around situations. What are the important constraints and how do they interact to shape performance? For this reason, the issue of fidelity can only be addressed iteratively. In general, it is best to start with as much fidelity as you can practically afford and to assume that it is not enough! The performance observed in synthetic tasks needs to be skeptically evaluated relative to generalizations to natural domains. In our view, to be effective, a program of synthetic task research should be tightly coupled to naturalistic field studies. The patterns observed in the laboratory need to be compared to patterns observed in naturalistic settings. In this way, it may be possible to titrate down to identify critical constraints and interactions. The synthetic task observations allow more rigorous control and more precise measurement. But there is always the possibility that the patterns observed in the synthetic task are a result of your simulation and that they are not representative of the natural domain of interest. Ideally, however, synthetic task environments can improve our ability to see and quantify patterns during more naturalistic observations.

It is also important to note that questions of fidelity should not be framed simply in terms of the simulation device. Consideration must be given to the participants of the research. Are they representative of the people who do this work in natural settings, in terms of knowledge, skill, motivation, and so on? Consideration also must be given to the task scenarios. Are the tasks representative of the work in the natural context in terms of probability of events, consequences, and organization? (For more on the design of complex task scenarios, see Chapters 13–15.) But more importantly, are the experiences of the participants representative of experiences in the real work domain (for example, in terms of stress)?

In order to bridge the gap between laboratory research and cognition in the wild, synthetic task research will be most effective when the questions are driven by field observations of natural environments and when the multiple nested measures are motivated by 1) the values of the problem owners; 2) by normative models of the work (for example, information theory, control theory, queuing theory); and 3) by basic theories of cognition. Currently, each of these three motivations has its champions and there seems to be a debate over which of these motivations is optimal. In our view, all three motivations are critical and none of these motivations alone will meet our aspirations for a science of cognition. With respect to these three motivations, the synthetic task environment may provide a common ground to facilitate more productive coordination between the disparate constituencies across the basic and applied fields of cognitive science.

Finally, it is important to recognize the inherent limits on any controlled scientific observation. The results will depend in part on properties of the phenomenon of interest and in part on the choices we make in designing the synthetic task environment. It is important to resist the temptation to become infatuated with a particular experimental paradigm (whether micro-task or specific synthetic task environment). It is important to leave ultimate control to Nature!

Conclusion

> Nature does exist apart from Man, and anyone who gives too much weight to any specific [ruler]. . . lets the study of Nature be dominated by Man, either through his typical yardstick size or his highly variable technical reach. If coastlines are ever to become an object of scientific inquiry, the uncertainty concerning their lengths cannot be legislated away. In one manner or another, the concept of geographic length is not as inoffensive as it seems. It is not entirely 'objective.' The observer invariably intervenes in its definition. (Mandelbrot, 1983, p. 27)

The quote from Mandelbrot reflects the difficulty in measuring a natural coastline—as the size of the ruler get smaller, the 'length' of the coastline can grow to infinity. If a simple attribute like 'length of a coastline' creates this difficulty for measurement, how much more difficult is the problem when the nature that we are trying to measure involves humans themselves. In our view, perhaps, this might be the biggest differentiator between micro- and macro-approaches to cognition. The micro-approach clings to the classical idea that it is possible to stand outside of ourselves to 'objectively' measure cognition, work, or situation awareness. The macro-approach believes that this is a myth.

The macro-approach understands that measurement is not neutral. There is no privileged measure or privileged level of description! Every measure, every level of description, every perspective offers an opportunity to see some facet of nature, but hides other facets. Thus, understanding requires multiple measures, multiple levels of description, and/or multiple perspectives. In this respect, the Abstraction Hierarchy or Measurement Hierarchy is simply a way to be explicit about the need to measure at multiple levels and a framework to guide the search for patterns that are invariant over multiple perspectives. In other words, the only way to eliminate or unconfound the invariant of a specific measurement perspective from an invariant of nature is to measure from multiple perspectives. One is more confident in attributing an invariant to nature, when that invariant is preserved over many changes of observation point.

Note that this is not a special requirement for studying humans or cognition. This will be true for any complex phenomenon in nature (for example, weather systems or coastlines). By complex, we simply mean a phenomenon that involves many interacting dimensions or degrees of freedom.

It is humbling to realize that nature/cognition cannot be reduced to reaction time and percent correct; to realize that the convenient measures (in terms of experimental control or in terms of analytic models) will not yield a complete picture; to realize that measures that work within the constraints of the ideals of Euclidean geometry do not do justice to the curves of natural coastlines. We get a distinct impression that the field of cognitive science is searching for a mythical holy grail—that is, a single framework (neuronets, neuroscience, chaos, and so on) and a specific measure (42, MRI, 1/f scaling, and so on) that will provide the key to the puzzle. We are skeptical.

Complex systems are difficult. They require multiple levels of measurement. An attractive feature of synthetic task environments is that they allow many measures (from micro-measures specifying the complete time histories of activity and state change, to macro-measures specifying achievement relative to the intentions of operators and system designers). The problem is making sense of all this data, weeding through the data to discover the patterns that allow insight, prediction, deeper understanding, and generalization. Success in this search requires the intuitions available from normative systems theory (for example, information, signal-detection, and control theory, computational and normative logic, nonlinear systems, and complexity theory), from controlled laboratory research, and from naturalistic field observations. Again, none of these perspectives on research is privileged. We expect that if there are answers to be discovered, they will be found at the intersection of these multiple perspectives. Thus, the value of synthetic task environments is to create common ground at the intersection of these various perspectives where we can constructively debate and test alternative hypotheses about the nature of cognitive systems.

Acknowledgments

We thank the organizers of the Measurement Workshop, Emily Patterson and Janet Miller, for giving us a forum to express our rather eccentric views about cognition, measurement, and the implications for system design. These views have evolved over many years of research. This research has benefited from several sponsors, most significantly the Air Force Office of Scientific Research (AFOSR) and the Japan Atomic Energy Research Institute (JAERI).

References

Adams, D. (1979). *A hitchhiker's guide to the galaxy.* London, UK: Ballantine Books.

Amelink, H. J. M., Mulder, M., van Paasan, M. M., & Flach, J. M. (2005). Theoretical foundations for total energy-based perspective flight-path displays for aircraft guidance. *International Journal of Aviation Psychology, 15*(3), 205-231.

Brungess, J. R. (1994). *Setting the context. Suppression of enemy air defenses and joint war fighting in an uncertain world.* Maxwell, AFB, AL: Air University Press.

Erikson, K. A. & Charness, N. (1994). Expert performance. *American Psychologist, 49*(8), 725-747.

Flach, J. M., Jacques, P., Patrick, D., Amelink, M., van Paassen, M. M. & Mulder, M. (2003). A search for meaning: A case study of the approach-to-landing. In Erik Hollnagel (Ed.), *Handbook of cognitive task design* (pp. 171-191). Mahwah, NJ: Erlbaum.

Flach, J., Mulder, M., & van Paassen, M. M. (2004a). The concept of the 'situation' in psychology. In S. Banbury & S. Tremblay (Eds.), *A cognitive approach to situation awareness: Theory, measurement, and application* (pp. 42-60). Aldershot, UK: Ashgate Publishing.

Flach, J. M., Smith, M. R. H., Stanard, T., & Dittman, S. M. (2004b). Collisions: Getting them under control. In H. Hecht & G.J.P. Savelsbergh (Eds.), *Theories of time to contact.* Advances in Psychology Series (pp. 67-91) London: Elsevier.

Flach, J. M. & Rasmussen, J. (2000). Cognitive engineering: Designing for situation awareness. In N. Sarter & R. Amalberti (Eds.) *Cognitive engineering in the aviation domain* (pp. 153-179), Mahwah, NJ: Erlbaum.

Hick, W. E. (1952). On the rate of gain of information. *Quarterly Journal of Experimental Psychology, 4*, 11-26.

Hutchins, E. (1995). *Cognition in the wild.* Cambridge, MA: MIT Press.

Hyman, R. (1953). Stimulus information as a determinant of reaction time. *Journal of Experimental Psychology, 45*, 188-196.

Mandelbrot, B. B. (1983). *The fractal geometry of nature.* New York, NY: Freeman.

Marr, D. (1982). *Vision.* New York, NY: Freeman.

Marzolf, G. S. (2004). *Time-critical targeting: Predictive versus reactionary methods: An analysis for the future.* Maxwell, AFB, AL: Air University Press.

Newell, A. & Simon, H. A. (1972). *Human problem solving.* Englewood Cliffs, CA: Prentice Hall.

Rasmussen, J. (1986). *Information processing and human-machine interaction: An approach to cognitive engineering.* New York, NY: North-Holland.

Rasmussen, J. & Vicente, K. J. (1989). Coping with human errors through system design: Implications for ecological interface design. *International Journal of Man-Machine Studies, 31*(5), 517-534.

Vicente, K. J. (1999). *Cognitive work analysis.* Mahwah, NJ: Erlbaum.

Vicente, K. J. & Rasmussen, J. (1990). The ecology of human-machine systems II: Mediating 'direct perception' in complex work domains. *Ecological Psychology, 2*(3), 207-249.

Chapter 17

System Evaluation Using the Cognitive Performance Indicators

Sterling L. Wiggins and Donald A. Cox

Introduction: Problem

> Jana checks her email and sees an urgent message from her boss. It says, 'Hi Jana, there's a system design review meeting in the next few days for that HS29 project to support users in time constrained, high-stress environments. You've taken classes and read about human systems integration and usability, so take a look at this prototype and system documentation. A traditional usability analysis has already been performed. The program manager suggested we conduct a "human-centered" evaluation to see how well the system supports the user on the mission-critical tasks. Let's meet tomorrow morning to discuss.'

What should Jana do next? Who should she consult with to figure out what a 'human-centered' system evaluation looks like and what methods are available to do this? What elements of human-system performance should she focus on to evaluate how well the system supports the user in handling the challenging tasks? The purpose of this chapter is to describe a range of criteria that can be used to conduct a human-centered system evaluation.

Cognitive Systems Engineering (CSE) experts have developed an informal set of criteria (or indicators) to identify and describe how systems support and hinder cognitive performance in naturalistic settings. Specifically, CSE experts use these indicators to identify how systems help or hinder their users in making decisions, assessing situations, making plans and replanning, detecting problems, and managing uncertainty and risk.

We identified these indicators and documented them as a set of assessment criteria for Humans System Integration (HSI) practitioners. These practitioners, who may not be familiar with CSE research, are being asked to conduct human-centered system evaluations, but do not have a comprehensive set of tools to help them do this. We assembled the indicators to support HSI practitioners in conducting these assessments.

We have three goals for the use of the cognitive performance indicators. First, provide a framework for analysis that will allow evaluators to have more insight into how system design features will impact the cognitive work of the human early in the development cycle. Second, provide all stakeholders in the evaluation process a more nuanced language with which to discuss the positive and negative potential of systems as they move through the development lifecycle. Third, assist system developers in prioritizing which design features to focus resources on in subsequent development efforts.

The number of cognitive performance indicators and their role in system evaluation has evolved since their inception. First, we describe how the indicators were derived and then give examples of how they have been used to evaluate systems.

Initial Approach

Initially, we investigated quantitative methods of evaluation. Such methods claim to yield measurements of various aspects of cognition. However, cognition cannot be directly measured in the same ways that one can measure length or width of an object. Instead, the closest we can come to cognitive measurement is to assess the behaviors that are thought to reflect the cognition present. The literature on quantitative methods of evaluation talks about cognitive measurement by examining time, accuracy, workload, or situation awareness, usually at the level of individual rather than team-based cognition. The NAVSEA 03 Human Performance Technical Metrics Guide (Hart & Slaveland, 1988) offers a comprehensive summary of this family of metrics.

We began by generating a set of assessment criteria directly from numerous descriptive models that make up the macrocognition framework (Klein et al., 2003a; also see Preface) and called them Cognitive Impact Metrics (CIMs) (Moon et al., 2004). Specifically, we generated assessment criteria from the Recognitional Planning Model (Klein et al., 2003b) that were composed of a metric and measure. The metric was a short title that described the aspect or phenomenon that was to be observed and was associated with one or more corresponding measures with which that phenomenon would be evaluated.

Unfortunately, when HSI practitioners without a CSE background tried to use these assessment criteria, they had difficulty applying them. They had trouble interpreting and applying the model-based measures. Our hypothesis is that CIMs work for those who are familiar with the specific descriptive models from which they are derived. These do not address the needs of HSI practitioners who have much less exposure to the field and models of CSE.

Furthermore, there are a number of shortcomings to using a quantitative approach to the evaluation of system design. A critical one is that a system must at least have working interface prototypes in order to employ quantitative methods. This is a concern because, when a system is this far along in the development cycle, often it is too late to make many significant changes due to the increased costs associated with them (see Chapter 13 for a related discussion of the strengths and limitations of formative and summative evaluations). This concern stresses the importance of being able to intervene based on evaluations early in the development cycle. Quantitative approaches are not well-suited for that need.

However, the fields of usability (Nielsen, 1993), ergonomics (Tilley & Drefus, 2001), human factors (Stanton et al., 2005), and accessibility (World Wide Web Consortium, 1999) have established qualitative and quantitative criteria that practitioners reference when performing system assessments. Consequently, we decided to move away from a focus on model-based, quantitative approaches to an expert judgment-based approach that leveraged the system evaluation practices in the CSE community.

Solution

We conducted a pilot observation study to understand how two CSE experts from the Naturalistic Decision Making (NDM) community, Dr. Gary Klein (Klein, 1998) and Dr. David Woods (Woods & Hollnagel, 2006), evaluate systems. The purpose of the study was twofold. First, we wanted to uncover the indicators they used. Second, we wanted to see if the indicators they used were different from the criteria discussed in the fields of usability, ergonomics, human factors, and accessibility. Dr. Klein and Dr. Woods independently evaluated actual documentation (concept of operations, requirements documents, and human-computer interaction design specifications) for a military ship-board system to determine how the design of these engineering artifacts supported or hindered the cognitive work of its intended user. They looked for examples of how the designs helped or hindered the user with a variety of cognitive tasks, including making decisions, assessing situations, rapidly locating key cues in information presented, and projecting the future based on historic information. This pilot study resulted in an initial set of seven indicators that were different from existing criteria in the fields of usability, ergonomics, human factors, and accessibility. This pilot study gave us confidence that we could identify a set of CSE-specific indicators for determining whether a system supports users' cognitive performance in naturalistic settings.

To extend the seven initial indicators to better reflect the overall breadth of CSE research, we studied many approaches within the CSE field (Hoffman et al., 2002). We took a cross-disciplinary approach by identifying the similarities around how expert CSE practitioners describe and identify systems that support and hinder cognition. Table 17.1 describes a sample of the expert practitioners and approaches that we studied.

For each expert CSE practitioner, we reviewed their articles and books on the subject of system design and, for some, conducted interviews (Interview with Gary Klein 12 Oct 2006; Interview with David Woods, 26 Oct 2006). Some practitioners referenced existing criteria from other fields such as usability and human factors. For example, one expert practitioner talked about the importance of '*consistent use of automation.*' Consistency is one of Jacob Nielson's ten usability heuristics (Nielsen, 1993). We removed the overlapping criteria to create a list of CSE-specific cognitive indicators.

Table 17.1   Sample of the expert practitioners and various approaches studied

Practitioner	Approach
Klein Associates Gary Klein	Recognition-Primed Decisions Decision-Centered Design
The Ohio State University David Woods	Laws That Govern Cognitive Work The Substitution Myth
SA Technologies Mica Endsley	Situation Awareness-Oriented Design
Institute for Human & Machine Cognition Robert Hoffman	Human-Centered Computing
University of Toronto Kim Vicente	Cognitive Work Analysis
University of Waterloo Catherine Burns	Ecological Interface Design

Those familiar with usability engineering will see similarities between Nielsen's heuristics and the indicators. The indicators all represent different heuristics. Several people who have used the indicators compared them to Nielsen's heuristics and noted similarities and differences. However, Nielsen's heuristics and the cognitive performance indicators have differing origins and uses. Nielsen's heuristics were developed based on a thematic and factor analysis of a set of reports of usability problems on a limited number of relatively simple applications, as well as the consideration of existing principles and guidelines. The cognitive performance indicators were developed based on the pilot study by Drs. Klein and Hoffman, extended by a thematic analysis of the NDM and CSE literature, and refined by considerations of existing CSE practices. Depending on the type of system being evaluated, it may be more fruitful to use one or the other. If you are reviewing a time reporting system, Nielsen's heuristics will highlight the main usability problems with the system. If on the other hand, you are evaluating a system that uses advanced decision-support models to recommend potential courses of action to a Joint Forces Commander, the cognitive performance indicators will point more directly to what will make the system successful or a burden for the commander. This is not to say that Nielsen's heuristics could not highlight issues with decision support systems. The indicators provide a different perspective from the heuristics and guides practitioners to explore the impact of the design on cognition in specific ways.

The expertise and experience of the various NDM practitioners was distilled into an initial set of indicators; differences between NDM indicators and existing criteria in other fields were resolved. The indicators and supporting documentation were placed in a standardized format, and iteratively tested and refined. Next, we needed to find a viable format for sharing them with other practitioners for use when conducting their own system evaluations.

Finding a Usable Format

After specifying the initial list of cognitive indicators, we developed a job aid to be used by HSI practitioners when performing formative system evaluations. Two types of documents were developed to enhance HSI practitioners' abilities to identify and describe how systems support and hinder cognitive performance. The first document was a simplified job aid. The second document was an expanded version of the job aid that provided more detailed information, including assessment questions, and examples that showcased each cognitive performance indicator.

The simplified job aid is meant to support HSI practitioners in using the indicators to conduct an expert review-style evaluation of a system. (The expert review technique structures the input of one or more experts to provide a rigorous yet relatively inexpensive method of evaluating the status of a system design artifact.) The two goals of the aid are: 1) to remind practitioners of what to focus on when assessing a system, and 2) to give a rationale for why it is important for the system to support each indicator. To achieve these goals, each indicator is described using a standard three-part description format (see the list of cognitive performance indicators in Table 17.2).

First, each indicator is given a short title to help practitioners remember the indicator, talk about the indicator to others, and reference the indicator in documentation. Second, each indicator starts off with a description of what a system should do to support the cognitive performance. This statement gives practitioners a description of what to look for in a system design or description. Third, each 'should do' statement is followed by one to two sentences about how users' cognitive performance will be hindered if the indicator is not supported.

Using the Indicators

To assess the utility and comprehensibility of the cognitive performance indicators, we conducted two system evaluations with diverse audiences. Through these evaluations, we learned about the reach and impact of each of the indicators. Each evaluation team had four to five evaluators (a mix of subject matter experts, HSI practitioners, and software developers) and used the format below to conduct the system evaluations using the indicators. (Note: we use 'system' to indicate the technology and human working together, not a particular stage of maturity of the technology.) We found that following the steps below prepared the team to conduct more efficient and effective evaluations using the indicators. These steps represent a synthesis of our best practices for conducting expert reviews.

Table 17.2 The cognitive performance indicators

Cognitive Performance Indicator	Description
1. Option Workability	Systems should enable users to quickly determine if an option is workable. Systems that require users to generate or compare alternative options hinder users' ability to act in time-pressured and rapidly-changing situations. Experienced users evaluate options individually, focusing on imagining how an option would be carried out to determine if it is workable.
2. Cue Prominence	Systems should allow users to rapidly locate key cues from the information presented. Representing all information as equal and presenting as much of it as technologically possible on a display reduces users' ability to recognize patterns. To recognize patterns, users generally make use of only 5-10 key cues. Additional information competes with and reduces the visibility of these cues.
3. Direct Comprehension	Systems should allow users to directly view key cues rather than requiring users to manually calculate information to comprehend these cues. In real-world settings, users' attention and memory are often scarce resources. Systems that only present data in a stove-piped format force users to manually integrate individual pieces of data to comprehend key cues. This hinders users' ability to track and recognize patterns as the demands of their work increase.
4. Fine Distinctions	Systems should allow users to investigate or at least access unfiltered data. Systems that remove variances and "noise" from data representations hinder users' abilities to spot anomalies and detect fine distinctions in a situation. Experienced users pay attention to small changes, differences, or absences to recognize patterns. Users don't need to always see unfiltered data, but they want the opportunity to investigate it.
5. Transparency	A system should provide access to the data that it uses and show how it arrives at processed data. Making the workings of systems invisible hinders users' ability to understand how processes work. Users build mental models about how system processes are supposed to perform and what to expect from them in various situations. These mental models permit users to remember how a process was performed in the past or predict how it will perform in the future.

Table 17.2 *Continued*

Cognitive Performance Indicator	Description
6. Enabling Anticipation	Systems should provide information that allows users to anticipate the future states and functioning of systems. It is not enough for systems to inform a user about what it is doing and why. Users need to know what the system will do next and when, so they can form expectancies about what will occur in the future. It is only through forming expectancies that users can notice the absence of events that were expected to happen (i.e., expectancy violations). Expectancy violations allow users to detect problems and then use mental simulation to diagnose them.
7. Historic Information	Systems should capture and display historic information so that users can quickly interpret situations, diagnose problems, and project the future. Limiting historic information hinders users' ability to recover from problems and decide on a course of action. When faced with unexpected or unexplained situations, users rely on historic data to build a story about what is currently happening. To build this story, users examine historic information to interpret trends, understand data inter-relationships, compare data, and identify key cues such as shifts and anomalies in data.
8. Situation Assessment	Systems should help users form their own assessment of a situation rather than provide decisions and recommendations. Systems that provide decisions have been shown to increase decision times and errors. Decision times increase because users do not work independently of system recommendations. Instead, users treat recommendations as additional data points that need to be taken into account before making a decision. Errors increase because users become reliant on systems for what they should do and thus are more likely to follow system decisions that are incorrect or faulty. Users need to form their own assessment of a situation through pattern matching and mental simulation to make rapid and effective decisions.
9. Directability	Systems should support the directing and redirecting of system priorities and resources so that users can effectively adapt to changing situations. Users want to focus the computational power of systems on particular problems to assist them in their problem solving, especially when users have information that is not available to systems.

Table 17.2 *Concluded*

Cognitive Performance Indicator	Description
10. Flexibility in Procedures	Systems should allow users to modify the order of the steps in procedures as doctrine changes or situations call for flexibility. Systems that lock in procedures or exact harsh penalties for not carrying out procedures in the correct order force users to follow inappropriate or out-of-date procedures. In real-world settings, users face non-routine situations where modifications will be necessary. Experienced users know when steps have to be followed and when to make exceptions.
11. Adjustable Settings	Systems should allow users to refine and adjust settings as they learn more about a situation. Requiring users to decide on settings in advance and keep them in place makes it difficult to solve ill-defined problems. Ill-defined problems require users to change the way they study data as they learn more about a situation. Consequently, in real-world settings users often have to adjust settings rather than keep them constant.

1. Develop Evaluation Plan

Identify the system you will evaluate and the resources available to do the evaluation (paper or working prototype; number of evaluators; size and focus of the evaluation; amount of time to complete the evaluation). This will allow you to determine the scope of the effort. If the system is large or complex, assign different components of the system to the evaluators to reduce their workload. If you have two to three hours for the entire evaluation, we recommend using all of the indicators. If you have less than 1.5 hours, consider using a subset of the indicators.

2. Collect and Study Background Material

Collect background material related to the system, like technical reports, reviewing screen shots of the user interface, reading marketing information, or examining system development documentation (for example, user requirements documents, software specification documents, or even a concept of operations document). Study the material to understand the nature of the system, its users, and the envisioned context of use. If time permits, develop a user profile and identify the work they will do using the system being evaluated. Cognitive task analysis tools, like Applied Cognitive Task Analysis (Militello & Hutton, 1998), can be used to quickly collect information on the cognitive work the user will engage in.

3. Identify the 'Scenarios of Use' to be Used in the Evaluation

If standard 'scenarios of use' for the project exist, consider using them because they typically encompass the most important tasks the system was designed to support. Otherwise, you will have to identify prior incidents that were the 'tough cases' and use those to do the evaluation. The tough cases are the difficult and challenging decisions the system has to support and often have negative consequences if not completed successfully. ('Edge case' is a closely related concept. See Roth & Eggleston, Chapter 13, and Potter & Rousseau, Chapter 15, for a discussion of the importance of including them in an evaluation.) Cognitive task analysis techniques are helpful for identifying the tough cases; they can also be used to create new, cognitively authentic scenarios that mirror the cognitive challenges found in the work environment (Crandall et al., 2006). If scenarios are not available and you do not have enough time to create them, use the system demonstration (see the next step) as an opportunity to identify a few tough cases.

4. Get a Demonstration or Walkthrough of the System You are Evaluating

Ask one of the system designers or developers to give you a demonstration of the system if a prototype is available. It can be very effective for an incumbent user to demonstrate the system. System development artifacts such as storyboards or screen shots can be used to do the evaluation if a working prototype is unavailable.

Use the demonstration to gather the following information:

- identify the intended user(s) of the system;
- identify the tasks the system is supposed to help the user to accomplish;
- identify the 'tough cases' for the task the system will be used for;
- ask 'what if' questions about the system ('If you wanted to compare data from four different time periods instead of two, how do you make the system do this?') or the work domain ('If you only had half the time to complete this task, what would you do differently?' 'If you lost your primary information feed, how would you figure out the answer then?').

Review the background material and the system demonstration material to identify which indicators might be most productive to use for the evaluation. Decide if all of the indicators will be used to conduct the evaluation; if not, select a subset of the indicators for the evaluators to use.

5. Familiarize all Reviewers with Background, System, and Scenarios

This step is the transition into the actual review stage. Do this with each evaluator individually or as a group. The evaluators should reference the indicators during this step and note what they are noticing in the system and what things warrant further investigation. Each evaluator should walk through the entire system whenever possible. Evaluators need to ask clarifying questions and add to their record of assumptions during the walk through.

6. Conduct Evaluations Individually Using the Indicators

In addition to the familiarization pass, each evaluator makes another pass through the system using the indicators. This is the formal evaluation of the system. If you are using scenarios, work through the scenario using the interface. Record not only potential problems, but also key design decisions or tradeoffs that the design has made appropriately. If you are not following scenarios, be sure to systematically review the design. Continue the review until you are not seeing anything new. At this stage, it's more important to generate insights about the system than it is to use the correct indicator name that produced the insight; this will be negotiated in the next step.

7. Regroup to Examine Findings from the Evaluators

After completing their individual evaluations of the system, the evaluators meet to compare and contrast their findings. The group should look for similarities or trends in their findings. It's very helpful to bring the system artifact being evaluated to this meeting to anchor the discussions. The goal is to present a deep picture of what is going on with the design—to generate a list of the ways in which the system supports the user's cognitive work and ways in which it hinders it. Use the affinity process (Beyer & Holtzblatt, 1998) or similar method to create a coherent set of findings. Review the entire structure to ensure that it is coherent and consistent, that it puts forward the central issues, captures all the issues discovered, and acknowledges what is done well and what the central design tradeoffs seem to be.

8. Determine Where to Focus Resources for Subsequent Design Efforts

System reviews usually uncover more design issues than the project has resources to fix. Prioritize the list of 'hindrances' or negative issues from the previous step to point the development team to the issues that most disrupt the cognitive work of the user. Several methods are available for prioritizing the results of systems evaluations such as Quality Function Deployment (Beyer & Holtzblatt, 1998) or results synthesis.

The indicators are a fast, low-cost evaluation tool that can be used early and often in the development cycle. Therefore, the indicators can be used to evaluate the system after each round of development with very little impact to the project budget or schedule. Moreover, using the indicators early and often can help avoid the conundrum of identifying significant systems flaws, but leave the development team too little time to fix them.

Example Evaluations

The process described above was used to conduct several system evaluations. Two of these evaluations are described below to provide insight into how the indicators can be used. The first was a formative evaluation of a system developed by the US National Weather Service (NWS) to improve the process of issuing severe weather warnings. The system developers had several design concepts that were evaluated in varying stages of technological maturity. Some design concepts were in the system specification and risk reduction phase of the development life cycle, while others were in the system design, development, and testing phase of development.

The second was a summative evaluation of a recently-deployed system in a hospital in the eastern US to improve demand and capacity management to reduce overcrowding in the emergency department. The project sponsors were prepared to invest significant resources to improve the system, but needed first to determine why the system was not supporting its intended users.

In both examples, the system developers used the results of the cognitive performance indicators evaluation as input to prioritize which design issues to address in subsequent development efforts and how to improve those design features.

Example 1. National Weather Service

The NWS has a history of interest in and appreciation for how CSE can improve their work (Hahn et al., 2002). They asked us if we could help them evaluate a new system integrating a variety of technology concepts to improve their performance on issuing severe weather warnings. Issuing erroneous warnings can trigger people to evacuate the area and businesses to shutdown unnecessarily, causing economic losses. Not issuing warnings soon enough can put humans in grave danger and can increase property losses.

We used the process described above to evaluate the system, and added an iteration to train the NWS participants on the use of the indicators. Prior to our trip to the NWS, we explained the indicators to five NWS participants. We trained them how to use the indicators by having them evaluate the Hertz™ NeverLost™ in-car navigation system using the job aid for reference. We reviewed NWS's use of the indicators in their evaluation of the NeverLost™ system and provided feedback. The NWS team provided feedback on the utility and ease of use of the indicators. We learned that the NWS team easily used the indicators that support pattern matching (for example, Cue Prominence; Direct Comprehension; Fine Distinctions); they had more trouble using the indicators that support the mental simulation (for example, Enabling Anticipation).

Next, the NWS team gave us a system demonstration of the research prototype system they were developing. Design concepts from the research prototype were being transitioned to the production system. As a part of this process, the development team was identifying which features should be implemented immediately and which could be deferred for later versions or pending further research. The NWS team wanted to use the results from the indicators evaluation to support this process.

During the system demonstration, the NWS and Klein Associates teams discussed the design concepts in the context of cognitive performance indicators. The Klein Associates team captured the findings from this discussion, collected screen captures of the design concepts, and gathered background documents on the prototype system. The following week we conducted a formal evaluation of the prototype system using the cognitive performance indicators and documented our findings in a report.

Evaluation results were organized around features of the system. Fourteen prototype features were identified from the content of background documents and the system demonstration. Three features were added as a result of the evaluation. Evaluated features were grouped into three categories:

1. Features that removed technical limitations from the earlier software releases and allowed the forecaster to view the raw data in its most useful form.
2. Features that gathered together other features and applied advanced computing power to make the raw data easier to manage.
3. Features that showed promise, but we could not clearly define their value because of our limited understanding of the domain.

Figures 17.1 and 17.2 present our findings for one the features in Category 2. Figure 17.1 shows the part of the finding that includes the feature's name, location, and type of cognitive work it most greatly impacts (for example, decision making, planning, situation awareness). Figure 17.2 shows the part of the finding that identifies how the feature could help or hinder the cognitive work of the forecaster. In using this structure we acknowledge that it could be both.

The NWS team found the document useful because it highlighted how the system they were developing might help or hinder their forecasters in issuing accurate and timely severe weather warnings. The report demonstrated that using the indicators to conduct a formative evaluation of a system can deliver design-relevant information to system developers and has the potential to foster iterative development.

Example 2. Hospital Patient Flow System

The second evaluation example took place in a hospital that had already deployed a system designed to enhance communication and collaboration to achieve optimal patient flow. Improving patient flow can increase the quality of patient care and help the hospital control costs. The existing paper-based system was cumbersome and resulted in poor patient flow. The new, home-grown system aimed to help

Feature: Linking between panels in the four-panel display

Location: Research prototype

Cognitive Impact:

This feature will have the most impact on the macrocognitive functions of situation assessment and developing mental models. Currently, when forecasters pan and zoom in one panel of the research prototype, it shows the effect of the navigation in the other three panels. This feature is particularly useful because it allows the forecaster to:

- Quickly scan the weather data from multiple perspectives using different types of raw data
- See how changes in data in one view affect data in other views
- Improve his or her ability to see patterns in the weather and develop richer mental models regarding weather patterns by allowing integration of the different data types and views.

The essence of the forecaster's job is the process of taking and continuously updating weather data, identifying how the data confirms or refutes their mental model for the weather, and adjusting their mental models based on this fit. Linking the four display panels helps forecasters to more smoothly and efficiently go through the process of identifying how weather data confirms or violates their mental models.

(continued)

Figure 17.1   Finding from NWS system evaluation conducted by the Klein team

Feature: Linking between panels in the four-panel display (continued)

How Feature Could Support Cognitive Work:

- Cue Prominence – This feature can support better cue prominence as some key cues in the weather data (e.g., height of reflectivity) may be more recognizable in some displays than others.

- Direct Comprehension – This feature supports direct comprehension because forecasters don't have to cycle through each panel individually to see the effect of changing the data in each view. Forecasters do have to combine data from multiple sources as they build a mental picture of what the weather (storm) looks like.

How Feature Could Hinder Cognitive Work:

- Directability – If the forecaster cannot control what is displayed in the panels, the panels may not have the data relevant to the question the forecaster is exploring.

- Adjustable Settings – If the forecaster cannot adapt the linkage between panels, then displays may not support all the cognitive performance needed by a forecaster as he or she makes sense of the evolving situation.

Figure 17.2   Finding from NWS system evaluation conducted by the Klein team (cont.)

hospital staff match the patient demand and care-providing capacity within the hospital, but was not being used by the hospital staff. We were asked to identify why the system was not having the desired impact, to prioritize further system development, and to recommend non-technology changes that would improve the system's performance.

We conducted 1.5 days of observations and interviews with sponsors, system designers, and users to understand the system, its purpose, how it was being used, and how the people working with the system understood how it worked. Figure 17.3 opposite illustrates the Demand-Capacity Management System (DCMS) in use by the hospital. The system has three main parts. The grid in the left-hand pane summarizes the state of capacity management (red equals overloaded) for all the departments in the hospital. The middle pane presents the individual metrics and how the resources of the department are matching up to patient demands. The right-hand pane shows actions that can be taken by the department or other departments to help relieve the overload.

Based on the insight and information gained at the hospital, we conducted an assessment of the system using the indicators. We used the indicators to organize and explain our findings about why the system was not having the anticipated impact. Results were documented in a report delivered to the sponsor. This report detailed the cognitive performance issues with the system. An example finding from the evaluation (transparency) appears below:

Finding 12. Transparency (negative finding)

The emergency department staff is unaware of the process by which the summary grids were formed. The system, and its designers, failed to make clear how the grids are formed. This lack of transparency discourages the staff from using the system because they do not understand how it's *supposed* to perform and do not know how it will perform in various situations; therefore, the staff do not use the system.

A follow-on teleconference was held with the participants at the hospital to discuss final results. Based on this and similar findings using the indicators, we were able to tell the project sponsor why the system was not having the impact they hoped (it was not being used by the staff) and what they could change to encourage its use. We received positive feedback from the sponsor on the value the indicators provided, as well as the insights we had regarding the cognitive performance issues we uncovered. The sponsor believes that if they had access to the indicators prior to the development of the system, they would have developed a more effective system. Such feedback supports the idea that keeping user cognition in mind throughout the system development lifecycle translates into better systems that are developed more efficiently, effectively, and have higher user acceptance.

We further refined the indicators after conducting additional evaluations. In these evaluations, we observed how HSI practitioners used the indicators, we

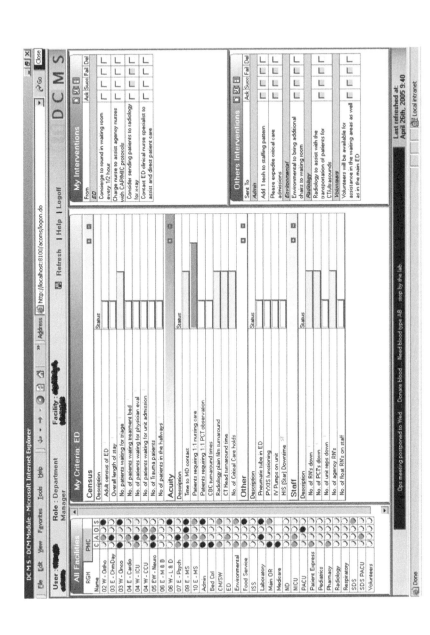

Figure 17.3 Demand-Capacity Management System

examined the analysis artifacts they generated using the indicators, and discussed with them their experience using the indicators. By doing this, we learned which indicators were used the most, which were consistently used correctly and incorrectly, and which were rarely used. Based on these findings, we reduced the original list of 13 indicators to 11.

Future of the Indicators

We believe the indicators provide a bridge to expertise. The indicators, or a future version of them, will enable less-experienced HSI practitioners to conduct assessments more like CSE experts. The indicators, as we have seen, can create a bridge between experienced practitioners and non-practitioners. The indicators provide a common language for discussing perceived faults in a system. We have had initial success in this regard, as discussed above, and have successfully used the indicators in a number of other projects and proposals. The indicators have great practical utility, but are not a completed work.

There are two concerns to be addressed in future work on the indicators. In our own use of the indicators we found a tendency for them to become slogans based on their names alone. While this can be seen as a testament to the power and reach of the concepts, our aim is for a rigorous system. Thus, the phenomenon of 'reading' more into an indicator than is there is should be discouraged. We recommend that evaluators regularly re-read the indicator descriptions during an evaluation to refresh their understanding of the purpose for the indicator and to prevent rater drift.

Our second concern is completeness. Looking at our own macrocognition framework, (Crandall et al., 2006), which describes the cognitive processes and functions that characterize how people think in naturalistic settings, the indicators do not provide equal coverage for all the components of the framework. In formulating his heuristics, Nielsen conducted a factor analysis (Nielsen, 1994) to claim that his heuristics accounted for a certain percentage of all the errors found in a sample of interfaces. We aim to develop a set of indicators that provides complete coverage and clear boundary conditions for their use.

Conclusion

We identified 11 indicators that CSE experts use to identify and describe how systems support and hinder cognitive performance in naturalistic settings. These cognitive indicators have proven successful in two system evaluations during this project and were well received by participants. We have used the indicators in other projects, and they now form an essential part of our conversation with HSI practitioners and system developers. While the indicators are not a complete solution, we believe they are an important step in making CSE expertise accessible.

They will help anyone wanting to improve the design and evaluation of systems for individuals and teams doing real-world work that is primarily cognitive in nature.

Succinctly, the indicators are a set of probes for inquiring about how a system is going to support or inhibit the cognitive performance of a person working with the system. We chose the qualifier 'indicator' with care. They are pointers to larger conversations amongst the design team and stakeholders, anchored by the concepts named by the indicators. These indicators provide new language that practitioners and others system development stakeholders can use to generate more insight during system evaluations about what works, why it works, and what might be done to make it work better.

It is not necessary to wait until major design review milestones are achieved to use the indicators. In fact, that is often too late to make substantive improvements. Using the cognitive performance indicators early and iteratively in the development life cycle is what we recommend. This will help to locate issues of how the system hinders the cognitive work of the user. Having conversations about these issues will help the development team retire risk sooner and promote incremental improvement of the system. Teams that have these conversations can markedly increase their ability to deliver high-quality systems that support real-world cognitive work.

Acknowledgements

The views, opinions, and findings contained in this article are the authors and should not be construed as official or reflecting the views of the Department of Defense, US Navy, or Applied Research Associates, Inc. We thank the Naval Surface Warfare Center, Dahlgren Division for their support through contracts N00178-04-C-1069 and N00178-04-C-3017. We specifically recognize our contracts monitor Owen Seely and topic author Patricia Hamburger.

References

Beyer, H., & Holtzblatt, K. (1998). *Contextual design: Defining customer-centered systems*. San Francisco, CA: Morgan Kaufmann Publishers.

Crandall, B., Klein, G., & Hoffman, R. R. (2006). *Working minds: A practitioner's guide to cognitive task analysis.* Cambridge, MA: The MIT Press.

Hahn, B. B., Rall, E., & Klinger, D. W. (2002). Cognitive Task Analysis of the Warning Forecaster Task, *Final Report for National Weather Service* (Office of Climate, Water, and Weather Service, Order No. RA1330-02-SE-0280). Fairborn, OH, Klein Associates Inc.

Hart, S. G., & Staveland, L. E. (1988). Development of a multi-dimensional workload rating scale: Results of empirical and theoretical research.

In P. A. Hancock & N. Meshkati (Eds.), *Human mental workload* (pp. 139-183). Amsterdam, The Netherlands: Elsevier.

Hoffman, R. R., Feltovich, P. J., Ford, K. M., Woods, D. D., Klein, G., & Feltovich, A. (2002). A rose by any other name would probably be given an acronym, *IEEE Intelligent Systems, 17*(4), 72-80

Klein, G. (1998). *Sources of power: How people make decisions*. Cambridge, MA: MIT Press.

Klein, G., Ross, K. G., Moon, B. M., Klein, D. E., Hoffman, R. R., & Hollnagel, E. (2003a). Macrocognition. *IEEE Intelligent Systems, 18*(3), 81-85.

Klein, G., Wiggins, S. L., & Lewis, W. R. (2003b) Replanning in the army brigade command post. In *Proceedings of the 2003 CTA Symposium* [CD-ROM].

Militello, L. G., & Hutton, R. J. B. (1998). Applied Cognitive Task Analysis (ACTA): A practitioner's toolkit for understanding cognitive task demands. *Ergonomics, Special Issue: Task Analysis, 41*(11), 1618-1641.

Moon, B. M., Wei, S., & Cox, D. A. (2004). Cognitive Impact Metrics: Applying Macrocognition During the Design of Complex Cognitive Systems. In *Human Factors and Ergonomics Society Annual Meeting Proceedings*, pp. 473-477.

Nielsen, J. (1993). *Usability engineering*. Boston, MA: Academic Press.

Nielsen, J. (1994). Enhancing the explanatory power of usability heuristics. In *Proceedings of the ACM CHI'94*, Boston, MA.

Stanton, N., Hedge, A., Brookhuis, K., Salas, E., & Hendrick., H. W. (2005). *Handbook of human factors and ergonomics methods*. Boca Raton, FL: CRC Press.

Tilley, A. R., & Henry Dreyfuss Associates. (2001). *The measure of man and woman: Human factors in design*. New York, NY: John Wiley and Sons, Inc.

Woods, D. D., & Hollnagel, E. (2006). *Joint cognitive system: Patterns in cognitive systems engineering*. Boca Raton, FL: CRC Press.

World Wide Web Consortium. (1999). Web Content Accessibility Guidelines 1.0. Retrieved from http://www.w3.org/TR/WCAG10/ on 30 March 2007.

Index

H

handoff, xxix, 137, 138, 139, 140, 141, 142, 143, 144, 145, 146, 147, 148, 149, 150, 151, 152
heuristics, 68, 70, 266, 287, 288, 300
Hierarchical Task Analysis, 5, 6
Hits, Errors, Accuracy, Time (HEAT), xxvii, xxi
human factors, 6, 12, 14, 16, 125, 132, 272
hypothesis exploration, xxvii, 68, 69, 73, 74, 75, 76, 77

I

information analysis, 65, 75, 77, 78
information processing, xxix, 65, 66, 67, 78, 271, 272, 275, 276
information search, 56, 68, 69, 72, 75, 76, 77
information synthesis, 32, 33, 68, 71, 72, 73, 74, 75, 76, 77
information validation, 68, 70, 71, 74, 75, 76, 77
intelligence analysis 20, 21, 65, 66, 77
intent, xxix, 17, 19, 50, 59, 67, 71, 109, 111, 127, 129, 168, 195, 216, 278, 279, 280, 282
internal validity, 207
internalized knowledge, 32, 34, 37, 40, 195, 207
interruptions, 140, 149, 246

J

joint cognitive system, 203, 204, 205, 207, 208, 211, 214, 215, 253, 254, 256, 257, 258

K

knowledge, xxvi, xxix, 3, 5, 19, 21, 22, 30, 31, 32, 33, 39, 65, 68, 90, 120, 141, 143, 144, 151, 161, 170, 179, 180, 181, 182, 183, 184, 185, 191, 193, 276, 280

L

learning, 19, 22, 118, 137, 138, 142, 149, 150, 181, 184, 189, 210

M

macrocognition, xxiii, xxiv, xxix, xxx, 29, 30, 41, 161, 162, 179, 180, 203, 210, 253, 271, 272, 286, 300
macrocognition functions, xxv, xxvi, xxix, xxx, 19, 47, 48, 54, 60, 65, 180, 203
measures, diagnostic, 207
memory, 19, 140, 275, 279, 290
mental model, xxv, 21, 42, 52, 53, 58, 101, 114, 116, 124, 161, 180, 181, 276, 290, 297
mental workload, 12, 14
model of support, 211, 212, 214, 247
model, xxv, 15, 21, 23, 26, 32, 47, 48, 52, 53, 54, 55, 58, 59, 60, 67, 71, 73, 77, 78, 97, 100, 101, 104, 116, 124, 125, 140, 151, 161, 168, 169, 180, 181, 195, 206, 208, 211, 212, 214, 215, 216, 247, 253, 254, 256, 276, 277, 278, 279, 281, 282, 286, 287, 288, 290, 297

N

narrative, 65, 76, 139, 140, 141, 142, 144, 149, 246
Naturalistic Decision Making (NDM), xxiv, 54, 188, 287
non-technical skills 125, 129

O

observability, 22, 125, 254, 256, 261, 262, 265, 267, 268, 274
outcome, xxvii, xxix, 29, 32, 34, 96, 102, 103, 112, 117, 118, 119, 138, 142, 143, 144, 170, 171, 174, 189, 193, 194, 246, 260, 276, 278, 279

P

paradigms, 48, 60, 203, 214, 215, 273, 279
pattern analysis, 166, 167
pattern recognition, 33, 39, 64
performance, team, xxiii, xxviii, 29, 56, 96, 123, 126, 127, 130, 131, 132, 162, 163, 170, 179